ACS SYMPOSIUM SERIES **394**

The Challenge of d and f Electrons

Theory and Computation

Dennis R. Salahub, EDITOR

Université de Montréal

Michael C. Zerner, EDITOR

University of Florida

Developed from a symposium sponsored
by the Divisions of Inorganic Chemistry and of Physical Chemistry
of the American Chemical Society
and the Division of Physical and Theoretical Chemistry
of the Canadian Society for Chemistry
at the Third Chemical Congress of North America
(195th National Meeting of the American Chemical Society),
Toronto, Ontario, Canada,
June 5–11, 1988

American Chemical Society, Washington, DC 1989

Library of Congress Cataloging-in-Publication Data

The Challenge of d and f electrons.

(ACS symposium series, ISSN 0065–6156; 394)

"Developed from a symposium sponsored by the Divisions of Inorganic Chemistry and of Physical Chemistry of the American Chemical Society and the Division of Physical and Theoretical Chemistry of the Canadian Society for Chemistry at the Third Chemical Congress of North America (195th National Meeting of the American Chemical Society), Toronto, Ontario, Canada, June 5–11, 1988."

Includes bibliographies and indexes.

1. Molecular orbitals—Congresses. 2. Quantum chemistry—Congresses.

I. Salahub, Dennis R., 1946–   . II. Zerner, Michael C. III. American Chemical Society. Division of Inorganic Chemistry. IV. American Chemical Society. Division of Physical Chemistry. V. Canadian Society for Chemistry. Division of Physical and Theoretical Chemistry. VI. Chemical Congress of North America (3rd: 1988: Toronto, Ont.) VII. American Chemical Society. Meeting (195th: 1988: Toronto, Ont.) VIII. Series.

GD461.C37   1989   541.2′2   89–6926
    ISBN 0–8412–1628–2
                                              CIP

Foreword

The ACS SYMPOSIUM SERIES was founded in 1974 to provide a medium for publishing symposia quickly in book form. The format of the Series parallels that of the continuing ADVANCES IN CHEMISTRY SERIES except that, in order to save time, the papers are not typeset but are reproduced as they are submitted by the authors in camera-ready form. Papers are reviewed under the supervision of the Editors with the assistance of the Series Advisory Board and are selected to maintain the integrity of the symposia; however, verbatim reproductions of previously published papers are not accepted. Both reviews and reports of research are acceptable, because symposia may embrace both types of presentation.

Contents

Preface

THE GOAL OF THEORETICAL CHEMISTRY is to predict chemistry, either with simple mathematical expressions or with the more complicated expressions that researchers must solve on a computer, rather than by working with test tube and beaker in the lab. Theorists have long dreamed of synthesizing and characterizing the properties of conceptualized materials and their stabilities and reactivities without early recourse to bench experiments. Computer experiments are already of considerable aid to the experimentalist in prescreening compounds of desired character. The day is fast approaching when organic chemists will routinely use computational procedures to examine targeted molecules for specific properties before attempting synthesis and characterization of these systems. Quantum chemical experiments already aid our understanding of chemical processes, and such computations are often included with experimental results in publications. The impact of such work is continually growing. Entire industries are being transformed at the most fundamental level, perhaps most strikingly so in the pharmaceutical industry.

Progress has been somewhat slower in the development of theoretical tools for transition metal systems. The localized nature of d and f electrons, for example, often not only makes molecular orbital calculations difficult but also makes the utility of such calculations uncertain. In addition, the chemistry of these systems requires consideration of large molecules, clusters, surfaces, and bulk systems. Phenomena as diverse as medicine, catalysis, and high-temperature superconductivity are complex, and they require the most modern techniques for their accurate study. Methods that are of proven value for the chemistry of hydrogen and the main group elements, methods that predict molecular structure with an accuracy rivaling experiment, are proving inadequate, even qualitatively, when applied to the more challenging metal-containing systems.

The symposium on which this book is based presented recent advances in the theory and computation of systems containing d and f electrons and applications to a number of the complex systems that have recently been examined using these techniques. Leading experimentalists presented work that would greatly benefit from advanced theory. The latest developments in molecular orbital theories, in correlation

ix

calculations, and in local and nonlocal spin density theories were discussed, as well as the effects that calculations using these techniques have had on our way of thinking about these systems. The foundations of future scientific, technological, and industrial revolutions are currently being laid in the laboratories and computer rooms dedicated to understanding d and f electrons.

This book contains contributions from most of the leading scientists in this area. It presents a snapshot of the state of the art as it existed in June 1988.

We take this opportunity to thank the sponsors of this symposium again, whose support helped make this meeting the dynamic forum it was. Financial support came from Eastman Kodak Company of Rochester, NY; IBM Corporation of Yorktown Heights, NY; Multiflow Computer Company of Branford, CT; and the Petroleum Research Fund of the American Chemical Society. We also take this opportunity to thank Tom Ellis in Montreal and Susanne Gaddy in Gainesville for their cheerful editorial assistance.

DENNIS R. SALAHUB
Université de Montréal
Montréal, Québec H3C 3J7, Canada

MICHAEL C. ZERNER
University of Florida
Gainesville, FL 32611

December 1, 1988

Chapter 1

Quantum Chemistry Throughout the Periodic Table

Dennis R. Salahub[1] and Michael C. Zerner[2]

[1]Département de Chimie, Université de Montréal, C.P. 6128, Succursale A, Montréal, Québec H3C 3J7, Canada
[2]Quantum Theory Project, Williamson Hall, University of Florida, Gainesville, FL 32611

An overview is presented of the state-of-the-art for quantum chemical calculations for d- and f- electron systems. The present role and the potential of ab-initio, density functional and semi-empirical methods are discussed with reference to contemporary developments in related experimental disciplines. Progress towards a true computational chemistry including the transition metals, lanthanides, and actinides is outlined with emphasis both on achievements and on the remaining barriers.

Imagine a chemist, a modern-day Rip Van Winkle, awake and refreshed following a twenty-year sleep and finding himself at the instrument exhibit at an ACS national meeting. Poor fellow! His search for rotating evaporators and simple bench-top IR's will not be an easy one. If there are any, they are lost in a sea of television screens; a dazzling display of rotating, vibrating, pulsating, dancing, reacting, technicolor molecules. A few well placed questions and Rip learns that, while he was sleeping the sleep of the just, there was quite a revolution going on, that now chemists and, it seems, in large numbers are using computer programs to simulate the molecules they will perhaps eventually make react. A bit more courage, and a few more questions and he learns that by-and-large the molecules dancing on the screen are organic - carbon, hydrogen, nitrogen, oxygen, with an odd halogen, or sulfur or phosphorus. Finding the colorful transition metals that he liked so much before he went to sleep turns out to be another difficult task; and when he finally finds a demonstrator to explain how they are handled, after much hemming and hawing, he learns that the methodological basis for the computations is not the same as it is for the lighter, s-p, elements.

The organic computational chemistry that so dazzled our allegorical eyes-wide-open chemist is based on the combined efforts of a large number of theoretical and experimental chemists over the last few decades. This combined experience has resulted, only recently, in a set of tools that can provide the entire chemistry community, specialist and non-specialist alike, with a manual and a set of "specs", in much the same way as any other instrument. The user, once he has read, or has been properly informed, that the instrument is appropriate for his problem (an organic system with N atoms and a required accuracy of x kcal/mole) can go ahead and use these enormously complex programs with the greatest of ease to help him in his work.

That the same is not true in general for problems involving transition metals or lanthanides or actinides can be attributed above all else to the difficulty of handling electron correlation for these systems. The d and f electrons are confined to a small volume and there are many of them, particularly if metal-metal bonds are included. This correlation problem will be at the core of this paper and indeed of the whole volume. But this is not the only problem when it comes to comparing d and f orbital systems with systems that do not contain these trouble makers. Transition metal systems are most often open-shell systems, unlike their organic cousins. The treatment of open-shell systems is more difficult both at the Hartree-Fock level, and at the post Hartree-Fock level. Of importance in transition metal complexes is a comparison of the relative energies of complexes of differing spin multiplicities, and this comparison is made very difficult. The Hartree-Fock procedure greatly favors states of higher multiplicity, and this bias is, again, only corrected by a high level correlated theory (see DAVIDSON, for example). In addition, the great wealth of states that lie near in energy for transition metal systems often make the SCF step required in most treatments much more troublesome. For organic systems, SCF is usually (but not always!) not a problem.

There are in general three common ways used today to go about examining the electronic structure of transition metal systems. All of these are concerned in one way or another in solving (or getting around) the correlation problem and we will give an overview of all three, underlining their respective strengths and weaknesses: "standard" ab initio techniques (the surest and often the best way to go for cases where the computational effort does not render them impractical or impossible), Density Functional methods where the correlation is handled through an electron gas model (still an "ab initio" framework but practicable for much larger systems) and semi-empirical where "nature's correlation", namely experimental data, are used to adjust parameters, resulting in very rapid methods that, when used judiciously, can provide accurate results for very large systems.

In this brief overview we can only touch on very few aspects of a vast and growing field. Our only goal here is to provide some structure so that newcomers, in particular, may more rapidly put the various questions and the various techniques into proper perspective, and obtain a first impression of the present state of the field and of where it is going, as we see it. Taken along with the other chapters of the book (referred to by the name of an author in capital letters) and the other references cited, we believe that this impression should be reasonably unbiased.

Before starting the methodological overview, it is worthwhile to remind ourselves briefly of the experimental context. The presence of d and f electrons, whose wave functions can be either core-like, essentially atomic, or more diffuse, and thus true valence like, or anything in between, does confer special, sometimes unique, properties. The fascination and the utility of the target systems for many of the chapters of this book are owed, in the final analysis, to the range of behavior that electrons in these wave functions demonstrate. The four "experimental" chapters of the book treat subjects ranging from ESR and other spectroscopies of clusters in matrices (WELTNER and VAN ZEE), to the thermodynamics and phase transitions of heavy fermions (OTT and FISK), to electron density maps (COPPENS), to energy conversion in photosynthesis (SMITH and GRAY). These represent only a very small sampling of the many challenges being offered to quantum chemistry by recent experimental advances. Other chapters will mention aspects of inorganic synthesis, of reaction mechanisms, of homogeneous, heterogeneous, and enzymatic catalysis, of cluster beams, of surface science, of superconductivity and magnetism, of NMR and Mossbaur spectroscopy, of biomolecules and much, much more. In fact this field provides a veritable smorgasbord of experimental data waiting for interpretation and guidance from quantum chemistry. A good smorgasbord is stocked with individual delicacies but also has a theme - ours has the d and f electrons to provide continuity, coherence and, if you will, the flavor that each of the authors of this volume so well enjoy.

Ab-Initio Calculations

The term "ab-initio" is often taken to include those methods which, given an initial choice for the general form of the N-electron wave function, (e.g. one or many determinants) attempt to solve the Schrodinger Equation

$$H\Psi = E\Psi$$

H=Kinetic energy operator + Potential energy operator

E= molecular energy

$$\Psi = \Psi(r1, r2, \ldots rN)$$

without introducing any empirical parameters. This does not mean that no approximations are made (indeed, the whole art is in the approximations) but only that recourse is not taken to experiment other than, perhaps, a-posteriori, that there results agreement with experiment. Table I gives a brief summary of some of the main classes of ab initio methods with their characterizing features. A good entry into the details of these methods is the book by Szabo and Ostlund (1). Recent issues of the Journal of Chemical Physics, Chemical Physics Letters, and the International Journal of Quantum Chemistry could also be consulted. Examples of applications will be found in the chapters of WILLIAMSON and HALL (GVB,CASSCF), DEDIEU and BRANCHADELL (CASSCF,CI), DAVIDSON (HF,MPPT,CI), NOVARRO

Table I.   Some Classes of Ab Initio Methods

Hartree–Fock HF (or SCF)	one (spin adapted) determinant $\Psi = \lvert \phi(1),\phi(2)....\rvert$ Det. built from molecular orbitals $[\phi(i)]$	work= N^4. $\phi(i)$'s adjusted iteratively to obtain minimum energy of this wave function through the SCF procedure. This energy is variational and size extensive.
Configuration Interaction CI	many determinants $\Phi = \sum_I \Psi(I)C(I)$	work= N^5 or greater. $C(I)$'s obtained by diagonalization of large matrix. The CI method is variational but converges slowly and is, in general, not size extensive.
Multi-Configuration SCF MCSCF	Φ = same as CI	Both $C(I)$'s and $\phi(i)$'s optimized simultaneously. More accurate than CI method with SCF $\phi(i)$ for a given CI expansion. Variational, not size extensive, convergence often difficult.
Complete Active Space SCF CAS SCF	Special case of MCSCF	all determinants from chosen set of "active" orbitals are included in the MCSCF. (i.e. Full CI) variational, size extensive; CI expansion list rapidly gets very large.
Generalized Valence Bond GVB	Special case of MCSCF	combinations of determinants represent VB functions are used to fix the CI coef. simplest is perfect-pairing
Many Body (Moller-Plesset) Perturbation Theory MP2,MP3, etc.	Φ = same as CI	$C(I)$'s determined by pert. theory. Most common is second order "MP2", but higher order is also common. Non-variational, size extensive.
Coupled Cluster CPMET CCD, CCSD, etc.	Φ = same as CI	$C(I)$ determined by exponential cluster expansion; non-variational, size extensive.
Quantum Monte Carlo QMC	quantum simulation HF or CI "guiding" function	so far, small systems only. difficult for excited states.

(HF,MPPT), KOKA and MOROKUMA (CI), PANAS, SIEGBAHN and WAHLGREN (CASSCF,CI), MESSMER et. al. (GVB), YANG and WHITTEN (HF), KOBER and HAY (HF,GVB), MALLI (HF,CI), CHRISTIANSEN (QMC), and NEWTON (HF,MPPT).

Most of the ab initio calculations presented at the meeting were performed within the HF-CI, CASSCF, or HF-GVB framework and this is a reasonable reflection of the overall use of these techniques for d and f systems. Nearly everyone agrees that, except for quick estimates or in "fortunate" situations, a straightforward Hartree-Fock calculation is unlikely to yield reliable results. Work on transition-metal atoms (2) is enlightening in this regard. On the other hand, the amount of correlation that is "actually" needed is a central question in the field and is very much an open one. The ultimate answer, at the moment, and quite unfortunately, seems to be system dependent.

Beyond the choice of a general method, further, technical choices have to be made and the utility of a calculation can depend critically on these. Perhaps the single most important one that is emerging is the choice of the basis set in which the molecular orbitals are expanded (3). Another important choice is the way in which many-electron correlation is to be included (1-2), and this is surveyed in the table above. Of some importance is whether or not to represent the core-electrons by an effective core potential, thereby greatly reducing the time for a calculation, but associated with the possibility of damaging the quality of the results (4). For the heavier elements, the incorporation of relativistic corrections (5) is important. Feasible ways in which to include correlation effects is, in no way, an agreed upon technology.

Although it is certainly dangerous to generalize in' this field, we will take the risk and offer a few rules of thumb in the hope that they will help the non-expert to situate, in something like the proper perspective, the various calculations that he may encounter in the literature. We INSIST that exceptions exist and advise prudence.

On the question of the one-electron (orbital) basis set it appears that the transition metals require significantly more flexibility than do the lighter elements. Whereas minimal, single-zeta, basis sets can often be used in organic computational chemistry to obtain a reasonable first idea of the properties, such an approach is inadequate for transition-metal chemistry. Errors of several electron-volts in the orbital energies result, incorrect orderings of spin states are predicted, and often geometries are in complete disagreement with experience. Even an incorrect charge distribution between ligand and metal can result. A "reasonable" basis set needed to describe the d functions is at the triple-zeta level, two "inner-loop" functions and a diffuse one, although double-zeta (including the diffuse function) can sometimes give useful results. Considerably less is known about the f orbitals active in the chemistry of the lanthanides and actinides. It is likely again that functions at least of triple-zeta quality will be required. As in the case of the d functions, these expansions must have the flexibility to reproduce the shape of orbitals representing the many possible oxidation, spin and valence states of the metal atom. The purist would correctly argue that even triple-zeta functions have not the flexibilty to represent the cobalt d orbitals with reported oxidation states from -2 to +6 and all the possible spin states associated with each of these oxidation states!

Rules of thumb for the many-electron wavefunction are even difficult to formulate, and the most efficient ways of including correlation are still under extensive theoretical review. But one observation is already clear. Methods that are effective for organic compounds are not necessarily effective for transition metal complexes. Much more experience is required. It might, however, be useful to have a rough idea of which types of system are proving the most demanding from the point of view of electronic correlation.

Transition-metal complexes with a single metal atom, although far from trivial, are the easiest to treat. If they are closed shell, fairly compact CI expansions, choosing determinants in about the same fashion as for organic molecules (double excitations from bonding to antibonding orbitals being perhaps the most important, (CID), can provide meaningful results. Unpaired electrons in the reference or ground state SCF description, generally complicate the situation, especially if comparisons between states of different multiplicities must be made. Coordinatively saturated bimetallic or cluster compounds with two or, perhaps three metal atoms, and a complement of ligands can also be treated, although these represent very computationally demanding situations. In systems containing several weakly coupled transition metals, either with or without coordinating ligands, there is a competition between a delocalized description, well described at zero'th order through molecular orbital theory, and a local orbital description, poorly described at zeroth order by m.o. theory, but described in a natural fashion through valence bond theories. In either case a complete description is obtained only after a great deal of configuration interaction is added to either starting description. In such cases, even transition metal diatomics, such as dichromium, can be considered state-of-the-art. Similar problems arise, for example, in the Creutz-Taube systems that show mixed valence behaviour. In these systems correlation becomes so important that the Hartree-Fock procedure itself "senses" problems, leading to instabilities that should not necessarily be assigned physical significance.

Recent work on the mechanisms of reactions involving transition-metal complexes represents the state-of-the-art of insightful ab initio calculations in this area. Beyond three metal atoms or so it is fair to say that ab initio approaches start to bog down. For such studies great care has to be taken in the choice of model system to be treated. For example, models involving clusters representing chemisorption sites and containing several transition-metal atoms can and have been treated with interesting results, but, more often than not, the metal chosen is nickel or copper where open shells are less of a problem and the details of d-electron correlation appear somewhat less crucial in describing the metal-adsorbate interactions. It is already quite apparent that even greater difficulties arise when considering metals more to the left in the periodic table.

The large number of chemically uninteresting core electrons for the heavier elements can usually be handled without great difficulty and with a high degree of accuracy by modern pseudopotential or effective core potential or model core potential techniques. Results corresponding closely with appropriate all-electron treatments can be obtained although this often requires that the so-called semi-core electrons (with principal quantum number one less than the valence shell) either be treated on the same footing as the valence electrons or that special core-polarization corrections be implemented. With these techniques, the number of electrons treated explicitly in the N-electron problem ranges from one to twenty or so depending on the element and on whether the semi-core electrons are included. The model potential operators include terms to represent the potential of the core electrons and also projection operators to ensure core-valence orthogonality. They are often represented in a parametric fashion (a gaussian form being convenient) and the parameters are adjusted to yield agreement with all-electron calculations for the atoms, typically using wave function and orbital energies as criteria, but sometimes including information on low-lying atomic excited states which are thought to be populated in the molecules. Some of the effects of relativity, notably the spacial contraction of the core electrons can be taken into account by adjusting the parameters on relativistic atomic calculations. This appears to work very well at least down to palladium or so. For the platinum row and beyond, the relativistic corrections (the correction to the mass due to the velocity, the Darwin correction and spin-orbit coupling, if a two-component, Pauli, treatment is adopted) are all of the same order of magnitude as the chemical bonding energies. It remains a subject of current research to find whether an indirect, relativistic pseudopotential approach will, in general, be adequate or whether the relativistic operators should be included directly in the valence shell Hamiltonian.

For the lanthanides and actinides the adequate treatment of relativistic corrections is absolutely essential. Indeed these atoms and their compounds may provide the ultimate testing ground for new theories of "many-electron" relativistic quantum chemistry. The question of how to simultaneously take account of electron exchange and correlation effects along with the effects of relativity remains one of the prime challenges of electronic structure theory. Of special interest here, for example, is the fact that spin-orbit splittings can be as large or larger than ligand field effects, and conventional electronic structure theories that consider these effects as perturbations (if at all) are likely to prove inappropriate.

So then, what is the role of these, sometimes discouraging, always ponderous, calculations? Does it make sense to expend so much time, effort, and money in return for these hard-earned results? Yes! Already the impact of these calculations on inorganic, or organometallic chemistry has been sizeable. How will, or should, this field evolve? If one looks to the history of organic computational chemistry then at least some partial answers to such questions can be formulated. The theory, the methodology, and the computer programs that implement both, will evolve, and

larger systems will become approachable with greater reliability
with this progress. Will ab-initio calculations on transition
metal systems ever become "black box" as they are now for some
organic systems? Probably, but there will never be a substitute
for careful thought and analysis, and this is going to be an
especially important part of this field for very many years to
come. The "incipient appeal" of just plain scalar numbers (the
total energy, for example) will have to be resisted in favor of
arguments based on sound chemical and physical principles.

In addition to the chemistry that may be addressed, ab-initio
calculations are needed to provide important guidance on which to
develop semi-empirical and empirical schemes that use, in fact, the
same formalism. There is an important interplay between all
methods, and the information that we gain from ab-initio
calculations is not only used to examine the chemistry of the model
systems, but also used to help develop semi-empirical methods. The
numbers obtained are often used to guide parameter choices where
these are not available from experiment. Most importantly,
discoveries in the underlying theory that we use are made in semi-
empirical, density functional and ab-initio studies, and these
improvements are adopted, perhaps in modified form, to affect all
theories used in computational chemistry.

Density Functional Methods

The methods classed under the general heading of Density Functional
Theory (6-10) strive to treat the electron correlation problem in a
somewhat different manner from that used in conventional ab initio
approaches. It is perhaps worth noting at the outset that these
methods are no less "ab-initio" than CI and the others, in that
they start from an exact formulation of the N-electron problem.
Only the approximations are different and it turns out that these
differences confer considerable advantages in terms of accuracy and
ease, in many circumstances. Although density functional
calculations have been performed since the days of the Thomas-Fermi
atomic theory, the field was put on a firm conceptual foundation in
1964 and 1965 when Hohenberg, Kohn and Sham presented their
remarkable theorems (6-7). Hohenberg and Kohn proved that a many-
electron ground state is completely determined by the electron
density. The energy functional depends on the density and it can,
in principle, be obtained variationally. Kohn and Sham showed, by
ingenious arguments (the paper is terrific fun to read, and
reread!) that Density Functional Theory can be reformulated in a
set of Hartree-like equations that can be solved self-
consistently. These equations are deceptively simple to write:

$$(-1/2 \ \nabla^2 \quad + v_{coul} \quad + v_{xc} \quad) \ \phi_i \ (1) = e_i \ \phi_i \ (1)$$

The final term in this equation, v_{xc}, the exchange-correlation
potential, contains all the intricacies of the many-electron
problem. It is defined as

$$v_{xc} \quad = \partial Exc/\partial \rho$$

where Exc is the exchange and correlation component of the energy functional. The various density functional methods can be classified according to the approximations adopted for Exc and also according to the techniques used to solve the Kohn-Sham equations. Tables II and III summarize some of the more common choices.

The cornerstone of the field (the "Hartree-Fock" of Density Functional Theory) is the Local Density Approximation (LDA) also called the Local (Spin) Density (LSD) method. Here the basic information on electron correlation, how electrons avoid each other, is taken from the uniform density electron gas. Essentially exact calculations exist for this system (the Quantum Monte Carlo work of Ceperley and Alder) and this information from the homogeneous model is folded into the inhomogeneous case through the energy integral:

$$Exc(LDA) = \int dr exc(\rho(\mathbf{r}))$$

To the extent that electrons in an atom or molecule do not behave too differently at the microscopic level from those in the uniform gas this approximation will yield meaningful results. In practice it works remarkably well. Various LDA band calculations have, for years, provided the theoretical and computational basis for many of the advances in solid state physics. Since the early days of Slater's Xα method, LDA approaches have occupied an ever increasing part of molecular and cluster physics and chemistry. Complex systems, for example clusters containing a dozen or more transition-metal atoms, can be treated with all of the techniques in Table III and reliable results are obtained for geometries, vibrational frequencies, photoelectron and optical spectra, and other properties which depend on the orbitals or on the density itself. The SW approach, although very rapid, is limited in that geometry variation is not generally reliable and also "non-compact" systems often suffer from rather large errors due to the muffin-tin form of the potential. The three "basis set" methods, LMTO, DVM, and LCGTO are all capable of yielding high quality results (the LSD limit) if proper attention is paid to the computational details (choice of the basis sets, sampling points, etc.). The relative advantages of each of these depend on the system, the accuracy needed, the computer machinery available (scalar, vector, parallel) and the evolution of the computer codes. Extensive comparisons of performance have not yet been made so it is too early to know which of these will survive (and in what form).

There has been some recent progress towards obtaining information relevant to dynamics using DF methods, notably in the area of surface diffusion. For these studies the local approximation appears to be reasonable. A crucial area where corrections to the LDA are needed is that of dissociation energies, which are usually seriously overestimated. Increasing attention is therefore being paid to methods that go beyond the LDA, primarily through terms involving the gradient of the density. Such quantitative treatment of many complex systems and properties. Whether and in what circumstances they will be adequate for the

Table II. Some Density Functionals

Kohn-Sham exchange only (Xα with α=2/3) (7) KS	$\rho^{1/3}$	exchange only for electron gas
Xα (11)	$\alpha \, \rho^{1/3}$	KS exchange weighted by $3\alpha/2$
VWN(Vosko, Wilk, Nusair)(12) PZ(Perdew, Zunger) (13)	$\rho^{1/3}$+correl.	electron gas exchange and correlation from accurate Quantum Monte Carlo calcs. of Ceperley and Alder (14)
LM(Langreth-Mehl) (15)	$\rho^{1/3}$+correl. +F($\nabla\rho$)	the first "modern" gradient corrected funct.
Becke (16) De Pristo-Kress (17) Perdew-Yue (18)	"	"semi-empirical" gradient function-als

----

Table III. Some Techniques for Solving the
             Kohn-Sham Equations

Scattered-Wave (SW) or Multiple Scattering (MS) (19)	Muffin-tin potential Partial-wave expansion	rapid good one-electron props., total energy unreliable
Linear Muffin-tin Orbital (LMTO) (20)	basis of MT eigenfuncs. linear expansion	quite rapid total energy ok basis set choice?
Discrete Variational Method (DVM) (21)	numerical sampling of Slater or numerical basis	quite rapid good energy requires fine grid
Linear Combinations of Gaussian Type Orbitals (LCGTO-LSD) (22)	Gaussian basis fits for density and exchange-correlation	most similar to "ab initio" analytical inte-grals accurate total energy possible

----

description of bond making and breaking will remain an area of intense investigation for the next few years. The general problem of treating excited states, and multiplets also requires attention as does the relationship of density functional methods to more conventional, wave function, techniques. A reasonable entry to the field should be provided by the references cited and the chapters of SONTUM, NOODLEMAN and CASE (Xa-SW), MALLI (relativistic Xa-SW), ZIEGLER, SNIJDERS, BAERENDS and RAVENEK (DVM-LSD + gradient), BECKE (numerical, gradient), NORMAN and FREEMAN (LSD bands), ANDZELM, WIMMER and SALAHUB (LCGTO-LSD), ROSCH et al. (LCGTO-LSD), KELLER and DE TERESA (LSD band), and STOLLHOFF and FULDE (LSD-quasiparticles).

For the future, in addition to research on the approximate Density Functional Hamiltonians, much work will be required on techniques and on computer programs if DFT is to live up to its full potential in terms of computational chemistry. By-and-large the programs (at least those with which the authors are familiar) are less highly developed than those used in more conventional quantum chemistry. More man-years of programming and testing effort will have to be invested before "user friendly" tools are widely available. Some of the work reported at the meeting is most encouraging in this regard. Some first attempts have been made to extract force fields from density functional calculations for systems involving metal-metal bonds, an area where DFT may prove to be particularly fruitful, either through parameterized force fields or even by direct calculations.

In a nutshell, methods based on Density Functional Theory are proving to be extremely powerful tools. Variants involving model core potentials and relativistic corrections allow the treatment of a wide variety of properties of polyatomic systems containing even the heaviest atoms. Work is in progress on writing and evaluating the computer codes to calculate energy gradients and various potential energy "walkers" and to make the programs more efficient, more flexible and more friendly. As this work progresses more and more chemists have come to appreciate that the DFT option provides a valuable tool in a wide variety of circumstances where other techniques are just not competitive.

Semi-Empirical Methods

Since the earliest days of quantum mechanics it has been clear that many, indeed the vast majority, of all interesting questions of chemistry that one might wish to examine through quantum calculations involved systems of such complexity that they were "just beyond" a straightforward solution of the Schrodinger equation. The resulting frustration was, is, and will continue to be the mother of nearly all developments in computational chemistry, and the reason for the fascination with semi-empirical methods. Surely we can use "exact solutions" of the Schrodinger equation - careful experimentation - to guide our approximate solutions.

Since the early work of Huckel on aromatic molecules, it has been realized that the basic and powerful structure of quantum mechanics could be combined with empirical elements to provide

techniques that furnish both numerical predictions and a framework
for explanations when applied judiciously to a selected set of
molecules and properties. The basic idea of these methods is to
take the equations from an established theory, and then to
approximate terms that appear in the formulation using experimental
information. This semi-empirical procedure has been adopted in all
areas of science to force a restricted application of that theory
to reproduce experimental results with greater accuracy. In the
case of quantum chemistry and condensed matter physics, the
electronic structure problem gives rise to equations of simple
appearance that are often nearly impossible to solve. The
Hamiltonian must therefore be simplified, and models, such as the
orbital model, are made, and yet even then the equations are
difficult to solve even on the most modern of computers. Empirical
parameters may then be introduced in the form of basis sets that
are justified a-posteriori, i.e., that they yield successful
results. Basis sets are even designed for specific purposes, spin
properties at the nucleus, excited state properties,
polarizabilities, to mention but a few. The introduction of such
parameters as described above, however, still renders the
description of the method ab-initio in the sense that all integrals
suggested by an exact implementation of the theory, be it
variational or perturbative, are still included.

The introduction of effective core potentials, as described
above to simplify the calculation, involves perhaps the most
rational deviation from an ab-initio framework. These potentials
are, in fact, often in the spirit of a true semi-empirical
approach. They not only can be designed to yield a more accurate
result than an exact application of the theory with a given basis
set (for example, they can often be designed to prevent significant
basis set borrowing and scalar relativistic effects), but they do,
in fact, significantly reduce the time required for a calculation,
making much large systems approachable.

Effective core potentials can be developed from an exact
application of perturbation theory. For this reason it is
sometimes controversial to call such methods "semi-empirical". In
electronic structure theory the term "semi-empirical" has been
pretty much reserved for modifications introduced into the
electronic Hamiltonian itself.

It is easiest to see this relationship by writing the
Hamiltonian in second quantized form:

$$H = \Sigma(i,j)\ h(i,j)\ i^{+}j\ \ + 1/2\ \Sigma(i,j,k,l)\ g\langle i,j//k,l\rangle\ i^{+}j^{+}l\ k$$

with $h(i,j)$ and $g\langle i,j//k,l\rangle$ parameters, and i^{+} and i the usual
fermion creation and annihilation operators respectively. If
$h(i,j)$ and $g\langle i,j//k,l\rangle$ are evaluated as the integrals

$$h(i,j) = \langle i/ -1/2\ \nabla^2\ - \Sigma\ (A)\ Z(A)/R(A)\ /j\rangle$$

$$g\langle i,j//k,l\rangle = \langle i,j//k,l\rangle = \langle i,j/k,l\rangle - \langle i,j/l,k\rangle$$

$$= \iint d\tau(1)d\tau(2)\phi_i(1)\phi_j(2)\ [1-P(12)]\ \phi_k(1)\phi_l(2)$$

then the method is generally classified as "ab-initio". It is important to stress here that all successful semi-empirical electronic structure Hamiltonians have this same form, and ab-initio and semi-empirical THEORIES are the same. The Table in the ab-initio section that describes theories, is also appropriate here: SCF, CI, MCSCF, CPMET, etc.

Semi-empirical theories generally parameterize h and g. Usually entire classes of integrals in g are neglected, and those that remain, along with h, are parameterized to compensate for this omission. The parameterization may be based on model ab-initio calculations, or directly on experiment, and often the proponents of these two different points of view don't talk to one another! A common goal, though, is that of reducing the N^4g integrals, where N is the size of the basis set, to one of N^3 or N^2.

For transition metal systems, it is probably fair to limit our attention to three methods, or rather three classes of methods since a number of variants exist for each. These are summarized in Table IV.

Table IV.   Some Semi-Empirical Methods Used
for d and f Electron Systems

Extended Huckel EH REX(Rel. EH)	h(i,i) related to ionization potentials, h(i,j) empirical param. set proportional to overlap (S). S included in secular Eq. $(\mathbf{H}-E(i)\mathbf{S})\mathbf{c}(i)=0$	pi and sigma elec. large systems not self-consist. relativistic vers.
Iterative EH IEH, SCCEH ITEREX	As above, iteration to charge consistency usually based on extrap. between I.P.'s of atomic ions.	As above. Relativistic version.
Fenske-Hall	As above: h(i,j) with terms proportional to S as well as kinetic energy terms.	Highly effective for complexes; iterative.
Neglect of Differential Overlap. CNDO, INDO NDDO	one center part of h(i,i) from IP's. Nuclear attract. included. CNDO contains all $\langle i,j/i,j\rangle$; INDO in add. all one-center $\langle i,j/k,l\rangle$; NDDO in add. all remaining two-center Coulomb type. h(i,j) usually proportional to S. Two-elec. ints often parameterized.	Large complexes CI, MBBT, MCSCF versions. Useful for geometry estimate and for electronic spectroscopy.

The Extended Huckel method, as it is known today, was first implemented by Lohr and Lipscomb in studies of boron hydrides (23) and later extended to organic systems by Hoffmann (24). It has been enormously fruitful, providing a framework for the analysis of orbital interactions which pervades modern organic chemistry. The application of the principles behind the Extended Huckel method, however, actually stem from application on transition metal complexes, initially by Wolfsberg and Helmholz, and then by Ballhausen, and Ballhausen and Gray (25). Iterative schemes for transition-metals compounds were devised by Ballhausen and Gray and by Zerner and Gouterman (26). More recently, primarily through the work of the school of Hoffmann, it has been adapted to solids and surfaces and again is providing insight that often cannot be obtained nearly so readily with other more elaborate techniques (27). Indeed, it could reasonably be argued that if one had to choose one technique to "do" chemistry (luckily one doesn't have to make this choice!) then EH, or its iterative versions, with its easy to interpret results, would probably be the best choice for a general overview of a wide variety of systems. It can provide invaluable guidance for questions that depend on orbital symmetry, orbital energies and orbital overlap. A good deal of understanding can be based only on these simple concepts! The Fenske-Hall method, which is in the spirit of the Extended Huckel Methods (28) but includes kinetic energy terms as well as much of the electrostatics, purports to greater accuracy, but qualitatively yields results similar to simple IEH.

For more quantitative, and more detailed treatments of, for example, multiplets and other aspects of spectroscopy, one must include most of the proper electrostatics. The methods described generically as Zero-Differential Ovelap (ZDO) types do this in a hierarchy of techniques; CNDO, Complete Neglect of Differential Overlap (29), INDO, Intermediate Neglect of Differential Overlap (30), and NDDO, Neglect of Differential Diatomic Overlap (31), all of which are generalizations of the earlier Pariser-Parr-Pople (PPP) method devised for pi electrons only (32). At present there is only one generally available program of this nature that includes transition metals, dubbed for un-acknowledged reasons ZINDO (33), although other interesting variations are being developed (34). An application of the INDO method is found in the chapter by LOEW, in which spin states of large porphinato Fe(III) complexes are examined, a task very difficult by ab-initio theories, as discussed in the chapter by DAVIDSON, and one still awaiting a theory of multiplet structure in DF methods.

Various CNDO and INDO schemes have also been proposed for transition-metal clusters and for chemisorption on them. Although there have been some successes, it remains true that both the level of theory being used and the parameterization are both very much experimental.

The semi-empirical methods continue to yield guidance to experimentalists and theorist alike, in a wide variety of fields. But in a sense the applications of these methods have a moving target. As ab-initio and density functional methods improve, becoming more rapid and applicable to a wider range of questions, those appropriate for semi-empirical studies become more and more complex. But there is also a fundamental question here. Any calculation that is performed on any system can be, and will be, performed at greater accuracy. At what level of theory and computation may the problem at hand be considered to be solved? Extended Huckel, the simplest of these theories, is apparently enough to describe a great deal of the actual chemistry of a complex; however, we still do not have ANY theory capable of reliably reproducing, for example, the ESR spectroscopy of these systems. Clearly the computational method of choice will depend on the problem at hand, and this will continue to be so.

Concluding Remarks

It has been impossible, in these few pages, to capture even a small fraction of the advances and of the excitement surrounding the treatment of d- and f- electron systems in contemporary quantum chemistry. The term quantum chemistry has been used advisedly, as opposed to either theoretical chemistry, with its connotations of dynamics and "things other than electronic structure calculations", or computational chemistry with its connotations of black-box programs and user-friendly graphics software. Very little has been done in either of these directions for the transition metals and even less for the lanthanides and actinides. Indeed, the excitement in the field is largely due to the fact that there is so much left to discover, and to the fact that the tools to allow pioneering discoveries are now just becoming available.

At the Toronto meeting we saw the first signs of progress towards both theoretical and computational chemistry for the d- and f- electron systems. Dynamics are definitely on the agenda for the next few years as are the development of convenient computer interfaces and the extraction of force fields and the like that will allow more and more experimentalists in inorganic chemistry to spend some of their time fruitfully at the computer keyboard.

Our friend Rip had better not go to sleep for another twenty years!

Literature Cited

1. Szabo, A.; Ostlund, N. S. Modern Quantum Chemistry: Introduction to Advanced Electronic Structure Theory MacMillan, New York, 1982.
2. Salahub, D. R. Adv. Chem. Phys. 1987, 69, 447.
3. e.g. Huzinaga, S.; Andzelm, J.; Klobukowski, M.; Radzio-Andzelm, E.; Sakai, Y.; Tatewaki, H. Gaussian Basis Sets for Molecular Calculations; Elsevier, Amsterdam, 1984 and references therein.
4. e.g. Huzinaga, S.; Klobukowski, M.; Sakai, Y. J. Phys. Chem. 1984, 88, 4880 and references therein.
5. Pyykko, P. Chem. Rev. 1988, 88, 1.
6. Hohenberg, P.; Kohn, W. Phys. Rev. 1964, 136, B864.
7. Kohn, W.; Sham, L. J. Phys. Rev. 1965, 140, A1133.
8. Lundqvist, S.; March, N. H., Eds. Theory of the Inhomogeneous Electron Gas; Plenum, New York, 1983.
9. Dahl, J. P.; Avery, J., Eds. Local Density Approximations in Quantum Chemistry and Solid State Physics; Plenum, New York, 1984.

10. Dreizler, R. M.; da Providencia, J., Eds. Density Functional Methods in Physics; Plenum, New York, 1984.
11. Slater, J. C. Adv. Quantum Chem. 1972, 6, 1; The Self-Consistent Field for Molecules and Solids; Vol. 4, McGraw-Hill, New York, 1974.
12. Vosko, S. H.; Wilk, L.; Nusair, M. Can. J. Phys. 1980, 58, 1200.
13. Perdew, J. P.; Zunger, A. Phys. Rev. B 1981, 23, 5048.
14. Ceperley, D. M.; Alder, B. J. Phys. Rev. Lett. 1980, 45, 566.
15. Langreth, D. C.; Mehl, M. J. Phys. Rev. B 1983, 28, 1809; erratum 1984, 29, 2310
16. Becke, A. D. J. Chem. Phys. 1986, 84, 4524.
17. DePristo, A. E.; Kress, J. D. J. Chem. Phys. 1987, 86, 1425.
18. Perdew, J. P.; Yue, W. Phys. Rev. B 1986, 33, 8800; Perdew, J. P. Phys. Rev. B 1986, 33, 8822.
19. Johnson, K. H. Adv. Quantum Chem. 1973, 7, 143.
20. Muller, J. E.; Jones, R. O.; Harris, J. J. Chem. Phys. 1983, 79, 1874 and references therein.
21. Delley, B.; Ellis, D. E.; Freeman, A. J.; Baerends, E. J.; Post, D. Phys. Rev. 1983, 27, 2132 and references therein.
22. Dunlap, B. I.; Connolly, J. W. D.; Sabin, J. R. J. Chem. Phys. 1979, 71, 3386, 4993.
23. Lohr, L. L.; Lipscomb, W. N. J. Chem. Phys. 1963, 38, 1604.
24. Hoffmann, R. J. Chem. Phys. 1964, 39, 1397: ibid. 40, 2047; 40, 2474; 40, 2480; 40, 2745.
25. Ballhausen, C. J; Gray, H. B. Inorg. Chem. 1962, 1, 111: ibid. Molecular Orbital Theory, Benjamin Press, New York, 1964.
26. Zerner, M.; Gouterman, M. Theoret. Chim. Acta. 1966, 4, 44.
27. Hoffmann, R. Rev. Mod. Phys. 1988, 60, 601.
28. Fenske, R.; Hall, M. B. Inorg. Chem. 1972, 11, 768.
29. Pople, J. A.; Santry, D. P.; Segal, G. A. J. Chem. Phys. 1965, 43, S129; Pople, J. A.; Segal, G. A. J. Chem. Phys. 1965, 43, S136; ibid. 1966, 44, 3289.
30. Pople, J. A; Beveridge, D. L.; Dobosh J. Chem. Phys. 1967, 47, 2026.
31. Pople, J. A.; Beveridge, D. L. Approximate Molecular Orbital Theory, McGraw Hill, New York, 1970.
32. Parr, R. G. Quantum Theory of Molecular Electronic Structure, Benjamin Press, New York, 1963.
33. Ridley, J. E.; Zerner, M. C. Theoret. Chim. Acta, 1973, 32, 111; Bacon, A.; Zerner, M. C. Theoret. Chim. Acta. 1979, 53, 21; Zerner, M. C.; Loew, G. H.; Kirchner, R. F.; Mueller-Westerhoff, U. T. J. Am. Chem. Soc. 1980, 102, 589.
34. Lipinski, J. Intern. J. Quantum Chem. 1988, 34, 423.

RECEIVED March 2, 1989

Chapter 2

Optimizations of the Geometry of Tetrahedral Ti(IV) Complexes

A Basis Set and Correlation Study of Tetrachlorotitanium and Trichloromethyltitanium

Rodney L. Williamson and Michael B. Hall

Department of Chemistry, Texas A&M University, College Station, TX 77843

Optimization of the geometry of $TiCl_4$ and $TiCl_3CH_3$ at the SCF level results in Ti-Cl bond lengths longer than the experimental values, even when d- and f-type polarization functions are added to the basis set. The bond lengths remain too long even as the Hartree-Fock limit is approached because the SCF level of theory over-estimates the noble-gas-like Cl···Cl repulsions, which hinder close Ti-Cl approach. The Ti-C-H angle of $TiCl_3CH_3$ is calculated to be close to tetrahedral geometry with little flattening of the hydrogen atoms, which apparently was observed in the electron diffraction. These same calculations do predict the anomalously low methyl-rocking frequency for the titanium complex in agreement with the experimental IR. This low methyl rocking frequency is due to stabilization of the Ti-C bond during the rocking motion by low lying empty d-orbitals on Ti. The large positive geminal hydrogen coupling constant observed in the NMR experiment is due primarily to the σ-donor and π-acceptor character of the $TiCl_3$ moiety and not to any flattening of the methyl group.

In the past few years, geometry optimizations of transition metal complexes have seen increased attention and improvement. Two minimal basis set studies (1-2) optimized metal-ligand bond lengths with errors of 0.05-0.25 Å. The errors decreased for most ligands in studies with moderate basis sets (3-8) with the exception of metal-cyclopentadienyl (Cp) bond lengths which showed errors ranging from 0.16-0.24 Å. Further work (9-12) has shown that the error in calculating the metal-Cp bond lengths with very large basis sets is primarily correlation error.

Two recent studies with 3-21G type basis sets have predicted the equilibrium geometries of some transition metal complexes with reasonable accuracy. In one of these studies (13), metal-carbonyl and metal-Cp distances were predicted to be 0.03 Å and 0.15 Å longer, respectively, than experimental values. It was shown that these long metal-Cp distances could be significantly reduced by including electron correlation. In the other study (14), the optimized metal-carbonyl distances averaged

0097–6156/89/0394–0017$06.00/0

0.11 Å longer than the experimental values. The optimized geometries for a large number of tetrahedral metal-halide complexes gave M-F bond lengths shorter than experimental values and M-Cl bond lengths longer than experimental values.

In order to show how different basis sets and electron correlation affect bond lengths and angles we report here the results of self-consistent-field (SCF) and generalized-valence-bond (GVB) geometry optimizations of two tetrahedral titanium(IV) chloride complexes, $TiCl_4$ and $TiCl_3CH_3$. For $TiCl_3CH_3$ we also did complete-active-space-self-consistent-field (CASSCF) geometry optimizations. We chose $TiCl_4$ because of its high symmetry, which greatly simplifies the calculation and interpretation, and we chose $TiCl_3CH_3$ as a second Ti(IV) complex because of its "unique" geometry. This geometry, which was reported by Berry et al. ([15]) from electron diffraction (ED), appeared to have a flattened methyl group due to three agostic hydrogens.

An agostic hydrogen is defined as a hydrogen atom covalently bonded simultaneously to both a carbon atom and to a transition metal atom ([16]). Agostic hydrogens have been reported for several titanium alkyl complexes ([16-19]) such as [$TiCl_3(Me_2PCH_2CH_2PMe_2)R$] (R=Et,Me). In these complexes, the geometry of the alkyl ligand containing the agostic hydrogen is distorted from the geometry it would have if it were bonded only to an organic substrate. In the methyl case this distortion is observed as a rocking of the methyl group such that one hydrogen atom moves toward the metal atom while the other two move away from the metal. The X-ray crystal structure of the above methyl complex ([17]) shows Ti-C-H angles of 70(2)°,105(4)°, and 117(3)°. A later neutron diffraction study ([19]) on the same complex showed Ti-C-H angles of 93.5(2)°, 118.4(2)° and 112.9(2)°. If the complex did not have an agostic hydrogen, one would expect the Ti-C-H angles to be close to 109.5°. An ab initio molecular orbital study ([20]) of the agostic hydrogen interaction in $Ti(CH_3)(PH_3)Cl_3$ reported direct interaction between the C-H σ-bond and an unoccupied Ti d-orbital. This type of interaction involving a single hydrogen bent towards the metal center is typical of most complexes with an agostic hydrogen.

However, Berry et al. ([15]) reported all three methyl hydrogens "flattened" toward the titanium atom, the first such complex with three agostic hydrogens. In addition to the ED results, they reported a large difference between the methyl rocking vibrational frequencies of $TiCl_3CH_3$ and $GeCl_3CH_3$ and explained this difference as resulting from the flattening of the methyl hydrogens. They also did ^{13}C and ^{1}H NMR studies and found the H,H coupling constant to have a large positive value, which they suggested was further evidence of hydrogen flattening.

We optimized the geometry of $TiCl_3CH_3$ to determine if the calculations would predict the symmetrical flattening of the methyl hydrogens. The model complexes TiH_3CH_3 and GeH_3CH_3 were used to calculate vibrational frequencies to determine if the calculation would predict the large difference between the rocking frequencies of the Ge and Ti complexes. Some of these results have been reported in a preliminary communication ([21]). And finally, we estimated the H,H coupling constant for the 1s orbitals of two methyl hydrogens on $TiCl_3CH_3$, CH_4, and CH_3Cl using the method of Pople and Santry ([22]). Although this method is known to have serious problems in predicting the sign and absolute magnitude of the H,H coupling constant ([23]), it does correctly predict the direction and size of the change of the coupling constant resulting from changes in the nature of the methyl substituent or changes in geometry of the compound.

Computational Details

We optimized the geometries of the Ti complexes using three different titanium basis sets. The first titanium basis set (I) is a (432-421-31) used in previous calculations (13), the second titanium basis set (II) is a Huzinaga (24) (5333-53-5) modified to a (533211-5211-3111) by splitting off the most diffuse s, p, and d functions and adding s, p, and d functions with exponents 1/3 the values of the split-off functions, and the third titanium basis set (III) is the titanium basis set II with an additional f-type polarization function (ξ=0.55) (25). For the main group atoms, basis set I is Cl(3321-321), C(321-21), H(21), basis set II is Cl(5321-521), C(421-31), H(31), basis set III is Cl(531111-4211), C(721-41), H(31), basis set IV is Cl(5321-521-1), C(421-31-1), H(31-1), basis V is Cl(533-53), basis VI is Cl(53111-5111), basis VII is Cl(533-53-1), basis VIII is Cl(533-53-11), and basis IX is Cl(5321-521-11). Basis sets I, II, and VI-IX are Huzinaga (24) basis sets which were all modified (expect for basis set V) by singly or doubly splitting the most diffuse s and p functions. Basis sets IV, and VII-IX were further modified by adding polarization functions. Basis set III is an unmodified Dunning-Hay-Huzinaga (26) basis set. The basis sets used for the geometry optimizations and force constant calculations of TiH_3CH_3 and GeH_3CH_3 are 3-21G type basis sets used in previous geometry optimization (13). The H,H coupling constant calculations were done using fully contracted titanium and chlorine basis sets I and carbon and hydrogen basis sets II.

The geometry of $TiCl_3CH_3$ and $TiCl_4$ were optimized in staggered C_{3v} and T_d symmetry, respectively. The GVB calculations (27) involve perfect-pairing for all seven sigma bonds for $TiCl_3CH_3$ and for the four Ti-Cl sigma bonds for $TiCl_4$. The first of two CASSCF calculations on $TiCl_3CH_3$ contains eight electrons in the eight orbitals (8/8) made up of four σ-bonding and four σ-antibonding Ti-C and C-H orbitals. The second contains eight electrons in eleven orbitals (8/11) made up of the eight orbitals in the 8/8 calculation plus two Ti-C π-bonding orbitals and one additional Ti-C σ-bonding orbital.

All of the calculations were done with GAMESS (Generalized Atomic and Molecular Electronic Structure Systems) except for the $TiCl_4$ calculations with f-type polarization functions which were done with QUEST (QUantum Electronic STructure). These programs were run on a CRAY X-MP at CRAY Research in Mendota Heights, an FPS-264 at the Cornell National Supercomputer Facility, an IBM 3090/200 at Texas A&M University, and the Department of Chemistry's VAX 11/780 and FPS-164.

Results and Discussion

$TiCl_4$ Geometry Optimization. The optimized Ti-Cl bond lengths (see Table I) for $TiCl_4$ with a variety of basis sets are all longer than the experimental value and differ from that by 0.016 to 0.051 Å. Splitting the Cl(533-53) basis set, which allows the orbitals freedom to expand or contract, only decreased the Ti-Cl bond length an average 0.005 Å. However, when we add d-type polarization functions to the chlorine basis set the Ti-Cl bond distance decreases an average 0.026 Å.

The effect of adding polarization functions is also seen in deformation density maps of $TiCl_4$. The deformation density of $TiCl_4$ without d-functions on Cl (Figure 1a) shows a buildup of density in the Ti-Cl bonding region and the Cl lone-pair regions. When a polarization function is added to Cl (Figure 1b) density in the bonding region increases and density in the chlorine lone-pair region decreases. The

a

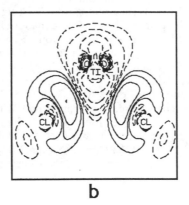

b

Figure 1. Electron deformation density plots of $TiCl_4$ in Cl-Ti-Cl plane:
a) deformation density without d-functions b) deformation density with
d-functions. Contours are geometric beginning at ±0.001 e^-au^{-3} and
incremented by doubling the previous contour value.

Table I. Ti-Cl Bond Distances for $TiCl_4$

	Method	Chlorine Basis	Titanium Basis Sets		
			I	II	III
1.	SCF	Cl(3321-321) (I)	2.197	2.209	
2.	SCF	Cl(533-53) (V)	2.203	2.221	
3.	SCF	Cl(5321-521) (II)	2.201	2.213	2.196
4.	SCF	Cl(53111-5111) (VI)	2.200	2.214	
5.	SCF	Cl(531111-4211) (III)	2.202	2.219	
6.	SCF	Cl(533-53-1) (VII)		2.186	
7.	SCF	Cl(5321-521-1) (IV)		2.187	2.181
8.	SCF	Cl(533-53-11) (VIII)		2.189	
9.	SCF	Cl(5321-521-11) (IX)		2.191	
10.	GVB	Cl(5321-521) (II)		2.234	
	Experimental		2.170(2)[a]		

[a]Reference 28, the error is reported at a confidence level of 2.5 σ.

origin of the shift in deformation density from the lone-pair region on Cl to the Ti-Cl bonding region is also seen in orbital plots (Figures 2a-d) of both the Ti-Cl σ-bonding and π-bonding molecular orbitals (MO). The value in the Ti-Cl bonding region of the a_1 MO with d-functions (Figure 2b) is greater than the value in the bonding region of the a_1 MO without d-functions (Figure 2a). The Cl t_2 π-orbitals also show a similar increase in the Ti-Cl bonding region (Figures 2e,f). Molecular orbital plots of the other orbitals (Figures 2c-d,g-j) do not show any difference between the plots with d-functions and the plots without d-functions.

As we improve the $TiCl_4$ wavefunction by adding f-type polarization functions to the titanium basis set the Ti-Cl bond distance shortens further. When a chlorine basis set without d-functions is used, the addition of an f-function on titanium shortens the Ti-Cl bond 0.017 Å. However, when a chlorine basis set with d-functions is used, the Ti-Cl bond only shortens 0.006 Å to 2.181 Å. Although this Ti-Cl distance is the shortest of all the optimized geometry calculations, it is still 0.011 Å longer than the experiment.

As the wavefunction approaches the Hartree-Fock limit one would expect the Ti-Cl bond distance to be shorter than the experiment because of the lack of bond-pair correlation. The bond-pair correlation added by the GVB wavefunction lengthened the Ti-Cl bond 0.021 Å, because the GVB wavefunction adds only limited left-right correlation and none of the dynamical correlation. For most A-B bonds, the calculated bond lengths at the SCF level are too short, and the correlation added by a GVB calculation accounts for a major portion of the non-dynamical correlation error in the SCF wavefunction. But for Ti-Cl bonds, both the SCF and GVB calculations predict too long a bond distance because they do not include necessary dynamical atomic correlation of the Cl atoms.

In $TiCl_4$, each Cl satisfies the octet rule by sharing a pair of electrons with Ti. Thus, to each other, the Cl atoms appear as noble gas atoms. As is well known (29-30), the Hartree-Fock wavefunction does not adequately describe the attraction of two noble-gas-type atoms. SCF level calculations (29) of the He_2, Ne_2, and Ar_2 potentials show the atoms in these dimers to be too strongly repulsive at close

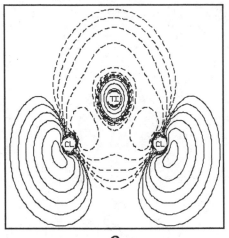

a

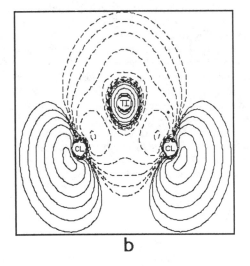

b

Figure 2. Wavefunction plots of $TiCl_4$ molecular orbitals in the Cl–Ti plane: a, a_1 without d functions; b, a_1 with d functions. Contours are geometric beginning at ± 0.001 $e^- au^{-3}$ and incremented by doubling the previous contour value. *Continued on next page.*

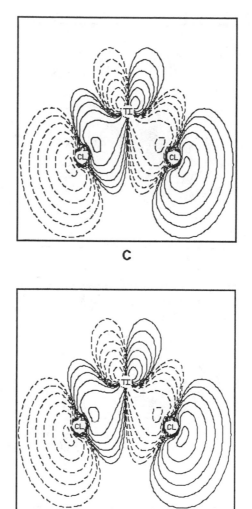

Figure 2. *Continued.* Wavefunction plots of TiCl$_4$ molecular orbitals in the Cl–Ti plane: c, t$_2$ σ without d functions; d, t$_2$ σ with d functions. Contours are geometric beginning at ± 0.001 e$^-$au^{-3} and incremented by doubling the previous contour value. *Continued on next page.*

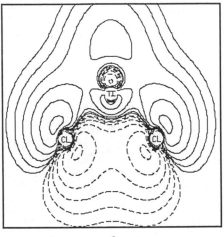

e

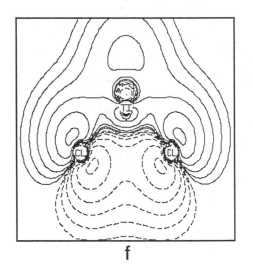

f

Figure 2. *Continued.* Wavefunction plots of TiCl$_4$ molecular orbitals in the Cl–Ti plane: e, t$_2$ π without d functions; f, t$_2$ π with d functions. Contours are geometric beginning at ±0.001 e$^-$au^{-3} and incremented by doubling the previous contour value. *Continued on next page.*

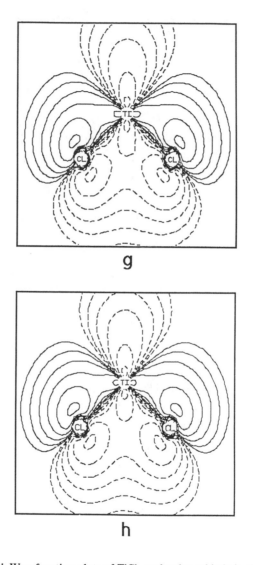

g

h

Figure 2. *Continued.* Wavefunction plots of $TiCl_4$ molecular orbitals in the Cl–Ti plane: g, e without d functions; h, e with d functions. Contours are geometric beginning at ±0.001 e^-au^{-3} and incremented by doubling the previous contour value. *Continued on next page.*

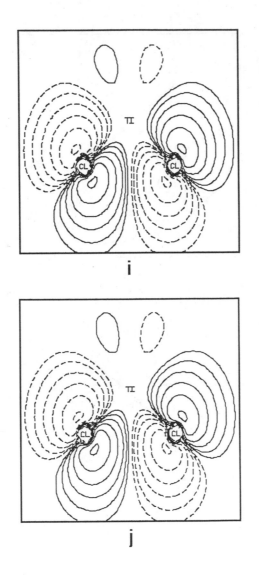

Figure 2. *Continued.* Wavefunction plots of $TiCl_4$ molecular orbitals in the Cl–Ti plane: i, t_1 without d functions; j, t_1 with d functions. Contours are geometric beginning at ± 0.001 $e^- au^{-3}$ and incremented by doubling the previous contour value.

distances and lacking a van der Waals minimum at long distances. Because the Cl atoms are noble-gas-like, the Cl··Cl interactions in $TiCl_4$ sterically hinder the Cl atoms from bonding close to the Ti atom. The accurate determination of the Ti-Cl bond distance will require a large configuration-interaction calculation.

$TiCl_3CH_3$ Geometry Optimizations. The results of complete SCF geometry optimizations on $TiCl_3CH_3$ using basis sets I-IV are shown in Table II. As the basis sets are improved, the Ti-C and C-H bond distances show the smallest changes of

Table II. SCF Geometries for Cl_3TiCH_3

	Basis Set		Bond Distances (Å)			Angles (°)	
	M	Ligand	Ti-Cl	Ti-C	C-H	Ti-C-H	Cl-Ti-C
1.	I	I	2.229	2.011	1.087	107.9	102.4
2.	II	II	2.251	2.013	1.092	108.3	103.4
3.	II	III	2.258	2.012	1.095	108.2	103.2
4.	II	IV	2.219	2.020	1.091	108.2	103.7
	ED Experiment[a]		2.185	2.042	1.158	101.0	105.2

[a] Reference 15

0.009 Å and 0.008 Å , respectively, and the Ti-Cl bond distance shows the largest change of 0.039 Å. The Ti-C-H and Cl-Ti-C angles vary only 0.4° and 1.3°, respectively. The geometry optimizations with Ti and ligand basis sets II give Ti-C and C-H bond distances 0.029 Å and 0.066 Å shorter and a Ti-Cl bond length 0.066 Å longer than the ED results. When polarization functions are added to the Cl, C and H functions (basis set IV), the differences between the ED and the calculated Ti-Cl and Ti-C bond lengths decrease to 0.034 Å and 0.022 Å, respectively, and the analogous difference for the C-H bond distance increases to 0.067 Å. For basis set II, the Cl-Ti-C and Ti-C-H angles are 1.8° smaller and 7.2° larger, respectively, than the ED angle. When polarization functions are added to all atoms but Ti, the differences decrease to 1.5° and 7.1°, respectively. Although most of these optimized geometric parameters are in good agreement (±0.03 Å, ±1.5°) with the ED results, the optimized Ti-C-H angle and C-H distance do not agree with the ED result.

In the GVB geometry (see Table III), the already long Ti-Cl distance increases 0.035 Å, the Ti-C bond lengthens 0.168 Å, and the Ti-C-H and Cl-Ti-C angles both decrease 4.4° and 3.9°, respectively. Previous work by Ditchfield and Seidman (31)

Table III. GVB Geometries for Cl_3TiCH_3

	Basis Set		Bond Distances (Å)			Angles (°)	
	M	Ligand	Ti-Cl	Ti-C	C-H	Ti-C-H	Cl-Ti-C
1.	III	III	2.296	2.180	1.108	103.8	99.3
2.	III	III	ED	ED	1.111	106.5	ED
3.	III	III	2.295	ED	1.109	105.5	100.5
	ED Experiment[a]		2.185	2.042	1.158	101.0	105.2

[a] Reference 15

has shown that the addition of electron correlation to the SCF wavefunction in AH_n (A=C,N,O,F) molecules usually has a small effect on bond lengths and angles. Although an increase in the bond length is not surprising for a GVB calculation, the magnitude of the increase in the Ti-C bond length is surprising. Because the GVB wavefunction overemphasizes dissociation, the Cl_3Ti and CH_3 moieties have too much radical-like $Cl_3Ti\cdot$ and $\cdot CH_3$ character. As the fragments become more radical-like they flatten toward their equilibrium planar geometry. Even with the long Ti-C bond distance, the Ti-C-H angle is still 2.8° larger than the ED angle.

If there were only steric repulsions between the hydrogens and titanium, the bond angles would increase as the fragments are forced back together. On the other hand, if there were attractive agostic hydrogen interactions, the hydrogens would bend towards the titanium atom as the fragments are brought closer, which would decrease the angle closer to the ED value. When we shorten the Ti-C bond distance to the ED bond length and optimize the remaining geometric parameters at the GVB level, the Ti-C-H and Cl-Ti-C angles increase 1.7° and 1.3°, respectively. When, in addition, we shorten the Ti-Cl bond length and the increase the Cl-Ti-C angle to the ED values, the Ti-C-H bond angle increases an additional 1.0°, to a value 5.5° larger than the reported ED angle. Thus, the bending of the hydrogen atoms away from the titanium atom as the Ti-C distance is shortened shows that the titanium-hydrogen interactions are repulsive and not attractive.

We improved the calculation further by using a CASSCF wavefunction with an active space of 8 electrons in 8 orbitals (8/8). The results of the geometry optimization (see Table IV) give a Ti-C bond length of 2.119 Å and a Ti-C-H angle of 105.4° (the Ti-Cl bond distance and Cl-Ti-C angle were frozen at the ED values). In

Table IV. CASSCF Geometries for Cl_3TiCH_3

Basis Set			Bond Distances (Å)			Angles (°)	
M	Ligand	Method	Ti-Cl	Ti-C	C-H	Ti-C-H	Cl-Ti-C
II	II	SCF	2.251	2.013	1.092	108.3	103.4
II	II	8/8 CAS	ED	2.119	1.112	105.4	ED
II	II	8/11 CAS	ED	2.106	1.112	106.2	ED
	ED Experiment[a]		2.185	2.042	1.158	101.0	105.2

[a] Reference 15

contrast to the SCF geometries, but like the GVB geometry, the CASSCF Ti-C bond distance is longer than the ED value and the Ti-C-H angle shows a slightly larger decrease from the tetrahedral angle. However, when we increase the size of the active space to 11 orbitals (8/11), with additional Ti-C σ and π bonding orbitals, the Ti-C bond shortens 0.013 Å and the Ti-C-H angle increases 0.8°. This CASSCF includes those virtual Ti $d\pi$-orbitals which are involved in the qualitative description of the agostic hydrogen interaction. The increase one observes in the Ti-C-H angle as the Ti-C bond distance shortens, which one also observed in the GVB geometry optimizations, shows that the Ti-C-H angle is very sensitive to changes in the Ti-C bond distance. If the wavefunction were improved by including even more correlation, one would expect the Ti-C bond distance to shorten further toward the ED value and as the Ti-C bond shortens, one would also expect the Ti-C-H angle to increase towards a tetrahedral angle.

Vibrational Frequencies Calculations. Berry *et al.* (15) reported that the methyl rocking frequency of $TiCl_3CH_3$ is much lower than the analogous frequency of $GeCl_3CH_3$. They presumed this anomalously low frequency of $TiCl_3CH_3$ to result from the "flattening" of the methyl hydrogens. We calculated the vibration frequencies of the hypothetical model complexes TiH_3CH_3 and GeH_3CH_3 after first optimizing their geometries. The results of the geometry optimizations (see Table V)

Table V. SCF Geometries for MH_3CH_3 (M = Ti, Ge)

	Metal	Basis	Bond Distances (Å)			Angles (°)	
			M-H	M-C	C-H	M-C-H	H-M-C
1.	Ti	I	1.710	2.035	1.092	109.9	108.3
2.	Ge	I	1.533	1.959	1.086	110.7	110.3

show the expected differences in geometry between the Ti and Ge complexes. The vibration frequencies were calculated by taking finite differences of energy gradients beginning at these optimized geometries.

The calculated frequencies, when compared with the experimental frequencies (see Table VI), show errors expected of calculations at the SCF level. However,

Table VI. MX_3CH_3 Vibrational Frequencies in cm^{-1}

	Mode	Calcul. (X=H)			Exper. (X=Cl)		
		Ge	Ti	Diff.	Ge[a]	Ti[a]	Diff
e	CH_3 rock	978	577	401	825	580	245
a_1	CH_3 deform.	1468	1360	108	1246	1052	194
e	CH_3 deform.	1611	1564	47	1403	1375	28
a	CH_3 stretch	3190	3122	68	2940	2894	46
e	CH_3 stretch	3270	3208	62	3019	2980	39

[a] Reference 15

when one compares the SCF differences between the Ti and Ge model complexes with the experimental differences between the Ti and Ge chloride complexes, the differences are in good agreement. The large difference in the methyl rocking frequency observed in the experiment is predicted by the SCF calculation, the same SCF calculation that predicts no flattening of the methyl hydrogen atoms. If the large difference between the Ti and Ge methyl rocking frequency is not the result of hydrogen flattening, what then is the cause of the observed difference in the methyl rocking frequencies?

To answer the above question we did point by point SCF calculations of the model complexes as the methyl group rocks. A plot of the resulting energies for each point calculated is shown in Figure 3 and, as expected, the titanium complex potential curve is flatter than the germanium complex potential curve. When we rock the methyl group on the model complexes +45°, the titanium complex is destabilized by

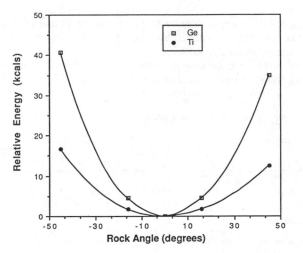

Figure 3. Relative energy of TiH$_3$CH$_3$ and GeH$_3$CH$_3$ for the methyl rock

12.9 kcal mol^{-1} and the germanium complex is destabilized by 34.9 kcal mol^{-1}. Eisenstein and Jean (32) using extended Hückel calculations on the titanium model complex also found the rocking motion to be weakly destabilizing.

At the equilibrium geometry the deformation density plot (Figure 4a) shows a large buildup of density aligned symmetrically along the Ti-C bond axis. As the methyl group rocks (Figures 4b and 4c) the density buildup is no longer symmetric about the Ti-C bond axis, but rather shifted to one side of the axis. Although the methyl group in Figures 4b and 4c has been tilted 45° from equilibrium the maximum density between the Ti and C atoms has only tilted an average 22°. As was seen in the titanium complex, the germanium complex at equilibrium geometry (Figure 5a) shows a similar symmetric buildup of density along the Ge-C bond axis. However, as the methyl group rocks 45° (Figures 5b and 5c) the density shifts to one side of the axis with an average tilt of 35°, 13° larger than the tilt in the Ti complex. It is also interesting to note that the density in the C-H bond region decreases slightly for both the Ge and Ti complexes as the hydrogen is tilted toward the metal. The density between the metal and the other hydrogen in the plot remains unchanged for the germanium complex but increases slightly for the titanium complex as the methyl group is rocked away from equilibrium.

Goddard, Hoffmann, and Jemmis (33) in a study of alkyl tantalum complexes have suggested that rehybridization of the metal-carbon bond with empty d-orbitals on the metal stabilizes the rocked methyl group. Titanium, like tantalum, has low-lying empty d-orbitals that allow facile rehybridization of the metal-carbon bond. Germanium, on the other hand, has used its s- and p-orbitals and the empty d-orbitals are at very high energy, therefore no empty orbitals are available for rehybridization. The difference in the ease of rehybridization explains the much lower methyl rocking frequency for the titanium complex than for the germanium complex. The deformation density plots do not contradict this explanation. Furthermore, this explanation is supported by the total metal-carbon overlap populations which actually increase for the Ti-methyl rock but decrease for the Ge-methyl rock. An increase in the titanium d_{xz}-carbon and p_x-carbon more than compensates for the decrease in the d_{z^2}-carbon overlap populations as the methyl group is rocked. The germanium complex however, shows no change in any germanium d-carbon overlap population and the very slight increase in the p_x-carbon overlap population does not compensate for the loss in the p_z-carbon overlap population as the methyl group is rocked.

Geminal H,H Coupling Constant. Using the method of Pople and Santry (22) we calculated the difference between the H,H coupling constant of ClCH$_3$ and HCH$_3$ to be 1.6 Hz, which is in good agreement with the experimentally determined value (34). The calculated H,H coupling constant for TiCl$_3$CH$_3$ at the optimized geometry is 8.0 Hz larger than the coupling constant for HCH$_3$. Although this change is smaller than the 23.7 Hz change observed experimentally (15,34), this simple model does predict a change which is both large and positive. When the optimized geometry of TiCl$_3$CH$_3$ is flattened to the ED geometry the H,H coupling constant increases only 3.6 Hz.

By comparing the increase in the coupling constant of XCH$_3$ as X changes from H to TiCl$_3$ with the increase in the coupling constant as the hydrogens in TiCl$_3$CH$_3$ are flattened, we see that the flattening of the methyl hydrogens has a relatively small effect on the coupling constant when compared with the large effect of the TiCl$_3$ substituent. However, Green and Payne (35) using extended Hückel calculations observed the effect of the substituent on the coupling constant to be only half of the

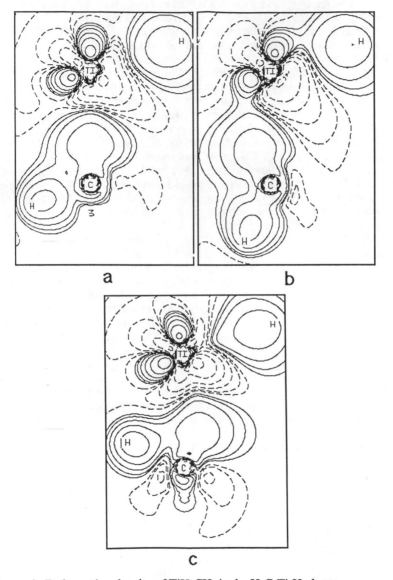

Figure 4. Deformation density of TiH_3CH_3 in the H-C-Ti-H plane:
a) equilibrium geometry b) methyl rocked 45° counterclockwise c) methyl
rocked 45° clockwise. Contours are geometric beginning at ±0.001 e‾au‾³ and
incremented by doubling the previous contour value.

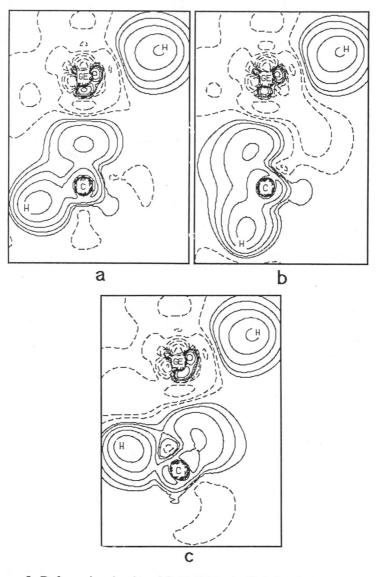

Figure 5. Deformation density of GeH_3CH_3 in the H-C-Ge-H plane:
a) equilibrium geometry b) methyl rocked 45° counterclockwise c) methyl
rocked 45° clockwise. Contours are geometric beginning at ±0.001 e⁻au⁻³ and
incremented by doubling the previous contour value.

effect of the geometry change on the coupling constant, which is opposite to what we predict using *ab initio* calculations. We believe that *ab initio* calculations describe the true nature of this titanium complex more accurately than the semi-empirical extended Hückel calculations. It is the strong σ-donor and weak π-acceptor character of the $TiCl_3$ substituent which results in the observed positive shift in the coupling constant. Thus, one can explain the coupling constant without postulating any geometric distortions.

Conclusions

Optimization of the geometry of $TiCl_4$ at the SCF level results in a Ti-Cl bond length which is longer than the experiment, even when d- and f-type polarization functions are added to the basis set. For covalently bonded systems one expects a wavefunction at the Hartree-Fock limit to give bond lengths shorter than the experiment if they are not sterically crowded. Because the Hartree-Fock wavefunction overestimates the Cl···Cl repulsions, the Ti-Cl bond distances remain long, even in large basis sets.

Geometry optimizations of $TiCl_3CH_3$ also show long Ti-Cl bond distances. The Ti-C-H angle is close to tetrahedral geometry with little, if any, flattening of the hydrogen atoms. Because of the known problem that electron diffraction has with determining the positions of hydrogen atoms, the large difference between the optimized value and the electron diffraction value for the Ti-C-H angle is not surprising. Our calculations correctly predict the anomalously low methyl-rocking frequency for the titanium complex without hydrogen flattening. This low methyl-rocking frequency is due to stabilization of the rocking motion by low-lying empty d-orbitals on Ti. The large positive geminal hydrogen coupling constant is primarily due to the σ-donor and π-acceptor character of the $TiCl_3$ moiety and not the flattening of the methyl hydrogens.

We do not believe agostic interactions are strong enough to symmetrically flatten a methyl group. The key to an agostic interaction is the low-energy rocking distortion. Thus, we would expect all agostic methyls to be rocked with the H-C-H angles close to those in similar organic compounds.

Acknowledgments

The authors would like to thank the National Science Foundation (Grant CHE 86-19420) and CRAY Research for support of the work. This research was conducted on an IBM-3090 and FPS-264 at the Cornell National Supercomputer Facility, a resource for the Center for Theory and Simulation in Science and Engineering at Cornell University, which is funded in part by the National Science Foundation, New York State, and the IBM Corporation, on a CRAY X-MP at CRAY Research, Mendota Heights, Minnesota, and on a VAX 11/780 and FPS-164 at the Chemistry Department of Texas A&M University. We would also like to thank Dr. Martyn F. Guest at SERC, Daresbury Laboratory, Warrington, UK. for making his version of GAMESS available and Drs. Thomas Dunning and Ron Shepard at Argonne National Laboratory for making QUEST available.

Literature Cited

1. Pietro, W.J.; Hehre, W.J. J. Comp. Chem. 1983, 4, 241.

2. Seijo, L.; Barandiaran, Z.; Klobukowski, M.; Huzinaga, S. Chem. Phys. Lett. 1985, 117, 151.
3. Faegri Jr., K. ; Almolf, J. Chem. Phys. Lett. 1984, 107, 121;
4. Luthi, H.P.; Ammeter, J.H.; Almolf, J.; Faegri Jr., K. J. Chem. Phys. 1982, 77, 2002.
5. Pitzer, R.M.; Goddard, J.D.; Schaefer, H.F. J. Am. Chem. Soc. 1981, 103, 5681.
6. Yates, J.H.; Pitzer, R.M. J. Chem. Phys. 1979, 70, 4049.
7. Kataura, K.; Sakaki, S.; Morokuma, K. Inorg. Chem. 1981, 20, 2292.
8. Spangler, D.; Wendoloski, J.J.; Dupuis, M.; Chen, M.M.L.; Schaefer, H.F. J. Am. Chem. Soc. 1981, 103, 3985.
9. Guest, M.F.; Hillier, I.H.; Vincent, M.; Rosi, M. J. Chem. Soc. Chem. Commun. 1986, 438.
10. Luthi, H.P.; Siegbahn, P.E.M.; Almolf, J. J. Phys. Chem. 1985, 89, 2156.
11. Luthi, H.P.; Siegbahn, P.E.M.; Almolf, J.; Faegri Jr., K.; Heiberg, A. Chem. Phys. Lett. 1984, 111, 1.
12. Almolf, J.; Faegri Jr., K.; Schilling, B.E.R.; Luthi, J.P. Chem. Phys. Lett. 1984, 106, 266.
13. Williamson, R.L.; Hall, M.B. Int. J. Quantum Chem., Quantum Chem. Symp. 21, 1987, 503.
14. Dobbs, K.D.; Hehre, W.J. J. Comput. Chem. 1987, 8, 861.
15. Berry, A.; Dawoodi, Z.; Derome, A. E.; Dickinson, J. M.; Downs, A. J.; Green, J. C.; Green, M. L. H.; Hare, P. M.; Payne, M. P.; Rankin, W. H.; Robertson, H. E. J. Chem. Soc. Chem. Commun. 1986, 519.
16. Brookhart, M.; Green, M. L. H. J. Organomet. Chem. 1983, 250, 395.
17. Dawoodi, Z.; Green, M. L. H.; Mtetwa, V. S. B.; Prout, K. J. Chem. Soc. Chem. Commun. 1982, 1410.
18. Dawoodi, Z.; Green, M. L. H.; Mtetwa, V. S. B.; Prout, K. J. Chem. Soc. Chem. Commun. 1982, 802.
19. Dawoodi, Z.; Green, M. L. H.; Mtetwa, V. S. B.; Prout, K.; Schultz, A. J.; Williams, J. M.; Koetzle, T. F. J. Chem. Soc. Dalton Trans. 1986, 1629.
20. Obara, S.; Koga, N. Morokuma, K. J. Organomet. Chem. 1984, 270, C33.
21. Williamson, R.L.; Hall, M.B. J. Am. Chem. Soc. 1988, 110, 4428.
22. Pople, J.A.; Santry, D.P. Mol. Phys. 1963, 8, 1.
23. Maciel, G.E.; McIver Jr., J.W. ; Ostlund, N.S.; Pople, J.A. J. Am. Chem. Soc. 1970, 92, 4151.
24. Gaussian Basis Sets for Molecular Calculations; Huzinaga, S., Ed.; Amsterdam: Elsevier, 1984
25. Bauschlicher Jr., C.W.; Seigbahn, P.E.M. Chem. Phys. Lett. 1984, 104, 331.
26. Dunning Jr., T.H.; Hay, P.J. In Methods of Electronic Structure Theory; Schaefer, H.F., Ed.; Plenum Press: New York, 1977; Vol. 4 Chapter 1.
27. Bobrowicz, F.W.; Goddard, W.A. In Methods of Electronic Structure Theory; Schaefer, H.F., Ed.; Plenum Press: New York, 1977; Vol. 4 Chapter 4.
28. Morino, Y.; Uehara, H. J. Chem. Phys. 1966, 45, 4543.
29. Gilbert, T.L.; Wahl, A.C. J. Chem. Phys. 1967, 47, 3425.
30. Schaefer, H.F.; McLaughlin, D.R.; Harris, F.E.; Alder, B.J. Phys. Rev. Lett. 1970, 25, 988.
31. Ditchfield, R.; Seidman, K. Chem. Phys. Lett. 1978, 54, 57.
32. Eisenstein, O.; Jean, Y. J. Am. Chem. Soc. 1985, 107, 1177.

33. Goddard, R.J.; Hoffmann, R.; Jemmis, E.D. J. Am. Chem. Soc. 1980, 102, 7667.
34. Pople, J.A.; Bothner-By, A.A. J. Chem. Phys. 1965, 42, 1339.
35. Green, J. C.; Payne, M.P. Magn. Res. Chem. 1987, 25, 544.

RECEIVED December 9, 1988

Chapter 3

LCGTO–Xα Study on the Agostic Interaction in Cl$_3$TiCH$_3$

N. Rösch and P. Knappe

Lehrstuhl für Theoretische Chemie, Technische Universität München, 8046 Garching, Federal Republic of Germany

The geometry and selected vibrational frequencies of Cl$_3$TiCH$_3$ have been calculated using the LCGTO–Xα method. The methyl group is found essentially undistorted, thus providing no indication for an agostic interaction.

We would like to comment on the findings of Williamson and Hall on tetrahedral Ti(IV) complexes ([1,2]) by reporting some results from our LCGTO–Xα calculations on the geometry of Cl$_3$TiCH$_3$ and of related titanium complexes (Knappe, P.; Rösch, N. J. Organometal. Chem., in press). Our study was prompted by the experimental finding ([3]) in Cl$_3$TiCH$_3$ of an unusual agostic interaction at an α carbon atom and by results of an Extended Hückel analysis ([4]) of the agostic interaction in tetrahedral and octahedral titanium complexes.

For TiCl$_4$ (in T$_d$ symmetry) perfect agreement was obtained between calculated and experimental values for the titanium chlorine bond length (2.17 Å ([5])). The geometry of Cl$_3$TiCH$_3$ was optimized in staggered configuration while retaining the threefold axis. As in previous LCGTO–Xα studies ([6]) we find satisfactory agreement between calculated and experimental values for the various bond lengths with the well–known underestimation of the transition metal to carbon bond characteristic for the local density approach ([7]). Essentially no distortion of the methyl group could be detected, the carbon to hydrogen bond length exhibiting a value falling in the usual range (Table I).

Table I. Comparison of experimental and calculated geometries for Cl$_3$TiCH$_3$

Method	Bond distances (Å)			Angles (degrees)	
	Ti–Cl	Ti–C	C–H	C–Ti–Cl	Ti–C–H
Exp[a]	2.19	2.04	1.16	105.2	101.0
Xα[b]	2.17	2.00	1.11	102.0	106.9
GVB[c]	2.30	2.04[d]	1.11	100.5	105.5
CAS[c]	2.19[d]	2.11	1.11	105.2[d]	106.2

a) Ref. ([3]);   b) LCGTO–Xα, this work;   c) Ref. ([1,2]);   d) Fixed value.

NOTE: LCGTO is an abbreviation for linear combination of Gaussian-type orbitals.

0097–6156/89/0394–0037$06.00/0

Table II. Selected vibrational frequencies (in cm^{-1}) of the methyl group
in Cl_3TiCH_3

Characteristic		Experimental[a]	Calculated	
			$X\alpha$[b]	Ab initio[c]
$\nu(C-H)_s$	(stretching)	2894	2855	3122
$\delta(C-H)_s$	(bending)	1052	990	1360
$\rho(C-H)$	(rocking)	580	625	577

[a] Ref. ([3]); [b] LCGTO–Xα, this work; [c] Ref. ([1]), results for H_3TiCH_3.

To corroborate the conclusions on the electronic structure of Cl_3TiCH_3 we have calculated selected vibrational frequencies whereby no coupling between the various motions was taken into account. The results for the symmetric C–H stretching mode, the symmetric bending mode and the rocking mode of the methyl group are presented in Table II together with values from experiment and from the *ab initio* calculation ([1,2]). Again, the agreement between experiment and available theoretical results is very satisfactory.

Summarizing we conclude – as has been done previously ([6,7]) – that LCGTO–Xα results on a transition metal compound are of a quality that is at least comparable to that of a Hartree–Fock calculation. For the title compound the agreement between the two rather different theoretical procedures (*ab initio* and local density method) is quite astonishing. It is therefore all the more disconcerting that both methods were unable to confirm the unusual geometry of the methyl ligand in Cl_3TiCH_3 as deduced from an electron diffraction experiment and attributed to agostic interaction ([3]).

<u>Literature Cited</u>

1. Williamson, R.L.; Hall, M.B. <u>J. Am. Chem. Soc.</u> 1988, <u>110</u>, 4429.
2. Williamson, R.L.; Hall, M.B. in <u>Computational Chemistry: The Challenge of d and f Electrons</u>; Salahub, D.R.; Zerner, M.C., Eds.; ACS Symposium Series No. 394; American Chemical Society: Washington, DC, 1989; p.
3. Berry, A.; Dawoodi, Z.; Derome, A.E.; Dickinson, J.M.; Dowens, A.J.; Green, J.C.; Green, M.H.L.; Hare, P.M.; Payne, M.P.; Rankin, D.W.H.; Robertson, H.E. <u>J. Chem. Soc. Chem. Comm.</u> 1986, 520.
4. Eisenstein, O.; Jean, Y. <u>J. Am. Chem. Soc.</u> 1985, <u>107</u>, 1177.
5. Krasnov, K.S.; Timoshin, V.S.; Danilova, T.G.; Khandozhko, S.V. <u>Handbook of Molecular Constants of Inorganic Compounds</u>; Israel Program for Scientific Translations: Jerusalem, 1970.
6. Rösch, N.; Knappe, P.; Sandl, P.; Görling, A.; Dunlap, B.I. in <u>Computational Chemistry: The Challenge of d and f Electrons</u>; Salahub, D.R.; Zerner, M.C., Eds.; ACS Symposium Series No.   ; American Chemical Society: Washington, DC, 1989; p.
7. Rösch, N.; Jörg, H.; Dunlap, B.I. in <u>Quantum Chemistry: The Challenge of Transition Metals and Coordination Chemistry</u>; Veillard, A., Ed.; NATO ASI Series C, Reidel: Dordrecht, 1986; Vol. 176, p 179.

RECEIVED November 22, 1988

Chapter 4

Chasing the Elusive d Electron with X-rays

Philip Coppens

Department of Chemistry, State University of New York at Buffalo, Buffalo, NY 14214

Models which allow interpretation of X-ray diffraction data in terms of orbital populations, radial dependence of the orbitals and LCAO coefficients are discussed. They are applied to experimental data on iron(II) phthalocyanine, iron(II) meso-tetraphenylporphyrin and its bis-pyridyl and bis-tetrahydrofurane derivatives. The diffraction studies indicate that the first two complexes are intermediate spin complexes with differing ground states while the last two are respectively low- and high-spin iron(II) compounds. A difference between the two intermediate spin complexes is thought to be related to the effect of the crystalline environment. This interpretation implies that the leading contributor to the ground state of the isolated complex is the $^3A_{2g}$ state. The more extensive formalism is applied to a set of theoretical charge and spin data on the iron(III) hexaaquo complex.

As d-electron complexes pose a special challenge for theoretical methods, alternative techniques capable of providing collaborating evidence are valuable. One such approach is the analysis of the elastic X-ray diffraction intensities of solids in terms of the one-electron density distribution. The method has been extensively tested on first-row atom molecular crystals, where it has shown to give reproducible results, in general agreement with densities obtained with the more advanced calculations. Much interesting chemistry concerns heavier elements, however, so that an extension to regions further down the periodic table is highly desirable. First-row transition metal complexes with organic ligands are especially suitable, as the relatively light core does not dominate the scattering, and effects such as X-ray absorption by the crystals are manageable.

The approach requires crystals of good quality and sufficient size (though that requirement is becoming less severe as more

0097–6156/89/0394–0039$06.00/0

intense sources are becoming available), and the absence of phase
transitions on cooling to temperatures suitable for accurate data
collection. Like any experimental technique, it is subject to
experimental error and limits in resolution, in particular in the
regions in the immediate environment of the nuclei. However, as the
interest is often in the angular distribution of the valence
electrons, which contains information on the relative population of
the orbitals, detailed knowledge of the region close to the nucleus
is not always essential. Theoretical energy-minimization
techniques, on the other hand, are very dependent on details in this
region. In this sense a complementarity exists between theoretical
calculations and the experimental diffraction method.
     In this paper we will comment on the background of the method,
and illustrate its application with a series of studies on iron
phthalocyanine and iron tetraphenylporphyrins.

Comparison of Theory and Experiment

The experimental sequence can be illustrated by the following
diagram:

$$
\begin{array}{c}
\text{electrostatic} \\
\text{properties} \\
\uparrow \\
\rho(r) \leftarrow \langle\rho(r)\rangle \ \leftarrow F(\underline{h}) \leftarrow I(\underline{h}) \\
\downarrow \qquad\quad \downarrow \\
\Delta\rho(r) \leftarrow \langle\Delta\rho(r)\rangle
\end{array}
\qquad (1)
$$

where the pointed brackets indicate the thermally averaged function,
and ρ, F and I are the electron density, the structure factor and
x-ray intensity respectively.
     The theoretical wavefunction can be used to extract the
theoretical equivalents of the experimental functions, provided that
a reasonable vibrational model is available for the averaging over
the modes of the crystal:

$$
\begin{array}{c}
\text{electrostatic} \\
\text{properties} \\
\uparrow \\
\Psi(1,2,\ldots n) \rightarrow \Gamma'(1,1) \rightarrow \rho(r) \quad \rightarrow \langle\rho(r)\rangle \rightarrow F(\underline{h}) \rightarrow I(\underline{h}) \\
\downarrow\sim \\
\Delta\rho(r) \rightarrow \langle\Delta\rho(r)\rangle
\end{array}
\qquad (2)
$$

     Though we can compare electron densities directly, there is
often a need for more condensed information. The missing link in
the experimental sequence are the steps from the electron density to
the one-particle density matrix $\Gamma'(1,1')$ to the wavefunction.
Essentially the difficulty is that the wavefunction is a function of
the 3n space coordinates of the electrons (and the n spin
coordinates), while the electron density is only a three-dimensional
function. Drastic assumptions must be introduced, such as the
description of the molecular orbitals by a limited basis set, and
the representation of the density by a single Slater-determinant, in
which case the idempotency constraint reduces the number of unknowns

(1). Even then no practical results have been obtained so far. But
in certain cases simplifications are possible, in particular if an
atom or fragment of a molecule can be identified which can be
treated separately from the remainder of the system. For 3d-
transition metal atoms in coordination complexes a major feature of
interest is the distribution of the electrons over the d-orbital
levels. On the other hand, the overlap between the metal and ligand
orbitals is relatively small, and represents a small fraction of the
total valence electron density. Therefore, in a first approximation
we can describe the d-electron density in terms of the population
P_{ij} of products of d-orbitals (d_i):

$$\rho_d = \sum_{i=1}^{5} \sum_{j \geq i}^{5} P_{ij} d_i d_j \qquad (3)$$

The most successful X-ray scattering model beyond the simple,
but widely used spherical-atom description consists of an expansion
of atom-centered spherical harmonic functions (2). The radial
dependence in this 'multipole model' is given flexibility by the
introduction of a radial parameter κ, which scales the radial
coordinate:

$$\rho_{atom}(\underline{r}) = \rho_{core}(\underline{r}) + P_v \kappa'^3 \rho_{valence}(\kappa' r) +$$
$$\sum_{\ell=0}^{\ell(max)} \kappa''^3 R_\ell(\kappa'' \zeta r) \sum_{m=0}^{\ell} \sum_{p} P_{\ell mp} y_{\ell mp}(\theta, \psi) \qquad (4)$$

where p is plus or minus for $m \geq 1$, P_v is the valence shell
population which multiplies the spherical valence density $\rho(r)$, and
the radial function R can have either a Slater type or a
Hartree-Fock radial dependence. The 'deformation terms' represented
by the summation include a spherical monopole term which for light
atoms is often omitted when the κ parameter is refined.
For covalently bonded atoms the overlap density is effectively
projected into the terms of the one-center expansion. Any attempt
to refine on an overlap population leads to large correlations
between parameters, except when the overlap population is related to
the one-center terms through an LCAO expansion as discussed in the
last section of this article. When the overlap population is very
small, the atomic multipole description reduces to the d-orbital
product formalism. The relation becomes evident when the products
of the spherical harmonic d-orbital functions are written as linear
combinations of spherical harmonics (3).

$$\rho_d = \sum_{i=1}^{5} \sum_{j \geq i}^{5} P_{ij} d_i d_j = \sum_{\ell=0}^{4} \kappa''^3 \{ R_\ell(\kappa'' \zeta r) \sum_{m=0}^{\ell} \sum_{p} P_{\ell mp} y_{\ell mp} \} \qquad (5)$$

In this case the second term in (4) describes the non-d (i.e
mainly 4s) part of the spherical valence density and the radial
depence of the terms in (5) is that appropriate for the d-orbitals
of the atom being considered.

In other words the multipole populations $P_{\ell mp}$ and the orbital populations are related through the matrix equation:

$$\underset{\sim}{P}_{\ell mp} = \underset{\sim}{M} P_{ij} \qquad (6)$$

where $\underset{\sim}{P}_{\ell mp}$ is a vector containing the coefficients of the 15 spherical harmonic functions with $\ell = 0, 2,$ or 4 which are generated by the products of d-orbitals. The matrix $\underset{\sim}{M}$ is a function of the Clebsch-Gordon coefficients, and the ratio of orbital and density function normalization coefficients ($\underline{4}$).

The d-orbital occupancies can be derived from the experimental multipole populations by the inverse expression:

$$\underset{\sim}{P}_{ij} = \underset{\sim}{M}^{-1} P_{\ell m \rho} \qquad (7)$$

The coefficients so obtained include the off-diagonal symmetry-allowed terms, which would not occur in the isolated atom. To some extent the d-σ overlap population will be projected into the one-center functions, but since the d-orbitals are relatively compact and the overlap is small this effect can be neglected in a first approximation.

In the following section we discuss application of the multipole formalism to a series of Fe(II) porphyrins. A next step towards derivation of the electronic wavefunction of a transition metal complex may be based on the LCAO formalism for the molecular orbitals. Test calculations with such a formalism, using a theoretical data set, are described in the final part of this article.

Application to Iron(II)-Porphyrins

In many respects the iron-porphyrin complexes offer favorable conditions for experimental study. The large organic component reduces the effect of absorption of X-rays by the crystals, which is sometimes difficult to correct for accurately, as well as the dominance of scattering by the core electrons, which may interfere in the charge density analysis of inorganic solids composed entirely of heavier atoms. The porphyrins exhibit a range of transition metal coordinations, spin states and ground state configurations. Because of the existence of closely-spaced electronic configurations, configuration interaction is essential in the theoretical treatment, in particular for the four-coordinate Fe(II) intermediate spin compounds, for which the theoretical results have been ambiguous. Of the four Fe(II) complexes we will discuss, the low-spin Fe(II)(bis-pyridyl)tetraphenylporphyrin has a well established ground state and serves as a suitable calibration of the method. Six-coordinate bis-tetrahydrofurane-FeTPP is the only six-coordinate high-spin Fe(II) compound known and shows an as yet unexplained temperature-dependence of the magnetic moment. The ground states of the two intermediate spin compounds Fe(II)phthalocyanine and Fe(II)tetraphenylporphyrin is not firmly established, a situation in which the independent information provided by the X-ray method is of considerable value.

Bis(pyridine)(meso-tetraphenylporphinato)iron(II) (BPyFeTPP). The deformation density in the plane of the porphyrin ring, shown in Figure 1, shows density accumulation around the iron atom in the directions bisecting the iron-pyrole nitrogen bonds, indicating preferential occupancy of the d_{xy} orbitals (5,6). The d-orbital population analysis was performed with both the harmonic and a more complete anharmonic thermal motion treatment. Results for the harmonic model are listed in Table I. In the anharmonic treatment the total d-electron population on the iron atom is somewhat reduced compared with the harmonic results, but the relative distribution of the electron over the d-levels as expressed by the percentage occupancies is rather invariant. The preferentially occupied orbitals are the t_{2g}-type orbitals of the pseudo-octahedral point group, which are almost equally occupied, with a slightly larger population of d_{xy}. Down to this detail the populations are in agreement with the Mulliken population analysis of the results an Extended Huckel (EH) calculation (Scheidt, unpublished results), listed in the third column of Table I. Reasonable agreement with EH is also obtained for d_{z^2}, but not for the destabilized $d_{x^2-y^2}$ orbital, for which the EH method gives a larger electron population. The occupancy of this orbital is due to the covalent interaction of the $d_{x^2-y^2}$ orbital with the ligand s and p-orbitals, which is clearly overestimated by the one-electron EH method. Nevertheless the agreement for the other orbitals is rather impressive given the approximate nature of the one-electron Hamiltonian in the EH method and the inadequacies of the Mulliken population analysis. The deviations from the spherical symmetry assumed in the standard crystallographic treatment (last column of Table I) are large.

Table I. d-Electron Orbital Populations in Bis(pyridine)(meso-phenylporphinato)iron(II)

Orbital	Experimental	Ext. Huckel	Spherical
$d_{x^2-y^2}$	0.35(4.8%)	0.81(11.0%)	1.2(20%)
d_{z^2}	1.05(14.4%)	0.724(9.8%)	1.2(20%)
d_{xz}	1.93(26.5%)	1.93(26.1%)	1.2(20%)
d_{yz}	1.93(26.5%)	1.93(26.1%)	1.2(20%)
d_{xy}	2.02(27.7%)	1.99(27.0%)	1.2(20%)
TOTAL	7.29	7.40	6.00

Bis(tetrahydrofurane)(mesotetraphenylporphinato)iron(II) (BTHFFeTPP). Though BTHFFeTPP is the only known six-coordinate high spin Fe(II)complex, its THF ligands are rather loosely bound. Crystals slowly loose THF when exposed to the atmosphere, while the

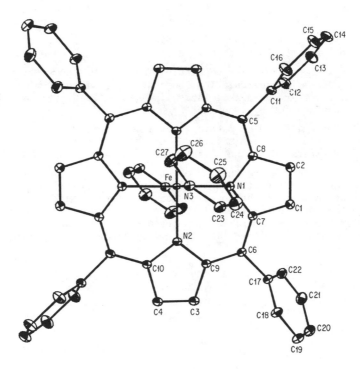

Figure 1a. Perspective drawing of bis(pyridine)-(meso-tetraphenylporphinato)iron(II).

CONTOUR INTERVAL = .10 E/A3

0 |1   2A

Figure 1b. Deformation electron density in the porphyrin plane in bis(pyridine)(meso-tetraphenylporphinato)iron(II). Contours at 0.10eÅ^{-3}. Negative contours broken. First positive contour at 0.05eÅ^{-3}. (Reproduced with permission from ref (6), Copyright 1988, International Union of Crystallography.)

iron is five-coordinate in a solution in benzene (7). But it is
surprising at first sight that the complex is high spin, while the
four-coordinate FeTPP obtained on detachment of the axial ligands is
in an intermediate spin state.

A further complication, not fully understood, is the
temperature dependence of the magnetic susceptibility (8), the
shrinkage of the axial Fe-O bonds on cooling from 1.35Å at room
temperature to about 1.29Å at nitrogen temperature, combined with
the absence of significant variations in the Mossbauer splitting.

The charge density analysis, performed at 100K, shows a pattern
typical for a high-spin complex in the tetraphenylporphyrin plane,
see Figure 2. The d-orbital peaks near the iron atom now occur in
the direction of the pyrole-nitrogen ligand atoms along the bonds,
indicating the effect of covalency (σ-donation) superimposed on the
cylindrical distribution of the d^6 ion. A very similar in-plane
deformation density pattern occurs in the theoretical map of the
(hexaaquo)iron(III) atom (19).

To our knowledge no theoretical calculation of the BTHFFeTPP
has been performed. Comparison of the orbital populations with the
idealized ionic states, see Table II, shows the best agreement with
the ^5E_g state. However the experiments indicate a depopulation of
$d_{xz,yz}$ and an excess population in $d_{x^2-y^2}$; both these effects are in
the direction expected when covalent interactions are taken into
account. A calculation of the quintet state of high-spin six-
coordinate bis-NH$_3$FeP (9), predicts a large depopulation of the π-
orbitals, as found here, thus supporting the validity of both
approaches.

Table II. Iron Atom d-Orbital Populations in
           Bis(tetrahydrofurane)(mesotetraphenylporphinato)iron(II)
           (The z axis is perpendicular to the molecular plane)

Term Symbol	$^5B_{2g}$	$^5A_{1g}$	$^5B_{1g}$	^5E_g	exp
$d_{x^2-y^2}$	1 (16.7%)	1 (16.7%)	2 (33.3%)	1 (16.7%)	1.42 (24%)
d_{z^2}	1 (16.7%)	2 (33.3%)	1 (16.7%)	1 (16.7%)	1.04 (17.5%)
$d_{xz,yz}$	2 (33.3%)	2 (33.3%)	2 (33.3%)	3 (50%)	2.52 (42.6%)
d_{xy}	2 (33.3%)	1 (16.7%)	1 (16.7%)	1 (16.7%)	0.93 (15.7%)
TOTAL	6	6	6	6	5.92

Iron(II) Phthalocyanine (FePc). The phthalocyanines and porphyrins
are closely related, and share the first- and second-nearest
neighbors of the centrally-located metal atom. While the S=1 nature
of their ground state is unambiguous, the exact term symbol has been
the subject of a large series of studies with often conflicting

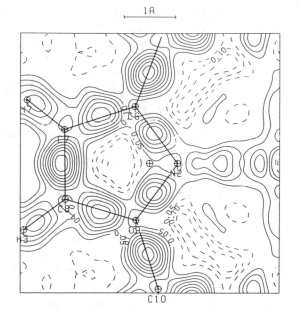

Figure 2. Deformation density in the plane through the iron atom and the pyrrole ring in bis(tetrahydrofurane)(mesotetraphenylporphinato)iron(II) after averaging over the molecular mmm symmetry. Contours at 0.5 e$Å^{-3}$. (Reproduced from ref. 8. Copyright 1986 American Chemical Society.)

results, ranging from from $^3B_{2g}$ based on magnetic data (10) to ^3E_g

and $^3A_{2g}$ from theoretical calculations on FeP, and from other
techniques such as NMR and circular dichroism measurements. Some of
the theoretical results are summarized in Table III. While the
results are contradictory, it is clear from the calculations that
the spacing of the $^3A_{2g}$ and ^3E_g levels is extremely small, of the
order of magnitude of a few tenths of an eV. While this is above
thermal energies it indicates that interaction of configurations
through spin-orbit coupling or other mechanisms will be important,
thus complicating calculations.

The experimental results from ref (16), obtained at 110K, are
given in Table IV together with values for the various ionic
configurations. The main difference between the competing ^3E_gA and

$^3A_{2g}$ states is a shift of one electron from the $d_{xz,yz}$ orbitals in
the former to the d_{z^2} orbital in the latter state. The experimental
populations are close to the almost 3:1 ratio of the $d_{xz,yz}/d_{z^2}$

populations predicted for ^3E_g and not compatible wth the other ionic

states. Compared with ^3E_g there is a depopulation of the $d_{xz,yz}$
and an excess population of $d_{x^2-y^2}$, a now recurring pattern
(compare BTHFFeTPP), that merits further quantitative analysis.

The distinction between the two states is very clear in the
density maps when a section containing the perpendicular to the
porphyrin plane through the iron atom is plotted. This is
demonstrated by theoretical maps of the two states (15) as shown in
Figure 3, in which there is a deficiency in the axial direction for
^3E_gA, and an excess for $^3A_{2g}$, resulting from the transfer of an
electron into the d_{z^2} orbital. The experimental map for FePc shows
the deficiency as expected from examination of the experimental d-
orbital populations (Figure 4a).

(Meso-tetraphenylporphinato)iron (II) (FeTPP). A first experimental
study of FeTPP did not provide conclusive results (17). Though a
large peak was observed in the axial direction in the deformation
density, the populations did not clearly distinguish between the
various ionic states, nor did they agree with the SCF calculation.
Since the experimental study was flawed by the use of two different
crystals in the data set collection, and by a larger than optimal
crystal size, a new analysis was undertaken (Li, Landrum and
Coppens, to be published). The new analysis confirms the peak in
the symmetry-related positions above and below the iron atom (Figure
4b) and gives d-orbital populations in reasonable agreement with the
$^3A_{2g}$, in particular with regard to the crucial d_{z^2} occupancy (Table
V).

Table III. Theoretical Results for the Ground State of Iron Porphyrin

Reference Number	Type	$\Delta E(^3E_g - {}^3A_{2g})$
11	EH	negative
12	SCF	0.32 eV
12	SCF after configuration mixing	0.08
13	SCF-Xα	0.2
14	SCF-CI	0.27
15	INDO	0.27
15	INDO-CI	0.03
9	SCF	0.29
9	SCF-CI	-0.47

Table. IV. d-Electron Orbital Population in Iron(II) Phthalocyanine

Term Symbol	^3E_gA	$^3A_{2g}$	$^3B_{2g}$	^3E_gB	X-ray
$d_{x^2-y^2}$					0.70(7) (12.9%)
d_{z^2}	1(17%)	2(33%)	1(17%)	2(33%)	0.93(6) (17.1%)
$d_{xz,yz}$	3(49%)	2(33%)	4(67%)	3(49%)	2.12(7) (39.1%)
d_{xy}	2(33%)	2(33%)	1(17%)	1(17%)	1.68(10)(30.9%)

Figure 3. Computed static deformation density map of FeP.
a, $^3A_{2g}$ in plane; contours at 0.10 eÅ^{-3}. b, ^3E_g in plane;
contours as in a. c, $^3A_{2g}$ bisecting plane; contours as in a.
d, ^3E_g bisecting plane. (Reproduced with permission from
ref. 14. Copyright 1985 Elsevier Science Publishers.)

Figure 4a. Deformation density sections perpendicular to the
molecular plane through the Fe and N atoms. Iron (II)
phthalocyanine Contours at 0.05 eÅ^{-3}, negative contours broken
(Reproduced with permission from ref. (16), Copyright 1984,
American Institute of Physics).

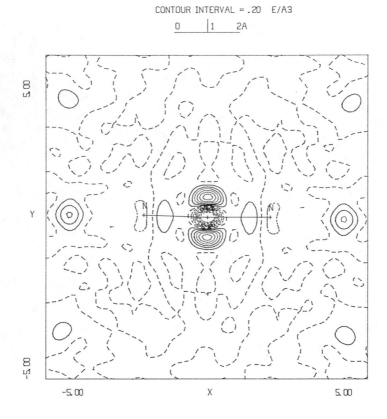

CONTOUR INTERVAL = .20   E/A3

Figure 4b. Deformation density sections perpendicular to the
molecular plane through the Fe and N atoms. Iron (II)(meso-
tetraphenylporphyrin). Contours at 0.1 e$Å^{-3}$ (Li, Coppens, and
Landrum,to be published).

Table V. d-Electron Orbital Populations in
(Meso-tetraphenylporphinato)iron(II)

| | Experimental | Theoretical([17]) | | |
		$^3A_{2g}$-SCF	^3E_gA-SCF	EH
$d_{x^2-y^2}$	0.24(15) (3.9%)	0.18 (2.9%)	0.18 (2.9%)	0.90(12.8%)
d_{z^2}	2.10(14)(33.7%)	1.94(31.7%)	0.99(16.1%)	1.07(15.3%)
$d_{xz,yz}$	2.28(18)(36.5%)	1.98(32.3%)	2.96(48.1%)	3.05(43.5%)
d_{xy}	1.52(15)(25.9%)	2.02(33.0%)	2.02(32.8%)	1.99(28.4%)
TOTAL	6.24(31)	6.12	6.15	7.02

The population of the $d_{x^2-y^2}$ orbital is small, in agreement with the
SCF-CI results, indicating relatively weak covalency in this
complex, in particular when the comparison is made with FePc. This
is not surprising as the Fe-N(pyrole) bond length is 1.927Å in FePc,
compared with 1.967Å in FeTPP. The d_{xy} population is significantly
lower than predicted by the calculation, an effect that requires
further analysis.

The conclusion that in the crystal at least the two four-
coordinate complexes have different groundstates is inescapable.
The explanation may be sought in differing Fe-N(pyrrole) bond
lengths and the differing ionicities, or in the effect of the
intermolecular interactions, as the molecular packing is quite
different in the two crystals. In monoclinic FePc nitrogen atoms in
the bridging (meso-) position of a neighboring molecule related by a
b-axis translation in the $P2_1/a$ space group, are located at 3.42Å
above and below the iron atoms, thus providing an axial "pseudo"-
ligand. In tetragonal FeTPP the molecules are aligned perpendicular
to the $\bar{4}$ axis of the space group I$\bar{4}$2d, and no such approach exists.
The sensitivity of the ground state of the prophyrins to axial
ligation is well known. A similar effect for non-bonded
interactions has been discussed by Mispelter, Momenteau and Lhoste
([18]), who show that small axial perturbations can induce a reversal
of the ground state in capped ferrous porphyrin derivatives. A
similar effect can occur because of crystalline packing and, in all
likelihood, due to the presence or absence of solvatation in
solutions of the porphyrins.

Thus, while the X-ray method of charge density analysis does
not suffer from some of the ambiguities of the theoretical
calculations, the measurements give information on a molecule in a
particular environment and not on the isolated state commonly
treated in theoretical work. Nevertheless the absence of strong
interactions in the FeTPP crystal leads to the conclusion that the
ground state of this molecule in its isolated state is close to the
$^3A_{2g}$ state.

Towards a Molecular Orbital Model: Analysis of Theoretical Structure Factors (19)

The X-ray scattering may be expressed directly in terms of a formalism related to the molecular orbitals of a transition metal complex, rather than to the charge distribution on the individual atoms, as is done in the multipole model. In such a formalism the overlap density is directly related to the orbital populations, and does not have to be projected into the one-center functions.

For a metal orbital ϕ and a ligand orbital χ the metal-ligand bonding and anti-bonding orbitals can be written as (20):

$$\psi = [1 + \gamma^2 + 2\gamma S]^{-1/2}\{\chi + \gamma\phi\}$$
$$\psi^* = [1 - 2\lambda S + \lambda^2]^{-1/2}\{\phi - \lambda\chi\} \tag{8}$$
$$\lambda = (\gamma + S)/(1 + \gamma S)$$

In the $d^1 o^2$ case of a singly-occupied metal d-orbital interacting with a doubly occupied ligand orbital (which in general will be a symmetry-adapted linear combination of atomic orbitals, see Table VI) the charge deformation can be written as (21):

$$\delta\rho^c = (S^2 + \gamma^2)\phi^2 + (S^2 - \gamma^2 - 2\gamma S)\chi^2 + 2(\gamma - S)\phi\chi - 2S^3\phi\chi -$$
$$2\gamma^3\phi\chi - 4\gamma^2 S\phi\chi \tag{9}$$

Thus the magnitude of the overlap term depends on the relative values of the mixing coefficient γ and the overlap integral S. For the spin deformation density we obtain with the same formalism, excluding spin polarization:

$$\delta s^c = (S^2 - \gamma^2)\phi^2 + (S^2 + \gamma^2 + 2\gamma S)\chi^2 - 2(\gamma + S)\phi\chi - 2\gamma^3\phi\chi$$
$$+ 4\gamma^2 S\phi\chi - 2S^3\phi\chi \tag{10}$$

In the spin deformation density the first-order overlap effect is dependent on $\gamma + S$, rather than on $\gamma - S$. It is therefore more pronounced than in the case of the charge density. It follows that a molecular orbital formalism is most likely to be productive if applied to a combination of X-ray data on the charge density and polarized neutron data on the magnetization density.

In order to test such an application we have calculated the spin and charge structure factors from a theoretical wave function of the iron(III)hexaaquo ion by Newton and coworkers (22). This wave function is of double zeta quality and assumes a frozen core. Since the distribution of the α and the β electrons over the components of the split basis set is different, the calculation goes beyond the RHF approximation. A crystal was simulated by placing the complex ion in a 10x10x10Å cubic unit cell. Atomic scattering factors appropriate for the radial dependence of the Gaussian basis set were calculated and used in the analysis.

For modelling purposes the molecular density is subdivided in a paramagnetic a part described by the molecular orbital formalism, and a diamagnetic b part for which the multipole formalism is retained (20). The rational for this separation is that the strong covalency and the atomic valence shell scattering in the water molecule (and in other molecules as well) are not well described by the minimal basis set based MO formalism applied to the paramagnetic molecular fragment.

In the final model tested the α and β densities are given by:

$$\rho_\alpha = \rho_\alpha{}^b + \sum_o \{|\psi_{o\alpha}|^2 + |\psi_{o\alpha}{}^*|^2\} + \sum_\pi \{|\psi_{\pi\alpha}|^2 + |\psi_{\pi\alpha}{}^*|^2\}$$

$$\rho_\beta = \rho_\beta{}^b + \sum_o \{|\psi_{o\beta}|^2 + \sum_\pi |\psi_{\pi\beta}|^2\} \tag{11}$$

with $\rho_\alpha{}^b = \rho_\beta{}^b$.

These functions account for 15 electrons (10 in the bonding and 5 in the antibonding orbitals), while the remaining 38 valence electrons occupy the multipole functions of the b part. The five β electrons are responsible for the bond polarization, as the 10 α electrons in the M.O 's equally occupy bonding and antibonding orbitals and give neither spin nor charge transfer. The molecular orbitals $\psi_{o,\pi}$ are defined in Table VI.

Table VI. Symmetry Orbitals

MO	NORMALIZATION	METAL d	LIGAND s^a	LIGAND $p(x,y,z)^b$
o_1	$(24)^{-1/2}$	d_{z^2}	$2s_1+2s_4-s_2-s_3-s_5-s_6$	$2z_4-2z_1+x_2-x_5$
o_2	$(8)^{-1/2}$	$d_{x^2-y^2}$	$s_2+s_5-s_3-s_6$	$x_5-x_2+y_3-y_6$
π_1	$1/2$	d_{xy}		$y_2-y_5+x_3-x_6$
π_2	$1/2$	d_{xz}		$z_2-z_5+x_1-x_4$
π_3	$1/2$	d_{yz}		$z_3-z_6+y_1-y_4$

a. s_k represents s - orbital on the kth oxygen atom.

b. x_k, y_k, z_k represent $2p_x$, $2p_y$, $2p_z$ orbitals respectively, on the kth oxygen atom.

The adjustable parameters to be 'retrieved' from the structure factors are the mixing coefficients λ and γ in (8), the 3d and 4s electron κ-parameters of both the α and β densities, and the parameters of the multipole model for the diamagnetic fragment.

Detailed results of the test refinements are given in ref.
(19). They show that the charge density can be fitted, and that the
orbital populations, net atomic charges and the bond-polarization of
the theoretical calculation can be retrieved with this 'minimal-
number-of-parameters' model.

This work is to be followed by an analysis of an experimental
set of intensity data obtained with both the X-ray and polarized
neutron techniques. This will allow application of the formalisms
described above, and a direct comparison of experimental charge
density parameters with theoretical results.

Concluding Remarks

The X-ray method is able to provide information on the detailed
electronic ground state of transition metal complexes. As the
agreement with EH results for the d-orbital populations of the
bipyridyl complex indicates, a quick calculation may often provide
answers which can only be obtained with much effort from the
experiment. However, as illustrated for the four-coordinated Fe
porphyrins and phthalocyanins, situations occur in which the
independent X-ray evidence is invaluable. The combination with
polarized neutron data is powerful, but requires the availability of
large, good quality crystals. With further improvements in neutron
source intensities this limitation should become less important.

Acknowledgments

Many of my collaborators have contributed to the work described
above. In particular, Naiyin Li, Eric Elkaim, Tibor Koritsanszky,
Claude Lecomte, and Liang Li made major contributions. Syntheses
were performed by John Landrum of Florida International University
and A. Tabard of the University of Dijon. Financial support of this
work was obtained from the National Institute of Health (HL23884)and
the National Science Foundation (CHE8711736).

Literature Cited

1. Frishberg, C.A.; Goldberg, M.J.; Massa, L.J. In Electron
    Distributions and the Chemical Bond; Coppens, P.; Hall, M.B.,
    Eds.; Plenum: New York, 1982; p 101.
2. Hansen, N.K.; Coppens, P. Acta Cryst. 1978, A34, 2336.
3. Holladay, A.; Leung, P.C.; Coppens, P. Acta Cryst. 1983, A39,
    377-387.
4. Paturle, A.; Coppens, P. Acta Cryst. 1988, A44, 6-7.
5. Li, N.; Landrum, J.; Coppens, P. Inorg. Chem. 1988, 27, 482-
    488.
6. Mallinson, P.R.; Koritsanszky, T.; Elkaim, E.; Li, N.; Coppens,
    P. Acta Cryst. 1988, A44, 336-342.
7. Reed, C.A.; Mashiko, T.; Scheidt, W.R.; Spartalian, K.; Lang,
    G. J. Am. Chem. Soc. 1980, 102, 2302-2306.
8. Lecomte, C.; Blessing, R.H.; Coppens, P.; Tabard, A. J. Am.
    Chem. Soc. 1986, 108, 6942-6950.
9. Rawlings, D.C.; Gouterman, M.; Davidson, E.R.; Feller, D. Int.
    J. Quantum Chem. 1986, 28, 733-796. Rawlings, D.C.; Gouterman,
    M.; Davidson, E.R.; Feller, D. Int. J. Quant. Chem. 1985, 28,
    797-822.

10. Barraclough, C.G.; Martin, R.L.; Mitra, S.; Sherwood, R.C. J. Chem. Phys. 1970, 53, 1643-1648.
11. Zerner, M.; Gouterman, M. Theor. Chim. Acta 1966, 4 44. Zerner, M.; Gouterman, M.; Kobayashi, H. Theor. Chim. Acta 1966, 6, 363.
12. Obara, S.; Kashiwagi, H. J. Chem. Phys. 1982, 77, 3155.
13. Sontum, S.F.; Case, D.A.; Karplus, M. J. Chem. Phys. 1983, 79 2881.
14. Rohmer, M.M. Chem. Phys. Lett. 1985, 116, 44.
15. Edwards, W.D.; Weiner, B.; Zerner, M.C. J. Am. Chem. Soc. 1986, 108, 2196-2204.
16. Coppens, P. and Li, L. J. Chem. Phys. 1984, 81, 1983-1993.
17. Tanaka, K.; Elkaim, E.; Liang, L.; Jue, Z.N.; Coppens, P.; Landrum, L. J. Chem. Phys. 1986, 84, 6969-6978.
18. Mispelter, J.; Momenteau, M.; Lhoste, J.M. J. Chem. Phys. 1980, 72, 1003-1012.
19. Koritsanszky, T., and Coppens, P., Proceedings of the Sagamore IX Conference on Charge, Spin and Momentum Densities, Coimbra, Portugal, 1988.
20. Becker, P.J., and Coppens, P., Acta Cryst. 1985, A41, 177-182.
21. Coppens, P., Koritsanszky, T., and Becker, P., Chem. Scripta 1986, 26, 463-467.
22. Logan, J.; Newton, M.D.; Noell, J.O. Int. J. Quantum Chem. 1984, 18, 213-235.

RECEIVED January 17, 1989

Chapter 5

Modeling the Structure and Reactivity of Transition Metal Hydride Complexes

Self-Consistent Field and Complete Active Space Self-Consistent Field Studies

A. Dedieu and V. Branchadell

Laboratoire de Chimie Quantique, ER 139 du Centre National de la Recherche Scientifique, Université Louis Pasteur, F–67000 Strasbourg, France

SCF and CAS SCF calculations on mono and bimetallic transition metal hydride complexes are reported. The importance of including the non dynamical correlation effects for the study of the cis-trans isomerism in dihydrido complexes and for the study of the CO insertion reaction into the metal hydride bond is stressed. The metal to metal hydrogen transfer in a class of bimetallic $d^6 - d^8$ hydride complexes is analyzed and the feasibility of the transfer discussed as a function of the coordination pattern around the two metal centers.

Transition metal hydride complexes are involved in many stoichiometric or catalytic transformation reactions of organic substrates mediated by transition metal complexes. In many instances a hydrogen transfer step, either intramolecular or intermolecular, is involved. They are of course various forms that this transfer may take: the ligand insertion reactions into the metal hydride bond may be, at least formally, regarded as hydride transfer from the metal to the ligand. Other reduction processes involve radical intermediates resulting from hydrogen atom transfer with or without previous electron transfer. Late transition metal hydrides can behave as acids in proton transfer reactions. Metal to metal hydrogen migrations which have been observed in several di- or polynuclear complexes form also another class of hydrogen transfer reactions that help to bridge the gap between homogeneous and heterogeneous catalysis.

Hence a detailed knowledge of the mechanisms of these transfers is of critical importance. One may expect that theoretical studies by focusing on the electronic structure of transition metal hydride complexes and on the electronic factors which control their hydrogen transfer ability will deepen our understanding of these processes. One may worry however about the criteria that the corresponding calculations should meet to achieve this goal. This is especially true for reactions in which a metal to hydrogen bond, which is essentially covalent, is broken. We report here the results of our own studies in this field, analyzing from this dual point of view some hydride transition metal carbonyl complexes and hydrogen transfer reactions. We shall first examine the influence of the non dynamical correlation on the structure and reactivity of hydrido transition metal complexes, focusing on the cis-trans isomerism in $H_2Fe(CO)_4$ and on the CO insertion reaction into the metal hydride bond. The feasibility of this latter reaction will be briefly discussed in the context

0097–6156/89/0394–0058$06.00/0

of other insertion reactions, such as the C_2H_4 or CO_2 insertion reaction. Our final concern will be with the hydrogen transfer between two metallic centers in $d^6 - d^8$ binuclear complexes.

Hydrido Transition Metal Carbonyls: SCF and CAS SCF Calculations.

The cis-trans isomerism problem of $H_2Fe(CO)_4$ arose in the course of a study of the relative thermodynamics of the CO insertion into the metal hydride and methyl bonds of the $H(CH_3)Fe(CO)_4$ system ([1]). To our surprise the trans isomer turned out to be, at the SCF level, more stable than the cis one. Further calculations showed that for related systems involving two σ donor ligands such as $Fe(CO)_4(CH_3)(COOH)$ or $H_2Fe(CO)_4$, SCF calculations invariably yielded the trans isomer as the most stable one, at odds with most experimental evidence: the electron diffraction structure of $H_2Fe(CO)_4$ ([2]) points to a cis disposition of the two hydrides, see 1. Infrared and Raman spectra ([3-5]) also indicate a ground state cis conformation, although ^{13}C NMR and ^1H -coupled ^{13}C NMR variable temperature spectra ([6]) suggest that an intramolecular carbonyl averaging process might involve the trans conformation 2. The SCF calculations carried out for both isomers 1 and 2 also failed to give Fe-C bond lengths in reasonable agreement with the experimental values obtained from the electron diffraction structure. This result is not peculiar to the $H_2Fe(CO)_4$ system, being also characteristic of other transition metal carbonyl complexes ([7-10]). More particularly Lüthi et al. have shown that the inclusion of electron correlation (especially non-dynamical correlation) greatly improves the description of the metal to carbonyl bonds ([10]). To what extent such an improvement influences the relative stability of both isomers has therefore to be assessed. It is interesting to note in this context that previous theoretical studies pertaining to H_2PtL_2 square planar platinum dihydrido complexes, having phosphine rather than carbonyl ancillary ligands, also ended up with the same conclusion, i.e. a thermodynamical preference for the trans isomer at the SCF level ([11-12]). The 23 kcal/mol computed energy difference for the $H_2Pt(PMe_3)_2$ system ([11]) is not in line with the observation of a cis-trans equilibrium involving 20% of the cis isomer in a toluene solution ([13]).The energy difference amounts to only a few kcal/mol in the $H_2Pt(PH_3)_2$ system ([11-12]) and the ordering of the two isomers is reversed when electron correlation is included through CI calculations ([12]). We therefore tried out in a comparative study ([14]) of 1 and 2 to understand why SCF calculations in these dihydrido complexes seemed to be systematically biased against the cis isomer and to find out the requirements to achieve a balanced description of both isomers.

As far as the basis set is concerned, increasing its quality from split valence to double zeta does not lead to any improvement of the situation: a slight increase in the energy difference was found on going from a (14,9,6/9,5/6) set of primitives contracted to $<6,4,3/3,2/3>$ for the iron atom, the first row atoms and the hydrogen atom respectively, to the (14,11,6/10,6/6) $<8,6,3/4,2/3>$ basis set ([14]). The addition of a p polarisation function on the hydrogen atom decreased this value somewhat, down to 1.8 kcal/mol, but in every case the trans isomer remained the most stable one ([14]).

CAS SCF calculations were therefore performed with the split valence basis set incremented by a p polarisation function on the hydrogen atoms. Two different sets of active orbitals were considered. The first one was designed to account for the $d_\pi \rightarrow \pi^*$ back donation and was therefore restricted to the π type valence orbitals. The three $3d_\pi$ orbitals, which are strongly occupied, were each correlated by two weakly occupied orbitals, owing to the mixed $4d_\pi$ and π^*_{CO} character of these weakly occupied orbitals. This $3 + 6$ set of active orbitals referred to as CAS SCF-6 is populated by 6 electrons. The second set, hereafter referred as CAS SCF-12, took into account both σ and π correlation effects. Twelve electrons were correlated and

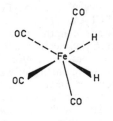

1

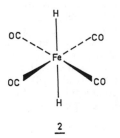

2

twelve orbitals were made active.   These orbitals correspond primarily to the bonding and antibonding orbitals of the Fe 3d and $4p_z$ orbitals with the appropriate symmetry combinations of the carbonyl and hydride ligands (14). In order to reduce the number of configurations the number of electrons in each symmetry was fixed. We checked (14) that this restriction is justified.

The improvement brought up by the inclusion of the near degeneracy correlation effects is quite substantial. As seen from Table I, the ordering is already

Table I. Total (in a.u.) and relative (in kcal/mol) SCF and CAS SCF Energies for both Isomers

	Cis	Trans	ΔE(Cis-Trans)
SCF	-1711.3795	-1711.3823	1.8
CAS SCF-6	-1711.4973	-1711.4959	-0.9
CAS SCF-12	-1711.6477	-1711.6378	-6.2
CAS SCF-12/opt/[c]	-1711.6500	-1711.6393	-6.7

[c] with Fe-C bond length optimized at the CAS SCF-12 level

inverted at the CAS SCF-6 level , but the improvement is quite small. The major effect arises from the inclusion in the active space of orbitals allowing for the correlation of σ bond electrons, the trans isomer now lying 6.2 kcal/mol below the cis isomer (or 6.7 kcal/mol when the Fe-C bond lengths are optimized). In fact, it is the correlation of the electrons involving the d'_σ and d''_σ orbitals which is necessary in order to obtain the correct ordering of the two isomers: CAS SCF calculations carried out with a set of 10 active orbitals (involving the bonding and antibonding orbitals between the d orbitals and the appropriate symmetry combinations of the ligand orbitals) yield the cis isomer more stable than the trans by 10.8 kcal/mol instead of 0.9 in the CAS SCF-6 calculations, where only the d_π correlation is taken into account. This energy difference of 10.8 kcal/mol is too large however. This is due to the fact that both the σ_g and σ_u combinations of the two hydrogen s orbitals are correlated in the cis isomer (by d'_σ and d''_σ respectively) , whereas only σ_g is correlated in the trans isomer (the CAS SCF-10 active space, being set up after the bonding and antibonding combinations made by the d orbitals, includes σ_g and d'_σ but not σ_u and its correlating counterpart $4p_z$). The CAS SCF-12 calculations correct for this inbalance. Our results therefore indicate that a proper description of the Fe-H σ bonds, which are essentially covalent in nature (15-16), is needed to account for the conformational isomerism problem. This is best achieved by including in the active space the orbitals accounting for the Fe-H σ^* bonds, i.e. the orbitals predominantly of d'_σ and d''_σ character in the cis isomer and those predominantly of d'_σ and $4p_z$ character in the trans isomer. The strongest correlating effect is provided by the orbitals involving the metal d AO's, as exemplified by the values of 0.074 and 0.038 found (14) for the respective occupations of the $d''_\sigma - \sigma_u$ orbital (in the cis isomer) and $4p_z - \sigma_u$ orbital (in the trans isomer).

An alternative but equivalent way of rationalizing is to consider that the two isomers 1 and 2 result from the interaction of two hydrogen atoms with the $Fe(CO)_4$ fragment in either a C_{2v} geometry (for 1) or a D_{4h} geometry (for 2), and not from the interaction of two hydrides with a d^6 $Fe(CO)_4^{2+}$ fragment. One has therefore to spin pair the two 1s hydrogen electrons with the two unpaired electrons of the corresponding $^3B_2$ (for the C_{2v} fragment) and $^3A_{2u}$ (for the D_{4h} fragment) triplet states. The lower energy computed for the $^3B_2$ triplet state , -1710.4322 a.u. vs. -1710.4071 a.u. for the $^3A_{2u}$ triplet state, accounts well for the preference of cis isomer (these values are obtained from CAS SCF calculations carried out with a set of 8 active orbitals populated by the 6 π electrons and the two unpaired electrons).

The inclusion of both σ and π correlation effects is also necessary to obtain a reasonable iron to carbonyl bond length, see Table II. The shortening observed on

Table II. SCF, CAS SCF-6 and CAS SCF-12 Optimized Fe-C Bond Length (in Å) for the Cis and Trans Isomers of $H_2Fe(CO)_4$

Calculation	Cis	Trans
SCF	1.97-2.04[a]	
CAS SCF-6	1.90	1.94
CAS SCF-12	1.87	1.85_5

[a] the axial and equatorial bond lengths were optimized independently at the SCF level

going from the SCF to the CAS SCF-6 level is traced, as in $Fe(CO)_5$ (10), to a better description of the π back donation at the CAS level. The further shortening obtained at the CAS SCF-12 level (which includes antibonding metal to ligand σ^* type orbitals) is traced to the involvement in the CAS SCF-12 wavefunction of configurations corresponding to cross excitations between σ and π type orbitals (14) which result in a marked increase in σ donation, again as in $Fe(CO)_5$ (10).

The above discussion therefore points to the covalent character of the metal hydride bond in $H_2Fe(CO)_4$. This has also been stressed by Goddard et al. for the square planar palladium and platinum dihydrides (15-16) and MH^+ systems (M = Ca through Zn) (17). Yet it is known that transition metal hydrides are not restricted to react through a radical mechanism but that they most often work as hydride donors or proton donors. Although this reactivity pattern is largely determined by the counter ion present, the solvent used and the stability of the conjugate base or acid left by the departure of a proton or of a hydride (18-20), it may be of interest to compare the electronic structures, especially the nature of the M-H bond, in a series of transition metal hydrides. We therefore extended our study of the $H_2Fe(CO)_4$ system to the $HMn(CO)_5$, $HCr(CO)_5^-$ and trans $HCr(CO)_4(NO)$ systems which are all isoelectronic and display the same octahedral geometry. Additional calculations were also performed on the $HFe(CO)_4^-$ system. The choice of the nitrosyl complex was dictated by the recent synthesis of trans-NO hydride complexes of tungsten (21) which, from spectroscopy and reactivity characteristics, were found to have a weak metal to hydride bond with a strong hydrido character for the hydrogen (22-23).

For the SCF and CAS SCF calculations the same basis set as for the $H_2Fe(CO)_4$ system was used. The CAS SCF calculations were of the CAS SCF-10 and CAS SCF-12 type described above and the number of configurations was again reduced by fixing the number of electrons in each symmetry. Since the CAS SCF-12 results do not differ significantly from the CAS SCF-10 results, as far as the metal to hydrogen bond is concerned, we report here only the results for the smallest of these two sets. We first note that on going from SCF to CAS SCF the s population of the hydrogen atom decreases.This is correlated by an increase in the $3d_{z^2}$ population (Table III). This merely results from a transfer in the CAS SCF wavefunction,

Table III. Selected SCF and CAS SCF Populations
for Hydrido Carbonyl Systems

	SCF		CAS SCF-10	
	s_H	$3d_{z^2}$	s_H	$3d_{z^2}$
$HCr(CO)_4(NO)$	0.827	0.780	0.832	0.854
$HCr(CO)_5^-$	1.002	0.701	0.945	0.826
$HFe(CO)_4^-$	1.115	0.785	0.960	1.029
$HMn(CO)_5$	0.802	0.814	0.739	1.022
$H_2Fe(CO)_4$	0.931	1.131	0.819	1.345

of electrons from an orbital more localized on the hydrogen atom ($s_H + \varepsilon 3d_{z^2}$) to an orbital more localized on the metal atom ($3d_{z^2} - \varepsilon s_H$). More interesting is the fact that the s_H and $3d_{z^2}$ populations do not show a regular trend along the series. This was not unexpected and has been noted by others (19-20). In particular the $HCr(CO)_4(NO)$ system does not appear, when using this criterion, to be much different from the $HMn(CO)_5$ system, especially at the SCF level, and does not display a strong hydridic character, at least something which would be comparable to $HCr(CO)_5^-$ as experimentally suggested (21-23). A better index of hydridic character is provided by the relative contribution (normalized to unity) of the s_H and $3d_{z^2}$ populations to the two active orbitals involved in the metal-hydrogen bond, the character of which is essentially ($s_H + \varepsilon 3d_{z^2}$) and ($3d_{z^2} - \varepsilon s_H$). One now finds (Table

Table IV. Hydrogen and Metal Atomic Orbital Contribution to the two Active
Orbitals Involved in the Metal Hydrogen Bond

	CAS SCF-10			
	s_H	$3d_{z^2}$	$4s$	$4p_z$
$HCr(CO)_4(NO)$	28.9	41.3	2.4	5.7
$HCr(CO)_5^-$	26.8	39.7	2.0	4.3
$HFe(CO)_4^-$	27.6	50.6	0.7	4.6
$HMn(CO)_5$	21.9	49.5	2.6	10.7
$H_2Fe(CO)_4$	13.1	64.0	3.8	0.0

IV) a close similarity between $HCr(CO)_5^-$ and $HCr(CO)_4(NO)$ (which is also reflected in the occupation number of the two natural orbitals involving s_H (Table V)). The relative high ratio between s_H and $3d_{z^2}$ in these two systems as compared to $HMn(CO)_5$ and $H_2Fe(CO)_4$ accounts for the tendency of the two Cr systems to behave as hydride donors. On the other hand the smaller value of the s_H / $3d_{z^2}$ ratio in the $HFe(CO)_4^-$, $HMn(CO)_5$ and $H_2Fe(CO)_4$ systems is in line with their known acidity. Interestingly the value of this ratio roughly follows the acidity scale of these three systems (24).

Another salient feature of these hydrido metal carbonyl systems is the underestimation, at the SCF level, of the $d_\pi \rightarrow \pi^*_{CO}$ back donation which we have already mentioned. This is not peculiar to the hydrido carbonyl complexes and has already been noted in many other carbonyl complexes (7-10,25,26). It is best evidenced by the orbital occupation number of the d_π orbitals (Table V) which range between 1.91 and 1.94e, with the noticeable exception of the $HCr(CO)_4(NO)$ system

Table V. Orbital Character and Occupation for the Hydrido Carbonyl
Complexes obtained from the CAS SCF-10 Calculations

	$HCr(CO)_4(NO)$	$HCr(CO)_5^-$	$HFe(CO)_4^-$	$HMn(CO)_5$	$H_2Fe(CO)_4$
$3d\pi_{ax}$	1.839	1.911	1.922	1.921	1.943
$3d\pi_{eq}$	1.918	1.915	1.910	1.921	1.933
$s_H + 3d'_\sigma$ [a]	1.961	1.960	1.934	1.942	1.927
$\sigma_L + 3d''_\sigma$ [a,b]	1.975	1.974	1.911	1.968	1.960
$3d'_\sigma - s_H$ [a]	0.039	0.040	0.069	0.057	0.072
$3d''_\sigma - \sigma_L$ [a,b]	0.025	0.026	0.089	0.033	0.040
$4d\pi_{ax}/\pi^*_{CO}$	0.161	0.089	0.077	0.079	0.057
$4d\pi_{eq}/\pi^*_{CO}$	0.082	0.085	0.089	0.079	0.067

[a] $3d'_\sigma$ and $3d''_\sigma$ are the axial ($3d_{z^2}$) and equatorial d_σ orbitals respectively
[b] σ_L stands for the symmetry combination of the CO ligands interacting with $3d''_\sigma$

where the strong π^* acceptor character of the NO ligand leads to an important decrease of the d_π population. That the $HCr(CO)_5^-$ system is not much different from the $HMn(CO)_5$ system, as far as the π back donation is concerned, may be surprizing since one would expect an anion to be much more efficient for the π back donation. But this agrees well with the experimentally known strong π back donation ability of manganese in Mn(I) complexes (27).

In all the compounds studied here the HF ground state wavefunction is the dominant configuration. Its weight in the CI expansion ranges from 80% in $HCr(CO)_4(NO)$ to 87% in $H_2Fe(CO)_4$, thus implying that the gross features of the reactivity of these complexes can be explained through the consideration of the HF wavefunction but that the energetics may be influenced by the inclusion of the non dynamical correlation. This is best exemplified by our results on the CO insertion reaction into the metal hydride bond.

The Ligand Insertion Reaction into the Metal Hydride Bond

The CO insertion reaction into the metal hydride bond is in fact a member of the class of ligand insertion reactions to which much theoretical work has been devoted (28,29-35). Some years ago we analyzed the ethylene insertion into the rhodium hydride bond of a Rh(III) hexacoordinated complex (36). We later focused our attention on the CO insertion reaction into the Mn-H bond of $HMn(CO)_5$ (37-39) and very recently we have undertaken the study of the CO_2 insertion reaction into the Cr-H bond of $HCr(CO)_5^-$ (C. Bo and A. Dedieu, Inorg. Chem., in press). We will concentrate here on the CO insertion reaction and compare it to the two other insertion reactions. The study of the reaction (1) was carried out at both the SCF and

$$HMn(CO)_5 \rightarrow Mn(CO)_4(CHO) \xrightarrow{+CO} Mn(CO)_5(CHO) \quad (1)$$

SD-CI levels within a limited set of active orbitals (37). From this study we were able to show that this reaction does indeed correspond to a hydride migration rather than to an actual CO insertion and this was rationalized (38) by considering orbital interactions in the vicinity of the transition state between the putative formyl ligand and a d^6 C_{2v} square pyramidal $Mn(CO)_4^+$ fragment. More specifically the HOMO, sketched in **3a**, was shown to correspond to a two electron stabilizing interaction between the doubly occupied $s_H + \pi_{CO}^* - 5\sigma_{CO}$ orbital of CHO^- (in fact its HOMO) and the empty $d_{y^2 - z^2}$ orbital of $Mn(CO)_4^+$. On the other hand a CO migration led (see **3b**) to a four electron destabilizing interaction between $s_H + \pi_{CO}^* - 5\sigma_{CO}$ and the occupied d_{yz} orbital. Replacing H by another σ donor ligand such as CH_3 would not modify appreciably this picture and this is why the CO insertion into the metal methyl bond also involves a methyl migration. The isolobal analogy (40) between d^6 C_{2v} ML_4 and d^8 C_{2v} ML_2 fragments also explains easily why the same feature was found by Koga and Morokuma in their study of the CO insertion into the metal methyl bond of $Pd(PH_3)(CH_3)(H)(CO)$ (31)

An interesting feature of the theoretical study of reaction (1) is the importance of the electron correlation, especially of the non dynamical type, on the energy profile. For the insertion step (2), truncated CI calculations led to an increase of the

$$HMn(CO)_5 \rightarrow Mn(CO)_4(CHO) \quad (2)$$

reaction energy from 10.5 kcal/mol at the SCF level to 38.4 kcal/mol at the CI level (37). This latter value was found to be in fair agreement with an estimated value of 39 kcal/mol obtained from gas phase data (38). It also agrees with the HFS- Xα value obtained later by Ziegler et al (30). SD-CI calculations carried out for the bare HMnCO and MnCHO systems, reaction (3), using the same geometries as in the

$$HMnCO \rightarrow MnCHO \quad (3)$$

molecular system and allowing for all single and double excitations from all occupied valence orbitals to all virtual orbitals led to a similar increase from 2.3 to 23.3 kcal/mol. A CAS SCF study of the reaction (3) suggested later that these results could be traced to the neglect of near degeneracy correlation effects (39). The CAS SCF energy curve was also highly endothermic and roughly parallel to the SD-CI curve. An analysis of the corresponding CAS SCF wavefunctions indicated (39) that the greater energy lowering obtained for HMnCO (as compared to MnCHO) arose essentially from an improved description of the $d_\pi \rightarrow \pi_{CO}^*$ back donation which is quite substantial for the hydrido carbonyl HMnCO, but much less effective for the formyl product MnCHO.

A similar analysis of the $(HPdCO)^+ \rightarrow (PdCHO)^+$ reaction (39) showed that in these systems, and in agreement with known experimental data (41), the $d_\pi \rightarrow \pi_{CO}^*$ back donation effects are rather weak and that they should not interfere significantly in the theoretical description of insertion reactions involving Pd(II) transition metal complexes. The calculations of Koga and Morakuma (31) did indeed exhibit a decrease of the endothermicity when going from the SCF to the MP2 level.

One should mention however that our conclusions have been very recently questionned by Axe and Marynick (42) who carried out calculations on the reaction (3) with various basis sets ranging from split valence to double zeta quality, with and without polarization functions on C, O and H atoms. They found a marked increase in the endothermicity value on going from the unpolarized basis sets (values ranging between 8.7 and 15.2 kcal/mol) to the polarized basis sets (with values between 19.5 and 25.2 kcal/mol, i.e. close to our SD-CI values). We have now carried out calculations adding to our original split valence basis set polarization functions on C, O and H. One polarized set includes the two sets of polarization functions ($\alpha_d = 0.920$ and 0.256 for the carbon atom, $\alpha_d = 1.324$ and 0.445 for the oxygen atom and $\alpha_p = 1.40$ and 0.26 on the hydrogen atom (43)) used by Axe and Marynick. The second one includes a single set of polarization functions (with $\alpha_d = 0.63$, 1.33 and $\alpha_p = 0.8$ for C,O and H respectively (44)). SD-CI calculations including again all single and double excitations from all occupied valence orbitals to all virtual orbitals were performed with this second polarized basis set. The results are summarized in Table VI. As found by Axe and Marynick there is indeed a basis set polarization effect,

Table VI. Total (in a.u.) and relative (in kcal/mol) SCF and SD-CI
Energies for the HMnCO and MnCHO Systems

	HMnCO	MnCHO	ΔE
SV[a]/SCF	-1260.5374	-1260.5338	2.3
SVP[b]/SCF	-1260.6073	-1260.5890	11.5
SVDP[c]/SCF	-1260.6210	-1260.5988	13.9
SV/SD-CI	-1260.8540	-1260.8169	23.3
SVP/SD-CI	-1261.0070	-1260.9633	27.4

[a] SV = split valence basis set; [b] SVP = SV basis set + one set of p and d polarization functions on H, C and O; [c] SVDP = SV basis set + two sets of p and d polarization functions on H, C and O

the SCF endothermicity increasing from 2.3 kcal/mol (with the unpolarized basis set) to 13.9 and 11.5 kcal/mol (with two and one polarization function respectively). One may wonder however, whether some artefact has not been introduced by an inbalance resulting from the lack of f polarization functions on the metal atom. Moreover the SD-CI calculations lead to an additional increase in the endothermicity (of about 16 kcal/mol) pointing again to the importance of the near degeneracy correlation effects. A recent study by Carter and Goddard (45) for the CH_2 insertion into the Ru-H bond of $ClRu(H)(CH_2)$ also stresses the importance of removing the HF deficiency in describing transition metal-ligand π bonds. Interestingly the calculations of Rappé on the $Cl_2ScH + CO \rightarrow Cl_2Sc(CHO)$ reaction (33) indicate that polarization functions on the ligand lower the energy of the reactants more than the energy of the product by 7 to 11 kcal/mol. and that the inclusion of electron correlation does not affect the results markedly. But this is understandable from the fact that in this formally d^0 metal complex there cannot be any substantial π back bonding.

That the wavefunction in these systems is nevertheless dominated by the HF wavefunction allows us to rationalize their chemical reactivity through the consideration of the molecular orbitals of the reactants, products and transition state. We have already seen that the shape of the HOMO at the transition state of reaction (1) was consistent with a hydride migration and not with a true CO insertion. An even better example of the use of the molecular orbital approach is provided by the comparison of the three insertion reactions which we have studied, the CO insertion into the Mn-H bond of $HMn(CO)_5$, the C_2H_4 insertion into the Rh-H bond of $H_2Rh(PH_3)_3(C_2H_4)$ and the CO_2 insertion into the Cr-H bond of $HCr(CO)_5^-$. These three reactions involve a d^6 hexacoordinated transition metal atom. But the CO insertion into the metal hydride bond is not experimentally observed (46), contrary to the two other insertion reactions which are facile (47,48). The corresponding HOMO's at the transition state or in the vicinity of the transition state are sketched in **3a** , **4** and **5** respectively. All three correspond to a two electron stabilizing interaction between the empty $d_{y^2 - z^2}$ orbital of the ML_4 C_{2v} metal complex fragment and the doubly occupied orbital of the ligand (either formyl, ethyl or formiate). This interaction is stronger however in **4** and **5** as the result of a better overlap with the upper lobe of the $d_{y^2 - z^2}$ orbital. It therefore accounts for the experimental findings.

The Hydrogen Transfer between two Metallic Centers

Despite the frequent observation of metal to metal hydrogen transfer in polymetallic hydrides, little is known about the actual mechanism of this process. Whether or not a rearrangement of the ligand framework occurs, is very often difficult to assess experimentally. It was not until very recently that both a bridged hydrido and a terminal hydrido coordination mode involved in a bridge-terminal equilibrium have been characterized by an X-ray crystal structure in a series of iron-platinum bimetallic hydride complexes (49). A similar process is also believed to occur in bimetallic $d^6 - d^8$ rhodium hydride complexes in which the two metal centers are linked by two bridging ligands (either H (50) or SR bridging ligands (51)). In this latter case however, the process goes until completion, a terminal hydrogen being transferred from one metal center to another one. Metal to metal hydrogen transfer is also a likely step of hydrogenation (50,52,53) or hydroformylation reactions (54,55) homogeneously catalyzed by these bimetallic complexes. In an attempt to assess the mechanistic details of the transfer, we chose the $H_4Rh_2(PH_3)_4$ system (see 6 of Figure 1, L= PH_3) as a prototype and tried to answer the following questions: (i) does the transfer proceed directly without major geometry relaxation, except the puckering of the inner core (path (A) of the Figure 1), which is quite easy since it is - from our calculations - slightly exothermic by 0.6 kcal/mol; or (ii) does it involve a rearrangement around the d^8 metal center - as suggested experimentally - from a square planar to a trigonal pyramid geometry (path (B) of Figure 1)? and (iii) how does the coordination of an extra ligand to the d^8 moiety affect the basic features of this transfer (path (C) of the Figure 2)? This is a rather important issue since the coordination of an alkene (or an alkyne) is supposed to occur in the catalytic reactions mentioned above.

SCF and CAS SCF calculations were therefore carried out on the $H_4Rh_2(PH_3)_4$ and $H_4Rh_2(PH_3)_5$ model systems for representative structures of the three pathways A-C (see 6-15 of the Figure 1, L= PH_3) (56). The use of a multiconfigurational approach was required in order to describe properly the metal hydrogen bond dissociation/bond formation processes, which not only occur simultaneously on both rhodium atoms but are in some instances associated with a diradical character of the resulting species (*vide infra*). Idealized geometries (the details of which may be found in the original publication (56)) rather than optimized ones were used and the geometry of the $H[Rh(\mu - H)(PH_3)_2]_2$ framework was frozen

<u>3a</u>

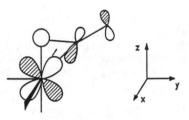

<u>3b</u>

<u>4</u>

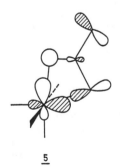

<u>5</u>

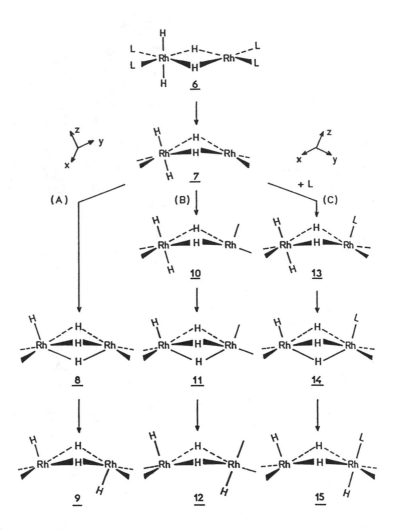

Figure 1. Three possible pathways for the hydrogen migration process in $\overline{H_2Rh_2(\mu - H)_2L_4}$, (L = PH_3). (Reproduced with permission from Ref. 56. Copyright 1988 Gauthier-Villars.)

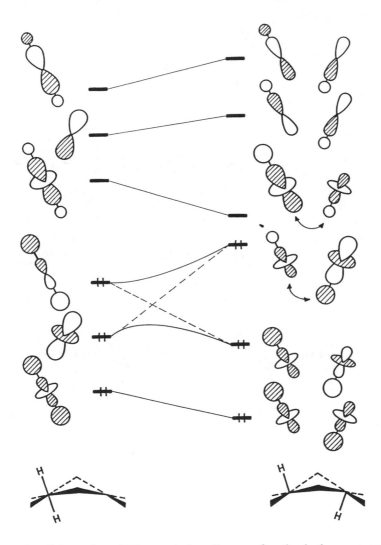

Figure 2.   Schematic orbital correlation diagram for the hydrogen transfer between 7 and **9**.   Only the in-plane valence orbitals have been represented. (Reproduced with permission from Ref. 56. Copyright 1988 Gauthier-Villars.)

in the puckered geometry. This is of course a limitation of our study, but optimization was precluded by the size of the systems and the necessity to carry out multiconfigurational type calculations. On the other hand, the CAS SCF calculations yield reasonable estimates of the energy profiles of the three pathways A, B and C of Figure 1. Our results should therefore be considered as semi quantitative.

The following basis set was used throughout the calculations: For the Rh atom a (15,10,8) set of primitives derived from a (15,9,8) set (57) by adding a diffuse p function of exponent 0.15 was contracted to $<6,4,4>$. For the PH_3 ligands the (11,7) and (4) sets of primitives (58,59) for the phosphorus and hydrogen atoms were contracted to $<4,3>$ and $<2>$. A somewhat more flexible basis set, namely (6) contracted to $<3>$ (59) was chosen for the hydrogen atoms bound to Rh.

The Pathway A. A clear picture of this pathway is provided by the corresponding orbital correlation diagram for the most important valence levels, see Figure 2. This diagram is characterized, in addition to an avoided crossing between the two highest occupied levels, by a near degeneracy of the product HOMO and the product LUMO which are both primarily of d_{z^2} character, but on a different metal center. One has here the molecular orbital representation of a diradical character (60) for **9** and one needs therefore an MC-SCF wavefunction based on the six orbitals sketched on Figure 2 to account properly for the reaction path between **7** and **9**. The residual interaction between the s orbital of the transferred hydrogen and the d_{z^2} orbital of the first rhodium atom (see the arrows on the correlation diagram) prevents a complete degeneracy of these two levels and hence a pure diradical character is not obtained. The partial diradical character of **9** is best illustrated by the value of 0.37 found for the CI expansion coefficient of the configuration corresponding to a double excitation from the HOMO to the LUMO (with a ground state coefficient amounting to 0.91). The residual interaction also explains why the second low energy covalent state, the $^3A'$ triplet state lies 13.2 kcal/mol above the singlet $^1A'$ state. One therefore needs to consider only the singlet state potential energy surface for this pathway.

The CAS SCF energy values obtained for the isomers **7**, **8** and **9** point to a continuous increase in energy, **8** and **9** being destabilized by 19.3 and 39.4 kcal/mol respectively (the corresponding total energies of **7**, **8** and **9** are -10722.2504, -10722.2197 and -10722.1877 a.u.). This strong endothermicity results from the existence of a doubly occupied d_{z^2} orbital on the attacked d^8 rhodium center (see the left part of the Figure 2) which repels the occupied s orbital of the migrating hydrogen. One may therefore conclude that pathway A is not likely to occur.

The Pathway B. That the d_{z^2} orbital on the d^8 rhodium center is doubly occupied results merely from the surrounding square planar coordination. From the preceeding analysis it is easy to forsee that an empty d_{z^2} orbital on the attacked rhodium atom should readily accomodate the s orbital of the migrating hydrogen. This is precisely what is obtained through an intramolecular rearrangement of the local coordination pattern, from a square plane (as in **7**) to a trigonal pyramid (as in **10**). This process is symmetry allowed but has to face avoided crossings. The energy demand is not too high, however: The computed endothermicity is 18.7 kcal/mol (at the CAS SCF level with an active space including the σ and σ^* combinations of the metal to hydrogen bonds) and there is almost no additional barrier (56). The important point is that, during the rearrangemement, the d_{xy} and d_{z^2} levels exchange; d_{z^2} becomes the highest in energy and therefore the empty level in a d^8 electron count. As a result the subsequent migration between **10** and **12** becomes exothermic by 5.6 kcal/mol, the halfway structure being destabilized by only 0.2 kcal/mol (the corresponding CAS SCF total energies are -10722.2363, -10722.2360, -10722.2453 a.u., computed from a set of 8 active orbitals, 4 strongly occupied orbitals and their correlating counterparts. These strongly occupied orbitals correspond to the two metal to hydrogen bonds and to two additional orbitals located on the trigonal pyramid moiety, namely the $\sigma_P + \varepsilon d_{z^2}$ orbital and the $d_{x^2-y^2}$ orbital which is the HOMO in the symmetric representation. The inclusion of these latter orbitals was found necessary for the stability of the CAS calculation along the reaction path.

The Pathway C. The addition of an extra ligand on the d^8 metal center has also quite interesting consequences on the feasibility of the hydrogen transfer and on the electronic structure of the system along the reaction pathway. This is because the transfer now occurs between two more or less symmetric entities. Let us first assume that one has a complete symmetrical situation. This will be the case for instance in the $H[Rh(\mu - H)(PH_3)_3]_2^+$ system where the non migrating H ligand has been replaced by PH_3. The corresponding correlation diagram for the hydrogen transfer, shown on the Figure 3, is characterized by an avoided crossing between the HOMO and the LUMO of A' symmetry, which are both primarily d_{z^2} on each center and which exchange their role in the transfer process. The reaction is athermic but one expects a high energy barrier, just as in a suprafacial [1,3] sigmatropic shift. Notice on the other hand that a system with two electron less would experience a stabilization of the halfway structure. There is indeed a variety of known $d^6 - d^6$ $H_3M_2L_6$ complexes with three bridging hydrides.

If one now replaces the phosphine ligand trans to the migrating ligand by a stronger σ donor ligand, as in 13 , the levels made of the $d_{z^2} - s_H$ combination, i.e. the LUMO in 13 and the HOMO in 15 , will be destabilized. The consequence of this destabilization is an endothermic process characterized by an avoided crossing occurring later and by some diradical character for the product. The CAS SCF calculations carried out with the orbitals sketched on Figure 3 as active orbitals (i.e. a set of 6 active orbitals populated by 6 electrons) yield an overall endothermicity of 33.8 kcal/mol with a halfway structure destabilized by 19.5 kcal/mol (the corresponding total energies are -11064.1907, -11064.1596 and -11064.1369 a.u.). It is also easy to understand that the greater the σ strength of the ligand trans to the migratory hydrogen, the greater the endothermicity will be. Conversely a reversal in the relative σ donor strength of these two apical ligands should lead to an earlier barrier (if any) and an exothermic process. One has therefore a way of tuning the hydrogen transfer through the relative σ donicity of the two apical ligands.

A second feature of this pathway is worth mentioning. For the complete symmetric situation, the halfway structure is of D_{3h} symmetry (or close to it) and the HOMO 16 becomes degenerate (or nearly degenerate) with the out of phase combination 17 of the d_{xy} rhodium orbitals. One therefore expects the corresponding triplet state to become more stable. The two compounds of this type which have been experimentally characterized are indeed paramagnetic (61-62). Although 14 is not strictly of D_{3h} symmetry, one may worry about the position of its triplet state ($^3A''$ in the actual C_s symmetry point group). CAS SCF calculations carried out by adding 17 in the active space indicate that this triplet state is indeed lower than the singlet state $^1A'$ by 19.6 kcal/mol (total energies of -11064.1938 and -11064.1625 a.u.). One therefore expects a crossing of the singlet and triplet potential energy surfaces during the transfer, the minimum on the triplet state potential energy surface being reached through intersystem crossing.

Experimental Implications. It is therefore clear from the above study that the metal to metal hydrogen transfer is indeed a facile process provided that a square plane to trigonal pyramid rearrangement at the attacked d^8 metal center occurs. A weak σ donor at the apex of the trigonal pyramid moiety, by keeping the empty accepting orbital at a low energy, should greatly help the transfer. Furthermore a strong σ donor trans to the migrating hydrogen in the octahedron will weaken the metal to hydrogen bond (63). In fact the location of the hydride ligand between the two metal

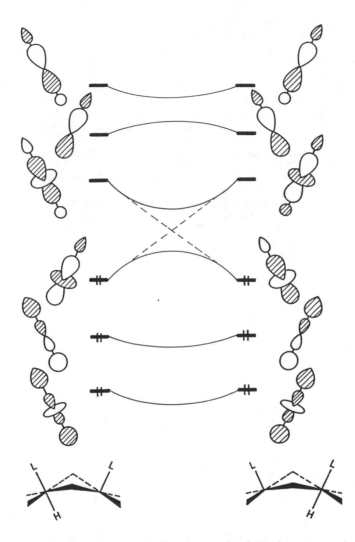

Figure 3. Schematic orbital correlation diagram for the hydrogen transfer in the symmetric $H[Rh(\mu - H)L_3]_2^+$ system. Only the in-plane valence orbitals have been represented. (Reproduced with permission from Ref. 56. Copyright 1988 Gauthier-Villars.)

16

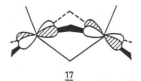

17

atoms depends on the balance between the respective bonding capabilities of the d_{z^2} orbitals on both centers, which in turn depend on the relative σ strengths of the ancillary ligands. This is why one also finds intermediate situations, such as $[(diphos)Rh(\mu - H)_3Ir(PEt_3)_3]^+$ which has a structure similar to **11** , with three bridging hydrogens (64).

Another way to induce the hydrogen migration in this type of dinuclear $d^6 - d^8$ complexes is to coordinate to the square planar d^8 metal center a ligand which has a greater σ donor strength than the ligand trans to the migrating hydrogen. The ease of the process will depend on the relative σ donor strength of the two apical ligands. The basal coordination of the olefin in the catalytic hydrogenation or hydroformylation reactions should for instance give rise to an easier transfer than the apical coordination.

Finally the reversal in the most favorable relative σ donor strength of the two apical ligands on going from the pathway B to the pathway C may be related to a different nature of the transferred hydrogen: The pathway B may be viewed as involving a transfer of a hydride, H^-, to a coordinatively unsaturated trigonal pyramid moiety acting as a Lewis acid. The pathway C, on the other hand may be viewed formally as an oxidative addition of H^+ on the coordinatively saturated square pyramidal d^8 center.

Acknowledgments

The calculations were carried out on the Cray-1 of the CCVR (Palaiseau) through a grant of computer time from the Conseil Scientifique du Centre de Calcul Vectoriel pour la Recherche, and on the IBM 3081 of the Centre de Calcul du CNRS in Strasbourg-Cronenbourg. VB is grateful to the CIRIT de la Generalitat de Catalunya for financial support.

Literature cited

1. Dedieu, A.; Nakamura, S. J. Organomet. Chem. 1984, 260, C60.
2. McNeill, E. A.; Scholer, F. R. J. Am.Chem.Soc. . 1977, 99, 6243.
3. Farmery, K.; Kilner, M. J. Chem.Soc.A 1970, 634.
4. Stobart, S. R. J. Chem.Soc.Dalton Trans. 1972, 2442.
5. Bradley, G.F.; Stobart, S.R. J. Chem. Soc. Chem. Comm. 1975, 325.
6. Vancea, L.; Graham, W. A. G. J. Organomet. Chem. 1977, 134, 219.
7. Faegri, Jr., K.; Almlof, J. Chem. Phys. Letters 1984, 107, 121.
8. McMichael Rohlfing, C.; Hay, P. J. J. Chem. Phys. 1985, 83, 4641.
9. Blomberg, M. R. A.; Brandemark, U. B.; Siegbahn, P. E. M.; Broch Mathisen K.; Karlstrom, K. J. Phys. Chem. 1985, 89, 2171.
10. Lüthi, H. P.; Siegbahn, P. E. M.; Almlof, J. J. Phys. Chem. 1985, 89, 2156.
11. Noell, J. O.; Hay, P. J. J. Am. Chem. Soc. 1982, 104, 4578.
12. Obara, S.; Kitaura, K.; Morokuma, K. J. Am. Chem. Soc. 1984, 106, 7482.
13. Paonessa, R. S.; Trogler, W. C. J. Am. Chem. Soc. 1982, 104, 1138..
14. Dedieu, A.; Nakamura, S.; Sheldon, J. C. Chem. Phys. Letters 1987, 141, 323.
15. Low, J. J.; Goddard III, W. A. J. Am. Chem. Soc. 1984, 106, 6928.
16. Low, J. J.; Goddard III, W. A. Organomet. 1986, 5, 609.
17. Schilling, J. B.; Goddard III, W. A.; Beauchamp, J.L. J. Am. Chem. Soc. 1986, 108, 582.
18. Darensbourg, M. Y.; Ash, C. E.; Kao, S. C.; Silva, R.; Springs, J. Pure Applied Chem. 1988, 60, 131 and references therein.
19. Bursten, B. E.; Gatter, M. G J. Am. Chem. Soc. 1984, 106, 2554.

20. Bursten, B. E.;   Gatter, M. G. Organomet.   1984, 3, 896.
21. Berke, H.;   Kundel, P. Z. Naturforsch.B 1986, 41, 527.
22. Berke, H.;   Kundel, P. J. Organomet. Chem. 1986, 314, C31.
23. Berke, H.;   Kundel, P. J. Organomet. Chem. 1987, 335, 353.
24. Cotton, F. A.;   Wilkinson, P. Advanced Inorganic Chemistry, 4th Edition, J. Wiley, New York, 1980; p 1066.
25. Bagus, P. S.;   Roos, B. O. J. Chem. Phys. 1981, 75, 5961.
26. Bauschlicher Jr., C. W.;   Bagus, P. S.;   Nelin, C. J.;   Roos, B. O. J. Chem. Phys. 1986, 85, 354.
27. Gross, R.;   Kaim, W. Angew. Chem. Int. Ed. Engl. 1984, 23, 614.
28. For a review see Dedieu, A. In Topics Physical Organometallic Chemistry, Gielen, M.F. Ed;Freund: London, 1985; pp 46-73.
29. Blyholder, G.;   Zhao, K. M.;   Lawless, M. Organomet. 1985, 43, 1371.
30. Ziegler, T.;   Versluis, L.;   Tschinke, V. J. Am. Chem. Soc. 1986, 108, 612.
31. Koga, N.;   Morokuma, K. J. Am. Chem. Soc. 1986, 108, 6136.
32. Axe, F. U.;   Marynick, D. S. Organomet. 1987, 6, 572.
33. Rappé, A. K. J. Am. Chem. Soc. 1987, 109, 5605.
34. Pacchioni, G.;   Fantucci, P.;   Koutecky, J.;   Ponec, V. J. Cat. 1988, 112, 34.
35. McKee, M. L.;   Dai, C. H.;   Worley, S. D. J. Phys. Chem. 1988, 92, 1056.
36. Dedieu, A. Inorg. Chem. 1981, 20, 2803.
37. Nakamura, S.;   Dedieu, A. Chem. Phys. Letters 1984, 111, 243.
38. Dedieu, A.;   Nakamura, S. In Quantum Chemistry,the Challenge of Transition Metals and Coordination Chemistry, Veillard, A., Ed.; NATO ASI Series C, No 176; D. Reidel, Dordrecht. 1986; p 277.
39. Dedieu, A.;   Sakaki, S.;   Strich, A.;   Siegbahn, P. Chem. Phys. Letters 1987, 133, 317.
40. Hoffmann, R. Angew. Chem. Int. Ed. Engl. 1982, 21, 711.
41. Calderazzo, F.;   Dell'Amico, D. B. Inorg. Chem. 1981, 20, 1310.
42. Axe, F. U.;   Marynick, D. S. Chem. Phys. Letters 1987, 141, 455.
43. Lie, G. C.;   Clementi, E. J. Chem. Phys. 1974, 60, 1275.
44. Roos, B.;   Siegbahn, P. Theor. Chim. Acta 1970, 17, 199.
45. Carter, E. A.;   Goddard III, W. A. Organomet. 1988, 7, 675.
46. Thorn, D. L.;   Roe, D. C. Organomet. 1987, 6, o17 and references therein.
47. Halpern, J.;   Okamoto, T.;   Zakhariev, A. J. Mol. Cat. 1976, 2, 65.
48. Darensbourg, D. J.;   Rokicki, A.;   Darensbourg, M. Y. J. Am. Chem. Soc. 1981, 103, 3223.
49. Powell, J.;   Gregg, M. R.;   Sawyer, J. F. J. Chem.Soc. Chem. Comm. 1987, 1029.
50. Sivak, A. J.;   Muetterties, E. L. J. Am. Chem. Soc. 1979, 101, 4878.
51. Bonnet, J. J.;   Thorez, A.;   Maisonnat, A.;   Galy, J.;   Poilblanc, R. J. Am. Chem. Soc. 1979, 101, 5940.
52. Meier, E.B.;   Burch, R. R.;   Muetterties, E. L.;   Day, V. W. J. Am. Chem. Soc. 1982, 104, 2661.
53. Burch, R. R.;   Shusterman, A. J.;   Muetterties, E. L.;   Teller, R. G.;   Williams, J. M. J. Am. Chem. Soc. 1983, 105, 3546.
54. Kalck, Ph.;   Frances, J.M.;   Pfister, M.;   Southern, T. G.;   Thorez, A. J. J. Chem. Soc. Chem. Comm. 1983, 510.
55. Dedieu A.;   Escaffre. P.;   Frances. J. M.;   Kalck, Ph.;   Thorez, A. Nouv. J. Chimie. 1986, 10, 631.
56. Branchadell, V.;   Dedieu A. In Recent Advances in Di and Polynuclear Chemistry,Braunstein, P. Ed; New J. Chem. 1988, 12, 443.
57. Veillard, A.;   Dedieu A. Theoret. Chim. Acta 1984, 65, 215.

58. Huzinaga, S. Approximate Atomic Functions, Technical Report, University
    of Alberta, Edmonton, 1971.;
59. Huzinaga, S. J. Chem. Phys. 1965, 42, 1293.
60. Salem, L.; Rowland, C. Angew. Chem. Int. Ed. Engl. 1972, 11, 92.
61. Bianchini, C.; Mealli, C.; Meli, A.; Sabat, M. J. Chem. Soc. Chem.
    Comm. 1986, 777.
62. Dapporto, P.; Midollini, S.; Sacconi, L. Inorg. Chem. 1975, 14, 1643.
63. Burdett, J. K.; Albright, T. A.; Sacconi, L. Inorg. Chem. 1979, 18, 2112.
64. Albinati, A.; Musco, A.; Naegeli, R.; Venanzi, L.M. Angew. Chem.
    Int.Ed. Engl. 1981, 20, 958.

RECEIVED October 24, 1988

Chapter 6

Potential Energy Surface of Olefin Hydrogenation by Wilkinson Catalyst

Comparison Between trans and cis Intermediates

Nobuaki Koga and Keiji Morokuma

Institute for Molecular Science, Myodaiji, Okazaki 444, Japan

The results of recent theoretical investigation with
the ab initio MO method on the full catalytic cycle is
presented. The catalytic cycle studied is for olefin
hydrogenation by the Wilkinson catalyst. We have
determined with the ab initio energy gradient method
the structures of the transition states as well as the
intermediates of the Halpern mechanism in which all
the intermediates have trans phosphines. The potential
energy profile thus obtained supports the Halpern
mechanism and gives evidence on the effectiveness of
the Wilkinson system as a catalyst. A new mechanism,
more recently proposed, considers that intermediates
with cis phosphines, in contrast to trans in the
Halpern mechanism, play an essential role.   Our
calculation indicates that energies of cis intermediates are high, and does not support the cis
mechanism for sterically unhindered olefins. However,
when steric effects inhibit reactions of trans
intermediates, the cis mechanism may become possible
and would exhibit the kinetics which is quite different from that of the Halpern mechanism.

Recent progress of methodology of quantum chemistry and technology
of electronic computer is making it possible for quantum chemists to
challenge the chemistry of d and/or f electrons. Now, such efforts
have covered a full catalytic cycle as well as structures of
complexes and intermediates and elementary organometallic reactions
(1).
  It has been our goal to design a catalytic system theoretically. To the end of this goal, we have so far analyzed the
organometallic reactions by using the ab initio MO calculations.
Recently, we have completed the theoretical study of the catalytic
cycle of hydrogenation by the Wilkinson catalyst (2), of which
mechanism has been proposed by Halpern (3). This catalytic cycle
shown in Scheme 1 consists of oxidative addition of H_2, coordination
of olefin, olefin insertion, isomerization, and reductive elimina-

0097–6156/89/0394–0077$06.00/0

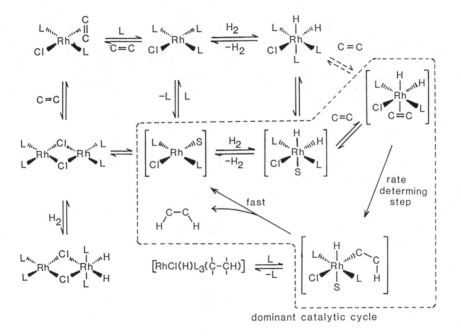

dominant catalytic cycle

Scheme 1

tion of alkane. The rate-determining step has been believed to be
the olefin insertion step. We used PH_3 instead of PR_3 and C_2H_4 as a
model of olefin. The model catalytic cycle we studied is shown in
Scheme 2. All the equilibrium and transition state structures were
optimized by the RHF energy gradient method.

   While two phosphines are always <u>trans</u> to each other in all the
intermediates in the Halpern mechanism, Brown et al. have very
recently proposed a different mechanism, in which the <u>cis</u> biphosphine intermediates play an essential role (<u>4</u>). Their molecular
modeling calculations where the van der Waals energy is calculated
between substituted olefins and the Rh fragment with bulky trans
phosphines have suggested that when the substituents on the olefin
are bulky, the steric repulsion is too large for the olefin to
coordinate. Their NMR experiments have shown the existence of the
following equilibrium (Equation 1) in which an intermediate with a
pair of cis phosphines can be formed. In fact, $H_2RhCl(PR_3)_3$ has
been detected in the catalytic system.

$$H-Rh\overset{\overset{H}{|}}{\underset{\underset{Ph_3\overset{a}{P}}{}}{}}\begin{smallmatrix}{}^{a}PPh_3\\{}^{b}PPh_3\\Cl\end{smallmatrix} \rightleftharpoons H-Rh\overset{\overset{H}{|}}{\underset{\underset{Cl}{}}{}}\begin{smallmatrix}{}^{a}PPh_3\\{}^{b}PPh_3\end{smallmatrix} + {}^{a}PPh_3 \qquad (1)$$

Based on these results, they have suggested that the intermediates
of the catalytic system have two cis phosphines. In Scheme 3 is
shown the new mechanism. In this Scheme, the key step is isomerization (Equation 2), trans-$H_2RhCl(PR_3)_2$ to cis-$H_2RhCl(PR_3)_2$ presumably through pseudorotation.

$$Cl-Rh\overset{\overset{PPh_3}{|}}{\underset{\underset{PPh_3}{|}}{}}\begin{smallmatrix}\backslash H\\ \\ H\end{smallmatrix} \rightleftharpoons H-Rh\overset{\overset{H}{|}}{\underset{\underset{Cl}{|}}{}}\begin{smallmatrix}PPh_3\\ \\PPh_3\end{smallmatrix} \qquad (2)$$

Then, the olefin insertion and the reductive elimination take place
from the resultant cis biphosphine complex.

   In this article, we will compare the energetics of the 'conventional' Halpern mechanism with that of the Brown mechanism. The
basis functions used are the 3-21G for ethylene and hydrides, the
STO-2G for 'spectator' ligands, PH_3 and Cl, and valence double zeta
basis functions for Rh with effective core potential replacing the
core electrons (up to 4p) (<u>5a,b,6</u>). In addition, we carried out the
MP2 calculations at selected, RHF-optimized structures with a larger
basis set, which consists of uncontracted (3s,3p,4d) functions from
the above valence DZ set for Rh, 4-31G for the ethyl group,
(10s,7p)/[3s2p] for P and Cl, and (4s)/[3s] for the hydrides (<u>5c-
e</u>). The basis set used is limited and the electron correlation
taken into account for a few critical steps is minimal. Therefore,

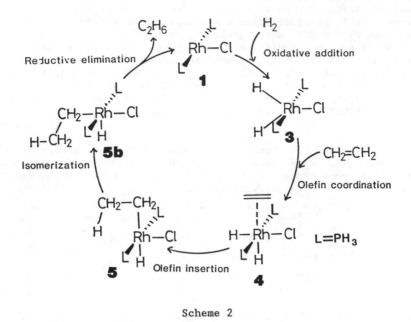

Scheme 2

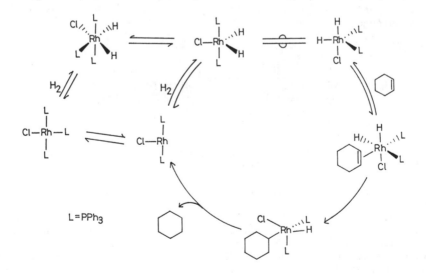

Scheme 3

the results presented here should be considered to be semi-
quantitative.

Halpern mechanism

Since our calculations on the Halpern mechanism have been published
(2), we will give a brief summary for comparison in a succeeding
section. The potential energy profile shown in Figure 1 is constructed from the energetics of the elementary reactions involved in
the Halpern mechanism. The optimized structures are shown in Figure
2.
 The first step of the H_2 oxidative addition is exothermic and
leads to the dihydride complex 3. During this step, there may be an
H_2 complex 2 from which oxidative addition takes place with almost
no activation barrier. The ethylene coordination that follows
requires no activation energy. The resultant ethylene dihydride
complex 4 is in the valley of the potential energy surface of the
catalytic cycle. Ethylene insertion requires a much higher activation energy of 18 kcal/mol and is endothermic by 16 kcal/mol at the
RHF level. The trans ethyl hydride complex, the direct product of
ethylene insertion, is unstable due to the cis effect of Cl to be
mentioned below and the trans effect of H and C_2H_5. Therefore
isomerization takes place to give more stable ethyl hydride complexes, which have ethyl and hydride cis to each other and are the
starting point of the final reductive elimination step. This
isomerization proceeds through hydride and chloride migration. The
final reductive elimination step requires a substantial energy
barrier of 15 kcal/mol.
 The potential energy profile is smooth without excessive
barriers and too stable intermediates which would break the sequence
of steps. The rate-determining step is found to be olefin insertion
followed by isomerization, supporting the Halpern mechanism.
Isomerization of the ethyl hydride complex is an important part of
the rate-determining step. These two reactions, exothermic overall,
has an overall barrier height of about 20 kcal/mol. The trans ethyl
hydride complex, the product of ethylene insertion, may not be a
local minimum (per MP2 calculation) and these two steps may well be
a combined single step.
 The activation barrier of reductive elimination, though
substantial, is smaller than that of the reverse of the rate-
determining step (isomerization and β-hydrogen elimination). This
is a very important requirement of a good olefin hydrogenation
catalyst. If this reverse reaction is easy, it would lead to
undesirable olefin isomerization. For instance, in the same hydrogenation cycle catalyzed by a Pt system, we found that the rate-
determining step is reductive elimination rather than olefin
insertion. This potential energy profile is expected to give olefin
isomerization through successive olefin insertion/β-hydrogen
elimination. Thus the Pt complex is not a good catalyst for olefin
hydrogenation. On the other hand, the potential profile of the
Wilkinson catalyst indicates an efficient hydrogenation without
isomerization.
 It is important, as mentioned above, that the olefin insertion
is rate-determining. Therefore, we have compared the reaction
energetics between $H_2RhCl(PH_3)_2(C_2H_4)$ and $H_2RhH(PH_3)_2(C_2H_4)$, and

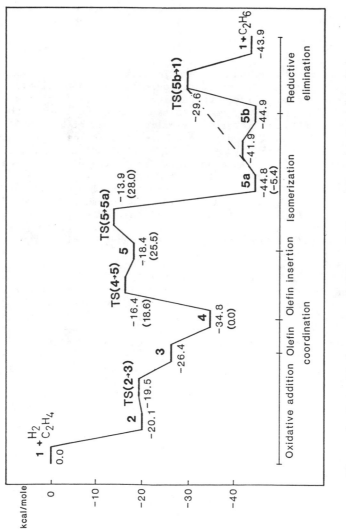

Figure 1. Potential energy profile of the entire catalytic cycle in the Halpern mechanism for olefin hydrogenation, in kcal/mol at the RHF level, relative to $1 + C_2H_4 + H_2$. Numbers in parentheses are the MP2 energy at the RHF optimized[2] geometries, relative to 4.

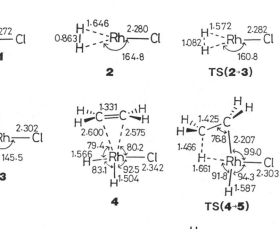

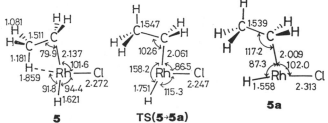

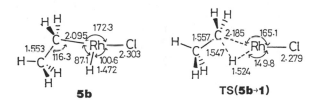

Figure 2. Optimized structures (in Å and deg) of some important species. TS(2→3), for instance, denotes the transition state connecting 2 and 3. Though practically all the geometrical parameters were optimized, only essential values are shown. Two PH3's, one above and one below the plane of paper, are omitted for clarity.

found that, in the former complex, the weak trans influence of Cl
makes the Rh-H bond to be broken stronger. In addition the strong Cl
cis effect makes the Rh-C bond to be formed weaker, resulting in
endothermic ethylene insertion. These two effects of Cl combined
appears to be essential to make ethylene insertion the rate-
determining step.

Brown mechanism

The first point of difference of the Brown mechanism from the
Halpern mechanism is isomerization of $H_2RhCl(PH_3)_2$. Therefore, we
have investigated the stability of isomers of $H_2RhCl(PH_3)_2$, 3. The
optimized structures of $H_2RhCl(PH_3)_2$ are shown in Figure 3. The most
stable isomer is found to be 3, the trans intermediate of the
Halpern mechanism. All the optimized structures but 3 are nearly
square pyramidal (though optimization was done without symmetry
restriction). The most stable square pyramidal isomer is 3a with
apical H and basal cis phosphines. 3b with apical phosphine and
basal cis hydrides is next. The remaining three isomers, 3c, 3d, and
3e are much more unstable; the energies relative to 3 are 33, 37 and
39 kcal/mol, respectively. Two hydrides with strong trans influence
are located trans to each other in 3c. This makes 3c 12 kcal/mol
less stable than 3b in which two hydrides are cis. The least stable
isomers, 3d and 3e, have apical Cl.
    Comparison of the stability among the isomers of 3 leads to
the order of apical preference: H>PH$_3$>Cl. H with the strongest
trans influence prefers the apical position that is trans to the
vacant site, and the most weakly trans-influencing Cl at the apical
position gives the most unstable isomers of 3d and 3e. 3b and 3c
are inbetween, in accord with the strength of PH$_3$ trans influence.
    Since 3a and 3b are low in energy, we have investigated
isomerization from 3 to 3a and 3b. Brown et al. have proposed that
the cis intermediate of the catalytic cycle is 3b in which one of
the bulky phosphines is trans to olefin and thus the vacant coordination site is less crowded. There are two possible pathways for
isomerization of 3 as shown in Scheme 4.
    The intermediates of the second pathway are unstable 3d and 3c,
and it is unlikely that isomerization takes place through them. The
easier isomerization pathway is through 3b to 3a. The transition
state for PH$_3$ migration connecting 3 and 3b have been located, as
shown in Figure 3, with the activation barrier for isomerization
from 3 to 3b of 27 kcal/mol (See Scheme 4). Therefore, one can
conclude isomerization from 3 to 3b or 3a is rather difficult (cf.
18 kcal/mol, the activation energy of the rate-determining step in
the Halpern mechanism at the same level of calculation).
    Setting aside this high activation barrier for a moment, the
remaining steps of the catalytic cycle in the Brown mechanism will
be discussed. As shown in the energy profile of the Halpern mechanism, the elementary reactions involved here are expected to be
very easy, and thus we have just determined the structures and
energies of intermediates but not of transition states (Figure 5).
The relative energies of the intermediates, shown in Figure 4, would
be a good indicator of the barrier of each elementary reaction
connecting them; an endothermic reaction requires a large activation
energy and the activation barrier of an exothermic step is low.

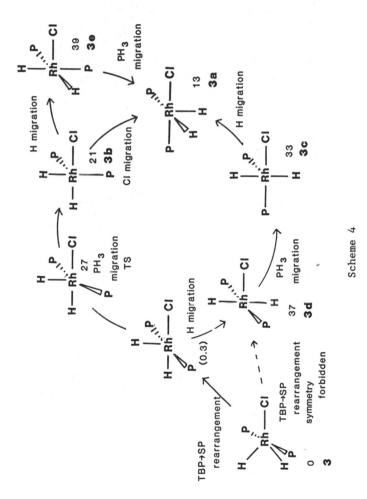

Scheme 4

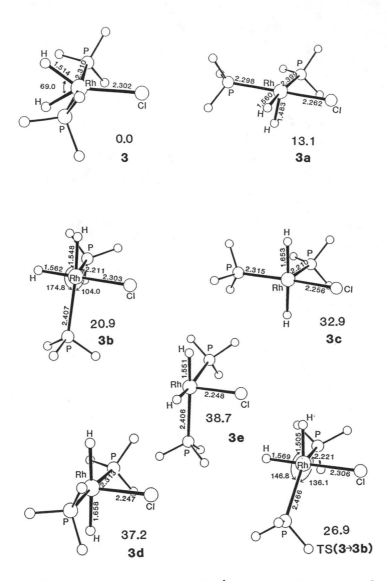

Figure 3. Optimized structures (in Å and deg) of isomers of
$H_2RhCl(PH_3)_2$ and the **3**→**3b** transition state, and their energies
(in kcal/mol) relative to **3**.

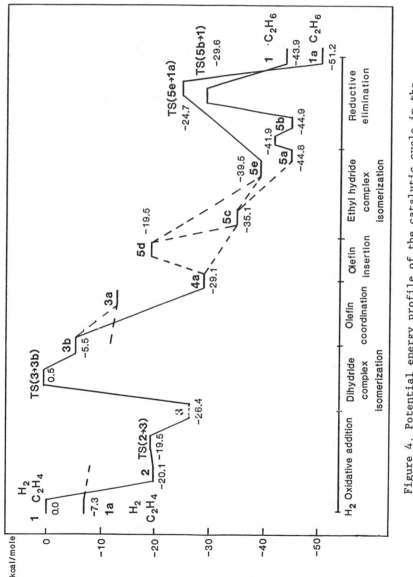

Figure 4. Potential energy profile of the catalytic cycle in the Brown mechanism for olefin hydrogenation, in kcal/mol at the RHF level, relative to $1 + C_2H_4 + H_2$.

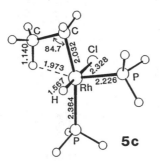

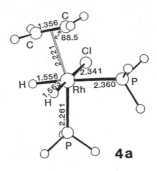

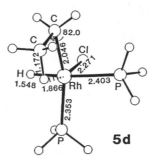

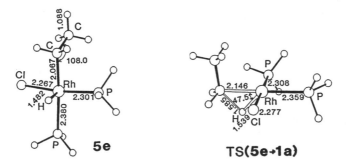

Figure 5. Optimized structures (in Å and deg) of some important species in the Brown mechanism.

Ethylene coordination to **3b**, the Brown's intermediate, gives **4a** which is higher in energy than **4** by 6 kcal/mol. Olefin insertion of **4a** can lead to **5c** or **5d**. Since **5c** is much more stable than **5d**, olefin insertion giving **5c** would take place exclusively. The instability of **5d** with an apical Cl is similar to that of **3d** and **3e** discussed above.

The ethyl group and the hydride in **5c** are cis to each other and thus reductive elimination might take place directly without isomerization. However, reductive elimination from a d^6 five-coordinate complex would favor a transition state where three ligands but C_2H_5 and H are in the same plane, as shown below.

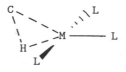

The reason for this preference is that the donation and back donation between a deformed alkane and a metal fragment shown below is expected to facilitate easy bond exchange.

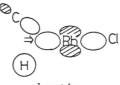   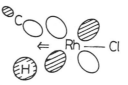

        donation   back-donation

Therefore, prior to reductive elimination, isomerization should take place from **5c** to ethyl hydride complexes which have H or C_2H_5 as an apical group, as shown below.

Ethyl migration from **5c** leads to **5c** itself, and hydride migration gives **5d**, an unstable intermediate; neither of these gives apical H or C_2H_5. The two remaining migrations give a more stable ethyl hydride complexes and they have either an apical H or C_2H_5; Cl migration leads to stable **5e** and PH_3 migration results in **5a** with trans phosphines, the intermediate of the Halpern mechanism. The transition state for reductive elimination of **5e** to give **1a** has been determined, as shown in Figure 5, and it has an activation energy of 14.8 kcal/mol. **5a** is more stable than **5e** by 5 kcal/mol and the activation energy for reductive elimination from **5a** to regenerate **1** is calculated to be about 15 kcal/mol, comparable with the **5e→1a** barrier. Therefore, the system is expected to return to the

Halpern mechanism, if it is not prevented for some reason (eg.
steric).

Comparison between two mechanisms

In the Brown mechanism, setting aside the high energy required for
isomerization from 3 to 3b, the final step of the reductive elimination would require the highest activation energy. Olefin insertion,
the rate-determining step in the Halpern mechanism, is an easy
process in this mechanism. The kinetics of the Brown mechanism is
thus expected to be completely different from that of the Halpern
mechanism. Therefore, in the cases in which olefin insertion has
been found to be rate-determining, the Halpern mechanism is clearly
more consistent and acceptable.

The above feature of the Brown mechanism that reductive
elimination is more difficult than olefin insertion may be related
to the nature of catalyst having a chelating bidentate ligand such
as DIPHOS. Halpern have also investigated the hydrogenation
(Equation 3) ($\underline{7}$), where isomerization from the trans- to cis-
biphosphine complex is not necessary.

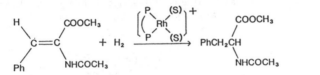

$$\tag{3}$$

They have found that at the low temperature, reductive elimination
is rate-limiting ($\Delta H = 17.0$ kcal/mol $-40°C$).

The present calculations using PH_3 as the phosphine and
ethylene as the olefin, however, does not exclude the possibility
of the cis mechanism completely, since the steric factor has not
been taken into account. The cis mechanism might be accessible in
the case where olefin is too bulky to coordinate to the trans
phosphine complex. Using the above calculations as a guide, here we
consider qualitatively what is expected to take place when a very
bulky olefin is hydrogenated. Let us assume that in such a case the
equilibrium (Equation 1) generates the cis intermediate. Then, the
reaction route will pass through 3b→4a→5c, each of which is sterically not too crowded. Isomerization of 5c is not allowed to lead
to 5a, which has the bulky alkyl group cis to two phosphines and is
overcrowded. Thus isomerization of 5c has to lead to 5e, which is
intrinsically (without steric effect) only slightly less stable but
sterically less crowded than 5a. Reductive elimination of 5e will
require an activation energy comparable to that of 5a→1 and generate
cis-RhCl(PH_3)$_2$, 1a.

There are two possibilities in the reactions of 1a. The first
is that 1a isomerizes to 1 due to the steric repulsion and that the
same reaction path 1→2→3→3b is followed. The second possibility is
that H_2 oxidative addition to 1a takes place to give directly 3a and
thus in the subsequent catalytic cycles intermediates always have
cis phosphines. Olefin coordination to 3a is prohibited because of
the steric repulsion between the bulky olefin and two bulky phosphines cis to the olefin. Thus 3a→3b isomerization has to take
place before the catalytic cycle proceeds.

One can consider that the Halpern mechanism and the cis mechanism are two extremes. The Halpern mechanism is most widely accepted and our ab initio MO calculations support this from the point of view of intrinsic electronic energy. However, there may be cases where the steric effect overshadows the electronic effect. There may also be cases where both effects are important. It may be, as Collman et al. said, that "this multistep reaction is very complicated. Like a chameleon, the dominant reaction mechanism changes when the nature of the catalyst, the ligands, or the substrate is altered" (8).

Conclusions

In this work, we have compared the potential energy profiles of the model catalytic cycle of olefin hydrogenation by the Wilkinson catalyst between the Halpern and the Brown mechanisms. The former is a well-accepted mechanism in which all the intermediates have trans phosphines, while in the latter, proposed very recently, phosphines are located cis to each other to reduce the steric repulsion between bulky olefin and phosphines. Our ab initio calculations on a sterically unhindered model catalytic cycle have shown that the profile for the Halpern mechanism is smooth without too stable intermediates and too high activation barrier. On the other hand, the key cis dihydride intermediate in the cis mechanism is electronically unstable and normally the sequence of elementary reactions would be broken. Possible sequences of reactions can be proposed from our calculation, if one assumes that steric effects of bulky olefin substituents prohibits some intermediates or reactions to be realized.

Literature Cited.

1. Koga, N.; Morokuma, K. Top.Phys.Organomet.Chem., in press.
2. (a) Koga, N.; Daniel, C.; Han, J.; Fu, X.Y.; Morokuma, K. J.Am.Chem.Soc., 1987, 109, 3455. (b) Daniel, C.; Koga, N.; Han, J.; Fu, X.Y.; Morokuma, K. J.Am.Chem.Soc., 1988, 110, 3773.
3. (a) Halpern, J.; Wong, C.S. J.Chem.Soc.Chem.Commun., 1973, 629. (b) Halpern, J. In Organotransition Metal Chemistry; Ishii, Y.; Tsutsui, M., Eds.; Plenum: New York, 1975; p109. (c) Halpern, J.; Okamoto, T.; Zakhariev, A. J.Mol.Catal., 1976, 2, 65.
4. Brown, J.M.; Evans, P.L.; Lucy, A.R. J. Chem. Soc. Perkin Trans. II, 1987, 1589.
5. (a) Binkley, J.S.; Pople, J.A.; Hehre, W.J. J.Am.Chem.Soc., 1980, 102, 939. (b) Hehre, W.J.; Stewart, R.F.; Pople, J.A. J.Chem.Phys., 1969, 51, 2657. (c) Ditchfield, R.; Hehre, W.J.; Pople, J.A. J.Chem. Phys., 1971, 54, 724. (d) Huzinaga, S.; Andzelm, J.; Kłobukowski, M.; Radzio-Andzelm, E.; Sakai, Y.; Tatewaki, H. Gaussian Basis Sets for Molecular Calculations; Elsevier: Amsterdam, 1984.
6. Hay, P.J.; Wadt, W.R. J. Chem. Phys., 1985, 82, 270.
7. Chan, A.S.C.; Halpern, J. J.Am.Chem.Soc., 1980, 102, 838.
8. Collman, J.P.; Hegedus, L.S.; Norton J.R.; Finke, R.G. Principles and Applications of Organotransition Metal Chemistry; University Science Books: Mill Valley, 1987; p.535.

RECEIVED January 18, 1989

Chapter 7

Ab Initio Studies of Transition Metal Dihydrogen Chemistry

Edward M. Kober and P. Jeffrey Hay

Los Alamos National Laboratory, Los Alamos, NM 87545

Examples of transition metal complexes containing dihydrogen ligands are investigated using *ab initio* electronic structure calculations employing effective core potentials. Calculated geometrical structures and relative energies of various forms of $WL_5(H_2)$ complexes (L = CO, PR_3) are reported, and the influence of the ligand on the relative stabilities of the dihydrogen and dihydride forms is studied. The possible intramolecular mechanisms for H–D scrambling involving $Cr(CO)_4(H_2)_2$ are investigated by examining the structures and energies of various polyhydride species.

The recent discoveries of a new class of metal complexes involving molecular hydrogen has spawned numerous experimental[1,2] and theoretical [3–5] investigations to understand the bonding and reactivity of these systems. In this paper we discuss recent theoretical calculations using *ab initio* electronic structure techniques. In the first part of this article we review calculations on W dihydrogen species of d^6 W complexes with emphasis on predictions of structures and energies of various chemical forms and on comparisons with available experimental information. In the second part a mechanistic problem involving scrambling of H_2/D_2 mixtures to HD by Cr dihydrogen complexes is addressed. In this section we hope to illustrate the role of for theory in helping to distinguish between various reaction mechanisms in transition metal chemistry.

Before proceeding to the specific examples of dihydrogen chemistry, it is worthwhile to summarize the particular challenges transition metal and actinide compounds present to this type of approach. There is a striking contrast to most compounds of first– and second–row main–group elements where reasonably accurate bond lengths and bond angles of stable species can be predicted at the SCF Hartree–Fock level with small basis sets and thermochemical quantities can be computed with reasonable accuracy by perturbative techniques to electron correlation (6). Accurate studies of molecular properties or transition states of chemical reactions are feasible using multi–configuration SCF (MC–SCF) and configuration interaction (CI) techniques. In contrast, for transition metal compounds one encounters a

0097–6156/89/0394–0092$06.00/0

situation where metal ligand distances are predicted to be 0.05–0.25 Å too long at the SCF level of calculation even with very accurate basis sets (7–10) and such approaches as Moller–Plesset perturbation theory can be unreliable for treating electron correlation effects (11,12). The relatively poor description of the ground state by a single configuration and the presence of numerous low–lying electronic states with very different electron correlation energies often requires sophisticated MC–SCF and CI treatments to obtain reliable molecular geometries and bond energies (13). In addition there are additional challenges arising from the sheer number of electrons in transition metal compounds and the relativistic effects which become increasingly important in second– and third–transition series compounds. These latter difficulties have been largely circumvented by the development of relativistic effective core potentials(14) to replace the chemically inert core electrons and to incorporate the relativistic effects on the valence electrons into the effective potential.

Tungsten Dihydrogen Complexes

In the molecular dihydrogen species $W(CO)_3(PR_3)_2(H_2)$ first characterized by Kubas *et al.* in both X–ray and neutron diffraction studies (1,2), the dihydrogen is bonded in η^2 sideways fashion to the d^6 metal center. Over 50 compounds involving many of the transition metals have since been synthesized by various workers. In addition, reanalysis(15) of existing hydrides such as $FeH_4(PR_3)_3$ have now been found to be formulated as molecular hydrogen complexes, i.e., as $Fe(H_2)(H)_2(PR_3)_3$.

In this section we review briefly our previous theoretical calculations(5) on two prototypical dihydrogen complexes $W(CO)_3(PH_3)_2(H_2)$ and $W(PH_3)_5(H_2)$. The calculations employed a relativistic effective core potential (ECP–1) to replace the inner [Xe] ($4f^{14}$) core on W and a nonrelativistic ECP on P with a flexible gaussian basis to describe the valence electrons of the system. Details of the calculation are given in Ref. 5.

Structures 1–3 in Fig. 1 exemplify the modes of H_2 bonding to a $W(CO)_3(PH_3)_2$ fragment: two sideways bonded (η^2–coordinated) forms (1 and 2) and the end–on bonded (η^2–coordinated) form (3). Using a rigid $W(CO)_3(PH_3)_2$ fragment, the geometries of these three forms of H_2 coordination have been optimized using Hartree–Fock wave functions. The sideways bonded species (Table I) are found to be stable with respect to the fragments 4 by 17 kcal/mol and more stable than the end–on form, which is bound by only 10 kcal/mol. Little difference in energy is observed between the two–sideways bonded form with the H_2 axis parallel either to the P–W–P axis or to the C–W–C axis. The former orientation is slightly favored, leading to a rotational barrier about the midpoint of the W–H_2 bond of 0.3 kcal/mol.

The calculated structure of the lower energy η^2 form (1) shows a slight lengthening (from 0.74 to 0.796 Å) of the H–H bond from uncomplexed H_2 with a W–H distance of 2.15 Å. Recent low–temperature neutron diffraction studies show two equal W–H bonds (1.89 ± 0.01 Å) and with the H_2 lying exactly parallel to the P–W–P axis as predicted by the present calculations and having a H–H separation of 0.82 ± 0.01 Å. Although the calculations have correctly described the preferred mode of H_2 binding, there remain some quantitative differences (Table I) between the theoretical and observed bond lengths. These differences are reduced considerably when one employs an effective core potential (ECP–2) which also treats the outermost 5s and 5p core orbitals of W as valence orbitals. The W–H and H–H distances are now calculated to be 1.93 and 0.81 Å, respectively, in much better agreement with the neutron diffraction results.

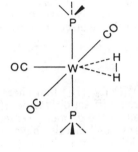

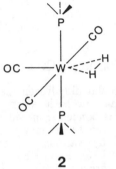

1   **2**

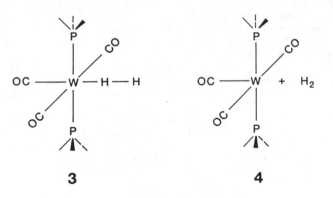

3   **4**

Fig. 1.   Structural forms of $W(CO)_3(PR_3)_2(H_2)$ species.

Table I. Calculated and Experimental W–H and H–H
Bond Lengths and Rotational Barriers for
$W(CO)_3(PR_3)_2(H_2)$ Species

Method	R	R(W–H),Å	R(H–H),Å	Rotational barrier, kcal/mole
SCF calc. (ECP–1 on W)	H	2.153	0.796	0.3
SCF calc (ECP–2 on W)	H	1.932	0.806	1.3
Low–temp X–ray diffraction	i–Pr	1.95±0.23	0.75±0.16	——
Low–temp neutron diffraction	i–Pr	1.89±0.01	0.82±0.01	2.4

Examination of the Mulliken population analyses for the fragment and the η^2 complex reveals an overall increase of 0.12 e on the W atom upon complexation and the total charge on each hydrogen has decreased slightly from 1.00 to 0.98 e. The σ–bonding orbitals of the W atom (6s, $6p_z$, and $5d_{z^2}$) show a net increase of 0.13 e, while the π–bonding orbitals ($6p_y$ and $5d_{yz}$) undergo a net loss of 0.03 e. Although the other five ligands also influence the amount of charge on the metal, the above trends are consistent with a mechanism involving some σ–donation from the H_2 ligand and a lesser degree of π–back–donation from the metal.

Of the possible seven–coordinate dihydrides let us consider the least motion reaction in which the two W–L bonds originally parallel to the H–H axis bend back as two W–H bonds are formed. The energies of these species are compared with the η^2–dihydrogen forms and the fragments in Fig. 2. Both seven–coordinate dihydride species lie higher in energy than the dihydrogen from (17 and 11 kcal/mol, respectively) and are only slightly bound compared to $WL_5 + H_2$.

For the case of having all PH_3 ligands in the $W(PH_3)_5(H)_2$ complexes, a much different situation prevails concerning the oxidative addition reaction. In contrast to $W(CO)_3(PH_3)_2(H)_2$, the seven–coordinate dihydride $W(PH_3)_5(H)_2$ lies 3 kcal/mol below the η^2 complex! Replacing the CO ligands by PR_3 groups favors the oxidative addition reaction proceeding to completion rather than being arrested in the η^2–dihydrogen stage. This preference for η^2–coordination in $W(CO)_3(PH_3)_2(H)_2$ correlates with the overall stabilization of the 5d orbitals, and the $5d_{xz}$ orbital in particular, by the back–bonding CO ligands. When these ligands are replaced by the less stabilizing PH_3 groups, the dihydride is the most favored form. This is consistent with experimental observations that the dihydrogen–dihydride equilibrium can be shifted depending on the basicity of the ligands. For example, in a series of complexes, $Mo(CO)(R_2PC_2H_4PR_2)_2 H_2$, the coordination mode changes from dihydrogen from R=Ph to dihydride for the more basic R=Et(2).

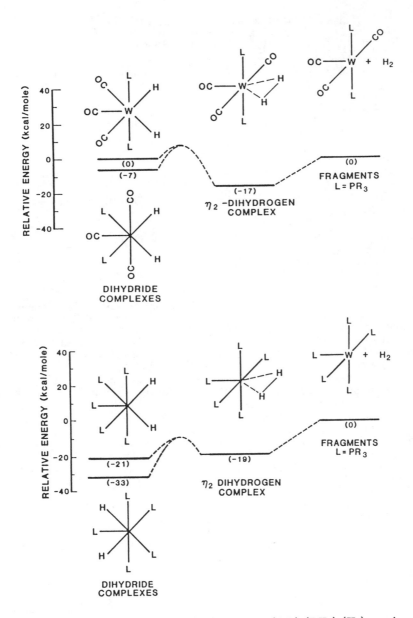

Fig. 2   Relative energies in kcal/mole of $W(CO)_3(PH_3)_2(H_2)$ species (above) compared to $W(PH_3)_5(H_2)$ species (below).

H–D Exchange Involving Cr(CO)$_4$Dihydrogen Species

Background. The gas phase reaction

$$H_2 + D_2 \rightarrow 2HD$$

occurs only under severe conditions such as in shock tubes with an activation energy (~100 kcal/mole) comparable to the H–H bond energy. In fact, the kinetics have been interpreted in terms of a free radical mechanism involving H atoms rather than the bimolecular process indicated in the above equation. By contrast, several cases of facile H$_2$–D$_2$ exchange have been observed under thermal or low temperature conditions involving dihydrogen complexes

$$L_n MH_2 + D_2 \rightarrow L_n MHD + HD$$

Upmacis, Poliakoff and Turner([16]) observed HD exchange for mixtures of Cr(CO)$_4$(H$_2$)$_2$ and D$_2$ but interestingly <u>not</u> for mixtures of Cr(CO)$_5$(H$_2$) and D$_2$. Kubas *et al.* ([1c]) observe a similar phenomenon where HD is produced from reacting W(CO)$_3$(PR$_3$)$_2$(H$_2$) with D$_2$ either in solution or in the solid state. In addition other cases of H–D scrambling occur readily with metal hydride complexes([17]) as in the case of Cp*ScH or Cp$_2$ZrH where Cp* = C$_5$Me$_5$.

Since the work of Upmacis *et al.* on Cr(CO$_4$)(H$_2$)$_2$ complexes is suggestive of an intramolecular mechanism, Burdett *et al.*([18]) have examined various polyhydride structures as possible intermediates in this process using extended Huckel theory. These studies have led us to pursue *ab initio* electronic structure calculations of Cr(CO)$_4$(H$_2$)$_2$ species and possible mechanisms leading to H$_2$–D$_2$ exchange. Implicit in these studies is the assumption that there is rapid equilibrium between

$$Cr(CO)_4(H_2)_2 + D_2 \rightarrow Cr(CO)_4(H_2)(D_2) + H_2$$

which subsequently undergoes intramolecular exchange

$$Cr(CO)_4(H_2)(D_2) \rightarrow Cr(CO)_4(HD)_2$$

although this is only inferred from the experimental studies. What is actually observed is Cr(CO)$_5$(HD) formation in a mixture of Cr(CO)$_5$(D$_2$) and Cr(CO)$_4$(D$_2$)$_2$ when reacted with H$_2$.

Results of *ab initio* calculations. The calculations have been carried out on stable structures of Cr(CO)$_4$(H$_2$)$_2$ or its fragments at the Hartree–Fock level, where the structures have been optimized using gradient techniques with the modified GAUSSIAN82([19]) or the MESA([20]) electronic structure codes. An effective core potential was used([14]) to replace the [Ne] core of Cr with a [3s 3p 2d] contracted Gaussian basis to describe the 3s, 4s, 3p, 4p and 3d orbitals. A flexible [3s 1p] bais was used for hydrogen and an STO–3G basis was employed for C and O. (Some results are presented using an unpolarized (3s) hydrogen basis.)

Of the possible forms for the parent molecule cis–Cr(CO)$_4$(H$_2$)$_2$ the lowest geometrical structure is found to have the H$_2$ molecules oriented in an upright position relative to the equatorial plane (Fig. 3). The structure with both H_2 ligands lying in the equatorial plane is 3.1 kcal/mole higher in energy, corresponding to a rotational barrier of 1.5 kcal/mole for the rotation of one H_2 about the metal–H_2 bond. Removal of one H_2 to form $Cr(CO)_4(H_2)$ requires 15 kcal/mole and removal of the second H_2 requires another 14 kcal/mole. At this level of calculation the dihydrogen–dihydride energies

$$Cr(CO)_4(H_2) \rightarrow Cr(CO)_4(H)_2$$

are comparable with the dihydride lower by 1 kcal/mole. The calculated Cr–H_2 and H–H bond lengths are 1.787 and 0.77 Å respectively for the upright form. (Fig. 4).

Some of the possible polyhydride forms—having either a square H_4 or a $H_3{}^+$–H^- species coordinated to $Cr(CO)_4$—are found to be very high in energy (at least 50 kcal/mole) and hence are unlikely intermediates in the H_2–D_2 exchange reaction.

Another pathway is shown schematically in Fig. 5, where the conversion of the bis–dihydrogen complex to a dihydrogen–dihydride complex. The process is symmetry allowed in the sense that the three relevant orbitals in the equatorial plane, the Cr $d_{x^2-y^2}$ and the two H_2 σ orbitals of the bis dihydrogen species transform into the Cr–H σ and H_2 σ bonds of the dihydrogen dihydride species. The diagram is oversimplified in the sense that the orbital characters change qualitatively through the course of the reaction. The $d_{x^2-y^2}$ orbital is empty in the dihydrogen dihydride complex and the H_2 bonding orbital actually descends from higher orbitals in the bis–dihydrogen complex. The dihydride form is calculated to lie 10 kcal/mole higher in energy with a Cr–H and Cr–H_2 bond lengths of 1.729 and 1.923 Å, respectively.

The transition state for the reaction

$$Cr(CO)_4(H_2)_2 \rightarrow Cr(CO)_4 H_2(H)_2$$

was located at the SCF level assuming C_{2v} symmetry and treating the H–Cr–H bond angle between the two central H atoms as the reaction coordinate. An activation barrier of 24 kcal/mole (see Fig. 6) was found for this process—a relatively low barrier compared to some of the other polyhydride species. A non–C_{2v}–pathway was also investigated where a similar barrier was also found.

Because of the electron correlation effects can have a significant effect on calculated energies for chemical reactions, calculations on several of the above $Cr(CO)_4(H_2)_2$ species were carried out using configuration interaction (CI) techniques. These particular calculations consisted of single and double excitations (SDCI) with respect to the single Hartree–Fock configuration employed in the above studies. The optimized geometries from the SCF calculations were used for the respective species. No excitations were allowed from the inner 16 orbitals corresponding to the carbon 1s, oxygen 1s and 2s, and chromium 3s and 3p core orbitals. This resulted in 140,642 spin eigenfunctions in C_{2v} symmetry for the SDCI calculations.

Cr (CO)$_4$ (H$_2$)$_2$ STRUCTURES AND FRAGMENTS

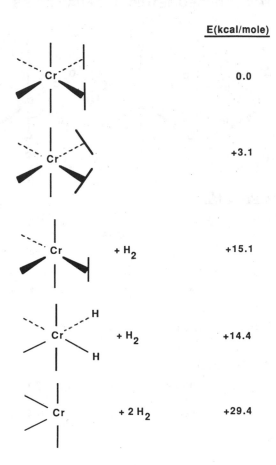

E(kcal/mole)

0.0

+3.1

+ H$_2$ +15.1

+ H$_2$ +14.4

+ 2 H$_2$ +29.4

Fig. 3.   Relative energies in kcal/mole of Cr(CO)$_4$(H$_2$)$_2$ species.

OPTIMIZED GEOMETRICAL PARAMETERS

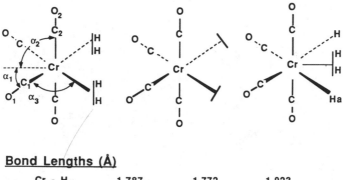

Bond Lengths (Å)

Cr - H$_2$	1.787	1.772	1.923
H - H	0.772	0.779	0.764
Cr - H$_a$			1.729
Cr - C$_1$	1.997	1.980	2.084
Cr - C$_2$	1.971	1.992	2.077
C$_1$ - O$_1$	1.147	1.148	1.143
Cl$_2$ - O$_2$	1.150	1.146	1.142

Bond Angles (deg)

α_1	46.9	44.9	42.1
α_2	90.8	93.6	94.2
α_3	90.3	88.9	67.4

Fig. 4.   Calculated structural parameters for $Cr(CO)_4(H_2)_2$ species.

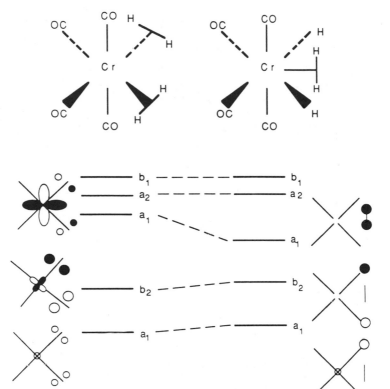

Fig. 5.   Correlation diagram for bis–dihydrogen to dihydrogen–dihydride forms of $Cr(CO)_4(H_2)_2$ species.

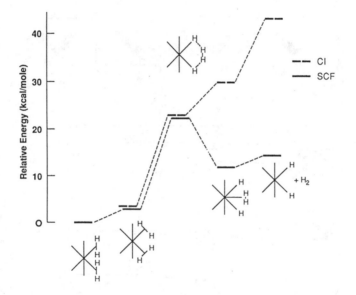

Fig. 6.   Calculated energies from Hartree–Fock (SCF) and configuration
            interaction (CI) calculations for one possible path for H–D
            exchange involving $Cr(CO)_4(H_2)_2$.

Table II. Relative Energies (kcal/mole) for
$Cr(CO)_4(H_2)_2$ and Related Species

	SCF	SDCI
$Cr(CO)_4(H_2)_2$—upright	0.0	0.0
$Cr(CO)_4(H_2)_2$—coplanar	3.1[a]	3.1[b]
$Cr(CO)_4(H_4)$—coplanar	22.5	22.5
$Cr(CO)_4H_2(H_2)$	11.8	29.7
$Cr(CO)_4H_2 + H_2$	14.4	43.3

[a]$E = -532.823307$ a.u.
[b]$E = -533.416149$ a.u.

The CI results are contrasted with the SCF calculations in Fig. 6 and in Table II. The bis–dihydrogen complex $Cr(CO)_4(H_2)_2$ is 27 kcal/mole and 40 kcal/mole more stable, respectively, than the dihydrogen–dihydride species $Cr(CO)_4H_2(H_2)$ and the $Cr(CO)_4H_2 + H_2$ fragment, respectively. These energies compare to 9 and 11 kcal/mole for the respective SCF calculations. The potential energy surface has also changed in that the $Cr(CO)_4(H_4)$ intermediate appeared to be a transition state at the SCF level, but actually lies below the energy of the $Cr(CO)_4(H_2)_2$ product at the CI level.

Finally we compare the results of the above *ab initio* calculations with the earlier extended Huckel theory (EHT) calculations of Burdett *et al.* (18). In this work, which helped to stimulate our own calculations, the authors emphasized that they were probing general trends and that reliable thermodynamic stabilities of MH_n species could not be obtained using this method. With these points in mind we compare the relative energies of tetrahydrogen species of $Cr(CO)_4$ in Table III.

Table III. Comparison of extended Huckel theory (EHT) results (18)
with SCF *ab initio* calculations on $Cr(CO)_4$ tetrahydrogen species

$Cr(CO)_4$ fragment		Rel. energy (kcal/mole)	
		EHT	SCF
$(H_2)_2$	cis–dihydrogen	0	0
$H_2(H_2)$	dihydrogen–dihydride	—	12
$H_2 + H_2$	dihydride + H_2	—	14
$(H_2) + H_2$	dihydrogen + H_2	—	15
(H_4)	planar H_4	17	23
$(H_3)(H)$	linear H_3	22	22
$(H_3)(H)$	triangular H_3	35	76
H_4	tetrahydride	36	65
(H_4)	square H_4	86	54
(H_4)	tetrahedral H_4	160	—

While the earlier study did not investigate the $H_2(H_2)$ species, the other intermediates involving "open " planar H_3 and H_4 moieties are actually placed at similar energies, while the intermediates involving closed species do not correspond with our findings. In particular, the $(H_3)(H)$ species previously described as having two—electron triangular H_3^+ and H^- ligands lies at considerably higher energy. There has, in fact, been considerable experimental activity to identify and isolate complexes containing H_3^+ and H_3^- ligands, but we find no evidence to support these forms for this particular class of complexes.

<u>Discussion of Mechanisms</u>. In summary, the above calculations have identified relatively low—energy (i.e., less than 30 kcal/mole) pathways for the conversion of H_2-D_2 into HD as exemplified by the dihydrogen—dihydride species originally formed according to the process

$$Cr(CO)_4(H_2)(D_2) \rightarrow Cr(CO)_4(H)(HD)(D)$$

as shown in Figs. 5 and 6. Once the incipient HD bond has begun to form, the reaction could proceed further by
  (a) rotation of HD about the Cr—HD bond followed by collapse to $Cr(CO)_4(HD)_2$,
  (b) dissociation of HD to form $Cr(CO)_4(H)(D)$ followed by insertion of H_2 or D_2, or
  (c) dissociation of HD followed by collapse to $Cr(CO)_4(HD)$, to mention some of the possibilities.
Despite the above low—energy pathways discussed above, it is not clear whether they are actually involved in the liquid xenon experiments of Upmacis *et al.* since relatively small barriers must be involved for any process at these temperatures. We would point out that the possibility of radical processes involving the presence of H atoms should also be examined thoroughly before these mechanisms are definitively understood. We observe that scrambling involving coordinated H_2 and H atoms is much more facile than processes involving two H_2 ligands. In cases such as

$$Cr(H_2)(H)(H) \rightarrow Cr(H\text{---}H\text{---}H)(H) \rightarrow Cr(H)(H_2)(H)$$

and

$$V(CO)_5(H_2)(H) \rightarrow V(CO)_5(H\text{---}H\text{---}H) \rightarrow V(CO)_5(H)(H_2)$$

the barriers involved in open H_3 intermediates are only 10—13 kcal/mole. A second possibility to be considered is that the $Cr(CO)_4H_2$ dihydride species actually possesses a triplet ground state much lower in energy than the singlet species discussed in Fig. 5 and Table II. Such a species could react with H_2 to form H—atom containing species.

In summary, the nature of some likely reaction intermediates involved in HD formation from H_2 and D_2 complexes of $Cr(CO)_4$ have been identified here. Unraveling the further details of the mechanisms involved in these fascinating complexes will require more extensive experimental and theoretical studies.

<u>Acknowledgment</u>

This work was carried out under the auspices of the U.S. Department of Energy.

Literature Cited

1. Kubas, G.J; Ryan, R.R.; Swanson, B.I.; Vergamini, P.J.; Wasserman, H. J. Am. Chem. Soc. 1984, 106, 451. (b) Kubas, G.J.; Ryan, R.R.; Wrobleski, D. ibid. 1986, 108, 1339. (c) Kubas, G.J.; Unkefer, G.J.; Swanson, B.I.; Fukishima, E. ibid. 1986, 108, 7000.
2. Kubas, G.J. Acc. Chem. Res. 1988, 21, 120 and references therein.
3. Saillard, J.–Y.; Hoffmann, R. J. Am. Chem. Soc. 1984, 106, 2006.
4. Jean, Y.; Eisenstein, O., Volatron, F.; Maouche, B.; Sefta, F. J. Am. Chem. Soc. 1986, 108, 6587.
5. Hay, P.J. J. Am. Chem. Soc. 1987, 109, 705.
6. Hehre, W.J.; Radom, L.; Schleyer, Paul v.R.; Pople, J. A. *Ab Initio Molecular Orbital Theory*. Wiley: New York, 1986.
7. Spangler, D.; Wendoloski, J.L.; Dupuis, M.; Chen M.M.L.; Schaefer III, H.F. J. Am. Chem. Soc. 1981, 103, 3985.
8. Faegri, K.; Almlof, J. Chem. Phys. Lett. 1984, 107 121.
9. Dobbs, K.D.; Hehre, W. J. J. Computational Chem. 1987, 8, 861.
10. Williamson, R.L.; Hall, M.B. Int. J. Quantum Chem. Symp. 1987, 21, 503.
11. Rohlfing, C.M.; Martin, R.L. Chem. Phys. Lett. 1985, 115,104.
12. Rohfling, C.M.; Hay, P.J. J. Chem. Phys. 1985, 83 4641.
13. Bauschlicher, Jr., C.W.; Walch, S.P.; Langhoff, S.R. Quantum Chemistry: The Challenge of Transition Metals and Coordination Chemistry; Veillard, A., Ed.; Reidel: Dordrecht, Holland, 1985; p. 15.
14. (a) Hay, P.J.; Wadt, W.R. J. Chem. Phys. 1985, 82, 270. (b) Wadt, W. R.; Hay, P.J. ibid. 1985, 82, 274. (c) Hay P. J.; Wadt, W.R. ibid. 1985, 82, 299.
15. Crabtree, R.H.; Hamilton, D.G. J. Am. Chem. Soc. 1986, 108, 3124.
16. Upmacis, R.K.; Poliakoff, M.; Turner, J.J. J. Am. Chem. Soc. 1986, 108, 3645.
17. Thompson, M.E.; Baxter, S.M.; Bulls, A.R.; Burger, B.J.; Nolan, M.C.; Santarsiero, B.D.; Schaefer, W.P.; Bercaw, J.E. J. Am. Chem. Soc. 1987, 109, 203.
18. Burdett, J.K.; Phillips, J.R.; Pourian, M.; Upmacis, R. Inorg. Chem. 1987, 26, 3061.
19. Modified GAUSSIAN82 Program: J.S. Binkley and R.L. Martin.
20. MESA program: P.W. Saxe and R.L. Martin.

RECEIVED December 9, 1988

Chapter 8

Activation of Small Molecules by Transition Metal Atoms
Theoretical Interpretation of Low-Temperature Experiments with Cu, Pd, and Pt Atoms

O. A. Novaro

Instituto de Física, Universidad Nacional Autónoma de México, 01000 México D.F., Mexico

Theoretical-experimental results on transition metal atom-small molecules systems are reported. The theoretical studies employ the PSHONDO-CIPSI sequence of programs that allow for variational and perturbational configuration-interaction calculations including up to 10^6 configurations for each ground and excited-states of the system. These theoretical results are contrasted with data from low-temperature matrix isolation experiments on these same systems supported by infrared, visible-ultraviolet, epr and other spectroscopic techniques. Interesting correlations between theory and experiment are found, including the following: the photoactivation of H_2 and methane by Cu atoms at low temperatures are rationalized from a theoretical standpoint; the theoretical prediction of H_2 activation by ground-state Palladium is verified experimentally and the preference of insertion over abstraction reactions and the formation of $Cu(N_2)_n$ complexes serve to explain some extraordinary isotopic effects found in experiment.

Matrix isolation experimental techniques [1-10] stand out among many other modern chemical research methods with regard to their ability to provide direct comparisons with quantum mechanical calculations. The use of photoexcitation methods to induce reactions [7-9] as well as the applications of multiple spectroscopic techniques to study such photochemical reactions allows for close control of the reaction parameters. Most of the high temperature and entropy effects, otherwise very large in thermochemical reactions, are therefore not present here and thus some of the limitations associated with applications of precise quantum mechanical calculations to kinetic processes disappear.

Specifically the low temperature studies which concern elementary interactions of small molecules and transition metal clusters or atoms isolated over "inert" solid matrices [5-10] are of high interest, especially now that the Schrödinger equation representing such interactions can be solved to relatively high precision using ab-initio configuration-interaction methods. Among such methods we could mention the CASSCF and CCI, GVB-CI, Monstergauss, PSHONDO and CIPSI [11-15] among other methods and programs, many of them mentioned and described in this book. The fact is that theoretical physicists and chemists in the recent past have developed very accurate methods for the study of d- and f-electron systems. Therefore, while several low temperature experiments concerning transition metal atoms or clusters and their interactions with small molecules have appeared in the literature, simultaneously many quantum mechanical calculations are appearing on the same type of systems. Rarely in the history of quan-

0097–6156/89/0394–0106$06.00/0

tum chemistry has a situation so tempting been found where theory can be compared with experiments concerning systems of great practical chemical interest and yet small enough and with such strict control of variables as to make direct comparisons feasible. However, actual theoretical experimental collaborations are not at all common. Some joint papers exist [16-18] and sometimes theoretical and a experimental papers on the same system are published back to back [19-20]. An example is that of Weltner and coworkers, who performed esr and endor studies of MnH supported over solid Argon at 4K [21] in order to compare with previous calculations on the same MnH molecules [22], among other similar efforts. The alternative situation of theoreticians directing their calculations of previous or current matrix isolation experiments is of course common also [22-31] but considering the potential benefits of following the experiments closely, we feel that in reality much more work should be done in this direction. In particular we believe that in order to be really relevant to the understanding of the kinetics, calculations should not only aim at obtaining the structure of a molecule but very importantly at determining reaction coordinates and activation barriers. This allows for predictions about feasibility and selectivity that may be contrasted with the experiments. In this chapter we shall review a few cases of such theoretical-experimental collaborations without pretending to be exhaustive but rather as examples of the mutual influence of theory and experiment.

Method

The method used in our calculations is the ab-initio pseudopotential method of Durand *et al.* [32-34]. We apply it for the Cu, Pd and Pt metal atoms whose pseudopotentials are also given in the literature: that of Cu in [35], that of Pd in [34], that of Pt in [36]. In every case all of the valence electrons as well as all the electrons of the outermost d-subshell are always treated explicitly and without restrictions. The basis sets used are always of double-zeta quality at least, those for Cu and H are given in [37], that of Pt in [36], that of C by Pacchioni *et al.* [38], that of Pd in [34], that of N by Daudey (Daudey, J.P. Preprint of the Laboratoire de Physique Quantique, Université Paul Sabatier, Toulouse, France, 1986). The convergence criterion of the SCF iteration energies was set at 10^{-6}. Basis set superposition errors were systematically tested and corrected for by following the counterpoise correction of Kołos [39] when necessary. The basis sets were selected by thoroughly testing their accuracy in the reproduction of the energy splittings between the ground and lowest excited states (*eg.* the ^2P-^2S and ^2D-^2S splittings in Cu, the ^3D-^1S splitting in Pt, etc.) but this is better seen by reading the original papers [36-37]. One limitation of these basis sets is the lack of f-polarization functions. In some instances it has been shown that their use introduces only marginal improvements in some calculations involving Cu [40] or Pt [41]. The use of f-functions is however an open subject of current interest as is exemplified by several chapters of this book.

We use the CIPSI algorithm [15] which introduces configuration interaction by perturbation with multiconfigurational wave functions selected by an iterative process using the Møller-Plesset barycentric values as proposed by Malrieu [42]. By this approach we then include many more configurations (typically of the order of a million or more) that interact effectively with the original reference states. These reference states correspond to more than one metal atom state, generally we take three or more states of each metal considered, say the ground and the lowest lying excited states. The importance of this will be evident when discussing the comparison of the specific results with the experimental data.

In every case a careful analysis of the configurations that are included in this large CI scheme is carried out trying to determine how their role during the process is to be understood. Such configurations represent very often a polarization of the d-subshell, which in many cases is closed or nearly closed, so that its relaxation substantially lowers the interaction energy between a transition metal atom and a small closed-shell molecule (*eg.* H_2, CH_4 or N_2). As interesting as this and also many other trends shown by the configurations participating in the CI schemes are, it is not adequate to describe them within the overall fashion in which we shall discuss the results on several different systems in the present paper. The reader is therefore referred to the original articles to review this important aspect [36-37,40,43-44].

Comparisons with other theoretical methods are important. Our Cu calculations are based on the pseudopotential of Péllisier [35], who applied it to the Cu_2 system. A controversy arose when CASSCF-CCI calculations on the same system seemed to imply [40] that the CIPSI calculations matched the experimental energy too well for a non-relativistic method. Péllisier [45] replied by showing his pseudopotential included relativistic effects.

On the other hand a recent CASSCF-CCI calculation on the $Pt+H_2$ reaction was published [46]. The author apparently was not aware of our previous work on the same system [36] and yet he obtained potential energy surfaces that are virtually identical to those obtained using CIPSI as is commented elsewhere [47]. He did make some comparisons with GVB-CI calculations [48] on the same PtH_2 systems concluding that both this method and his agreed well. From all this we must conclude that sometimes CIPSI, CASSCF-CCI and GVB calculations can lead to the same results and quite similar chemical pictures. Other methods also match well in their predictions. In one of our papers both the CIPSI and Monstergauss programs were used in the $CuH+H$ thermal reaction giving coinciding results in most aspects of the process [18].

Activation of Small Molecules by Closed-Shell Transition Metal Atom States

The matrix isolation experiments using epr, ir, uv-visible and other spectroscopic techniques on transition metal-olefin complexes [8,49] have naturally attracted the attention of theoretical chemists and calculations on the $Ni-C_2H_4$ system were reported in one of the first theoretical-experimental papers mentioned in the introduction [16]. These results were later supplemented with a larger (double-zeta) basis set [50] and also [51] extended for a $Ni(C_2H_4)_2$ system. The main conclusions are that a net charge transfer of almost 1/5 of an electron from the metal to the ethylene is evident and that a donation and back donation mechanism consistent with a classical Dewar-Chatt-Duncanson model exists. The Ni-ethylene binding energy is 12.8 kcal/mol.

Another system that has been studied theoretically is $Cu-C_2H_4$ where, in contrast [54], it was suggested that a weak charge transfer from the olefin to the metal ($0.164e$) without the participation of the carbons and the unpaired electron remaining in the $4s - 4p$ *hybridized* orbital exists. This is indeed very far from the Dewar-Chatt-Duncanson model [52-53]. We shall now report our results on $Pd-C_2H_4$ which are much more in coincidence with those of Ozin *et al.* [16] and Siegbahn and coworkers [50-51].

Palladium-Ethylene Interaction

The case of the interaction of Pd with C_2H_4 is interesting because the Palladium ground state has a closed shell, d^{10}, configuration. This notwithstanding, the existence of a stable PdC_2H_4 complex was established experimentally [8] and a net charge transfer from Palladium to the olefin carbons was reported. They also showed that the π-bonded $Pd-C_2H_4$ complex had very similar stretching modes to those observed for ethylene adsorption on Palladium surfaces [55], thus concluding that the complex is an acceptable model for this adsorbed species. This system is then interesting enough to justify a theoretical study and this was done by us using the methods described above [56-57]. The main conclusions of the matrix isolation experiments were confirmed by us as depicted in Fig. 1 where the $Pd+C_2H_4$ interaction energy curve as well as the geometrical and charge transfer properties of the complex are given. The binding energy of our complex, 47 KJoules/mole, was close to the value (54 KJoules/mole) of the desorption energy on Pd surfaces reported in [55], fulfilling the expectations of Huber,

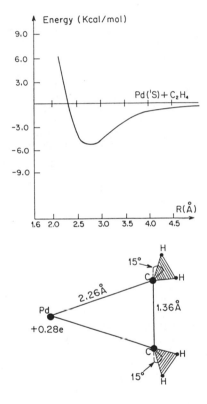

Figure 1. Potential energy curve and geometrical and charge transfer parameters of the Pd-ethylene complex.

Ozin and Power [8] that the Pd-C_2H_4 complex is an acceptable model of the adsorbed olefin species. We also see from Fig. 1 that the C-C distance is lengthened and the C-H_2 planes are rotated by $15°$ to the original plane of the ethylene, thus explaining the red shifts in the respective vibrational modes reported in [8]. Also, their ultraviolet results show a charge transfer from the metal to the olefin, which is also evident in Fig. 1. The value of this donation is similar to that reported for the Ni-C_2H_4 complex [50] and we also find that a reasonable consistency of the Dewar-Chatt-Duncanson model exists for Palladium as for Nickel [50] but apparently not for CuC_2H_4 [54] although the fact that the latter study included the d-electron subshell in the pseudopotential may completely falsify this aspect of their results. At least our own studies of Cu reactions, to be described later on, systematically showed the need of having a flexible and explicit description of the d-subshell in order to obtain the very important avoided crossings and activation barriers and in general the multiple-well potential energy surfaces that will be discussed later on.

H_2 Capture by Palladium Atoms

Several calculations on the PdH_2 system exist [58-60] showing that a weakly bound $^1A_1(C_{2v})$ Palladium dihydride can be stable. However no attempts to depart from this C_{2v} symmetry were carried out except for some analyses of the H_2 positions around a Palladium dimer [61]. We have studied both the side-on (C_{2v}) and the head-on approaches of Pd to H_2 showing that both present attractive curves without any activation barriers [19]. More recently we have studied a substantial part of the potential energy surface of the $Pd(4d^{10}5s^0, {}^1S) + H_2({}^1\Sigma_g^+) \rightleftharpoons PdH_2$ reaction. These results are partially reported in Figs. 2 and 3, including portions of the potential energy surface as well as the whole rotation coordinate between the stable PdH_2 head-on and side-on complexes. It should be noted that there is no rotation barrier between these two extreme situations, *i.e.* that one can shift smoothly from side-on to the head-on structures and viceversa. Bear this in mind for the following discussion.

When our first results [19] were obtained it seemed adequate that we propose an experiment because our energy curves clearly indicated that under the conditions of matrix isolation experiments Pd atoms deposited on rare gas matrices would be capable of capturing H_2 molecules, and at perhaps different geometries due to the relatively smooth potential energy surface connecting the various minimal energy structures. Consequently a proposal was made to G. Ozin to carry out such experiments with J. García-Prieto of our group, in his laboratory. The results were highly rewarding and they obtained the infrared band structure of PdH_2, it clearly had $^1A_1(C_{2v})$ [20] symmetry which was established by using isotopic mixture of H_2, HD and D_2 in the experiments. However the existence of other symmetries for PdH_2 was also established when further experiments were done substituting the Xe matrix for another of solid Krypton. For the Kr matrix the infrared spectra showed additional splitting for the lines of the HD reactant, a result that implies the existence of two clearly inequivalent hydrogen sites for the bound HD molecule when attached to the Pd atom. When the D atom occupies one type of site or the other the lines are shifted. This necessarily leads one to postulate a non-C_{2v} structure that must correspond to our calculated head-on geometry or perhaps a structure intermediate between the side-on (C_{2v}) and the head-on structures as described in Figs. 2 and 3. In these figures we have seen that a barrierless shift going from the head-on to the side-on and including all intermediate structures implies easy rearrangements from one to the other. This property of the PdH_2 surface that allows for different orientations of the Pd-H-H system may perhaps be related to the chemisorption (for a cluster model of this see [62]) bulk-absorption and catalytic properties of Pd solids which are vastly different from the properties of Platinum, for instance, but this shall be discussed elsewhere (Novaro, O; Poulain, E. and Ruiz, M.E., to be published). Here we only want to advance that from the PdH_2 potential energy surface one may conclude that in the matrix isolation experiments the selectivity of one

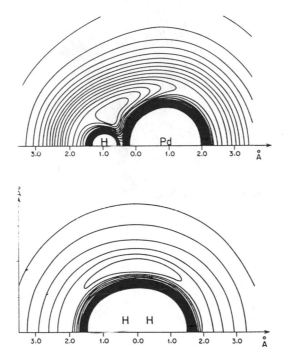

Figure 2. Two sections of the Pd-H_2 potential energy surface, the upper part shows the movement of an H moiety over a fixed PdH moiety, the lower the movement of the Pd moiety around a fixed H_2. Every line represent changes of about 5.75 kcal/mol.

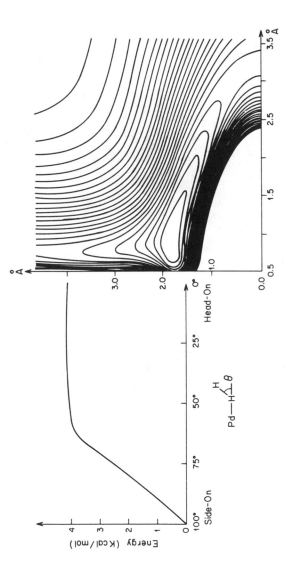

Figure 3. Energy changes of PdH$_2$ through the shifts from a linear head-on structure to a side-on (near C$_{2v}$) geometry. To the right another sector of the three dimensional PdH$_2$ potential energy surface, related to perpendicular motions of the Pd and the H moieties, is given.

geometrical structure or the other was mostly controlled by the solid matrix used in the experiments. The different minima in this surface shared the common characteristic that they do **not** relax the H-H distance, so that not only hydrogen scission is not favored by its interaction with Pd [19] but that the only difference in the spectroscopy of the various experiments is the appearance of additional splittings of infrared bands that depend on the solid matrix used. Our conclusions have been reinforced by more recent experiments (Ozin, G.A., personal communication 1987) where, for a single rare gas matrix, one actually may observe both C_{2v} and non-C_{2v} PdH$_2$ structures simultaneously.

Excited Platinum Atom States and their Interactions

The case of Pt apparently is expected to have some important differences with Pd. After all its ground state is an open-shell triplet. Recent studies [36] of the Pt+H$_2$ reaction have shown however that the ground state of Pt$(3d^94s^1)^3$D gives rise to repulsive interactions with H$_2$. In the case of C_{2v} symmetry the Pt$(d^9s^1)^3$D+H$_2$ $\rightleftharpoons$ Pt+H$_2$ reaction gives rise to two $^3A_1$ and $^3B_2$ states in which the metal does not capture H$_2$. However the Pt$(d^{10}s^0)^1$S closed-shell excited state lying 17.84 kcal/mol above the ground state energy (this energy separation is the one obtained by the CIPSI calculations, the experimental separation is however very similar: 17.56 kcal/mol) does have an attractive interaction. Another open-shell singlet state still higher in energy also presents an attraction in the Pt+H$_2$ reaction as is shown in Fig. 4. Both of these excited states correspond to $^1A_1$ configurations in the C_{2v} geometry, a crucial fact because as is evident in Fig. 5 it is the avoided crossing that these equal symmetry states present at a Pt-H$_2$ (midpoint) distance of approximately 4.4 a.u. that makes the $^1A_1$ closed-shell configuration of PdH$_2$ have the deep well evidenced in Fig. 4. Interestingly this state, which is an excited state of Pt, has exactly the same configuration as the ground state of Palladium (*i.e.* $(d^{10}s^0)^1$S). Thus we see that the H$_2$ capture by both these metals stems from the interaction of two closed shells ($^1\Sigma_g$ of the H$_2$ molecule and ^{1}S for Pd and Pt). One important difference in the behaviour of Pd and Pt towards H$_2$ capture and activation is apparent by remembering that all attempts to see if the H-H bond is labilized by its interaction with Pd were negative. In all studies of the PdH$_2$ system [19, 58-60] the H-H distance is basically unmodified, it is lengthened by a mere 0.1Å [19, 59]. In the case of Pt in the other hand the H-H bond is substantially weakened as Fig. 4 shows, where several Pt states allow for H-H distance relaxation. Thus Pt has the capacity of provoking an important labilization of the H$_2$ bond. For the closed-shell $^1A_1$ state of C_{2v} symmetry the H-H distance is almost doubled (1.38Å), as this curve has an absolute minimum implying that PtH$_2$ reaches an energy below not only the ^{1}S excited state asymptotic limit but also below the ground state Pt and H$_2$ energy limit. This has been the basis of our rationalization of the results of Low and Goddard [63] and others on catalytic Pt-bis-phosphine complexes as has been reported elsewhere [36]. Very recent results on the head-on approach of H$_2$ to Pt show also attractive interactions but less capacity to activate the H$_2$ bound [47].

In conclusion we have seen that just as for Pd, for Pt a closed-shell state is responsible for H$_2$ capture. However, in the case of Platinum this is an excited state albeit with exactly the same electron configuration as the Pd ground state. Pt is furthermore predicted to be more effective in activation of the H$_2$ bond. Unfortunately in this case, contrary to Pd, no experimental studies adequate for comparison exist. On the other hand, as has been mentioned the mutual support of these different techniques (GVB, CAS-CCI and CISPSI) on the PtH$_2$ system [36, 46, 48] should confirm the validity of our conclusions, pending experimental confirmation such as was obtained for PdH$_2$ [20].

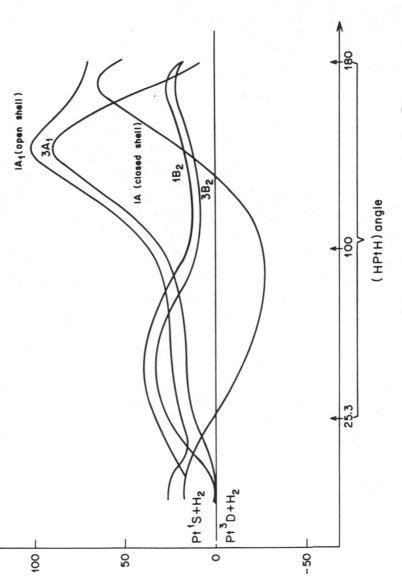

Figure 4. Potential energy curves for the H_2 capture and activation of various Pt-atom states.

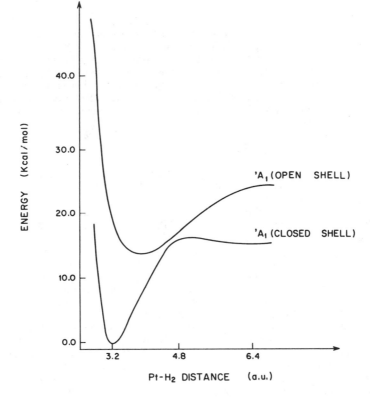

Figure 5. Representation of the avoided crossing of the $^1A_1$ closed- shell and open-shell curves for the approach of an unrelaxed H_2 molecule to the Pt site.

Different Aspects of H_2 Activation by Copper

A series of theoretical studies on the photoexcitation

$$Cu^* + H_2 \rightleftharpoons CuH + H \tag{1}$$

have appeared in the literature [37, 43-44]. Before referring to this specific photochemical process we shall mention other results on the inverse thermal process studied by Ozin and Gracie [64]:

$$CuH + H \rightarrow Cu + H_2 \tag{2}$$

which produced high yields of ground state Copper atoms and hydrogen. Theoretical calculations using both the Monstergauss and CIPSI programs showed an energetically downhill reaction coordinate for the H+CuH addition reaction [18]. The addition reaction implies a linear approach of the H atom towards the Cu moiety of CuH. Also an abstraction reaction was studied, which implied a linear approach of H now towards the H moiety of HCu. The latter process does have an activation barrier of less than 7 kcal/mol [18]. The CuH_2 complex formed by either of these reactions spontaneously dissociates into the final products (H_2 and ground state Copper atoms) thus explaining the experimental results of [64].

Photoexcited Cu Activation of H_2

In their study of the activation of H_2 by photoexcitation of Cu atoms deposited in rare gas matrices, Ozing and coworkers [65-66] irradiated with 320nm photons to produce the $3d^{10}4p^1(^2P) \leftarrow 3d^{10}4s^1(^2S)$ transition. This was sufficient and necessary for the reaction of eq. (1) to take place. The efficiency of this photochemical process was originally attributed by them [66] to the radiationless transition of the ^2P state to the lower excited ^2D state which hypothetically was the one that reacted with H_2. Our first calculations showed that the ^2P state itself is also capable [37] of capturing H_2 effectively. Furthermore it was established that while the photoactivation by the 320nm was a *sine qua non* condition for the process to be triggered (without it Cu simply does not activate H_2) right after it the process of eq. (1) proceeds *regardless* of whether Cu^* suffers a radiationless transition to the lower excited state ^2D or eventually to the ground state ^2S of Cu. This was demonstrated [43] from the relatively moderate ($\sim$ 20 kcal/mol) activation barriers that the potential energy curves of $Cu(^2S)$, $Cu(^2D)$ and $Cu(^2P)$ present to H_2 capture. These barriers are easily overcome by the great energy gain from the transition from the ^2P state (whose own potential curve is initially downhill in energy) to the lower ^2D and ^2S states. This process implies a very interesting mechanism involving Landau–Zener–Stückelberg transitions and Herzberg–Teller couplings between the main $^2A_1$ and $^2B_2$ symmetry potential energy surfaces of the $C_{2v}Cu^*+H_2$ system. These potential energy surfaces [44] are reproduced in Fig. 6 but the details of the reaction mechanisms envolving the three ^2S, ^2P and ^2D states of Copper in activating H_2 and eventually leading from the CuH_2 energy minima of Fig. 6 towards the final products of eq. (1) cannot, for reasons of space, be reproduced here. For this the reader is referred to the original papers [37, 43-44]. It is worth mentioning here however that the restriction to C_{2v} symmetry in Fig. 6 is not a limitation for the actual chemical process of eq. (1) because, as was shown in [43], any deviation from C_{2v} symmetry would only enhance the scission of H_2 and the liberation of the CuH+H products because the activation barrier ($\sim$ 20 kcal/mol) would necessarily be lowered even more considering that both the $^2A_1$ and the $^2B_2$ surfaces belong to a single representation $^2A'$ of the C_s group when C_{2v} symmetry is broken [37]. In conclusion: the activation of H_2 by Cu is only feasible if the $^2P \leftarrow ^2S$ photoexcitation is induced. After this transition however,

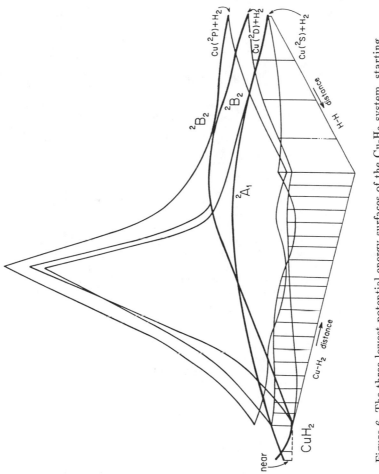

Figure 6. The three lowest potential energy surfaces of the Cu-H$_2$ system, starting from the reactants H$_2$ and (^2S), (^2P) and (^2D) copper at the right hand side and all three leading to the CuH$_2$ intermediate structures at the left hand side. The upper $^2B_2$ does this *via* a non-adiabatic transition at the point where this upper curve becomes thinner. CuH$_2$ has two minima: a linear form and, lying 2.6 kcal/mol lower in energy, a bent configuration.

the reaction proceeds to the valley of the products regardless of which of the surfaces of Fig. 6 we are in, and also regardless of any deviation from perfect side-on symmetry.

<u>Competing Effects for the Activation of H_2 by Cu; Presence of Methane</u>

One of the most important discoveries made by matrix isolation techniques in recent years is the photo-activation of natural gas by copper atoms at low temperature [67]. Beyond the possible technological consequences of this discovery, several fundamental questions were raised by it. Some of them were addressed in Ref. [17] concerning the reactants and products relative energies. In Ref. [68] in turn the activation barriers between the former and the latter were studied and they are depicted in Fig. 7.

A particularly challenging problem was posed by the observation (Ozin, G.A.; Gracie, C. and Mattar, S.M., unpublished, 1984) that the reaction:

$$Cu(^2P) + CH_4 + D_2 \rightarrow CH_3CuH + D_2 \qquad (3)$$

shows no formation of any CuD or D fragments! The same reaction using H_2 as reactant is harder to elucidate because of the secondary photochemical step:

$$CH_3CuH \begin{cases} \nearrow CH_3CuH + H \\ \searrow CH_3 + CuH \end{cases} \qquad (4)$$

which would make the presence of H or CuH to be inconclusive as to whether the origin of the hydrogen was from the H_2 reactant or from methane. With D_2 as reactant the complete absence of D or CuD fragments (only H and CuH were detected) meant that D_2 is **not** being activated by $Cu(^2P)$. Considering the results of the preceding subsection one must ask why D_2 does not —in the presence of methane— undergo the activation discussed there. Asked to elucidate this apparent paradox theoretically we have up to now obtained the potential curve for the insertion of $Cu(^2P)$ on a C-H bond in methane, which is completely downhill in energy forming a stable complex. We also have obtained the CIPSI results for the abstraction reaction corresponding to the linear approach of the H_2 molecule to the Cu molety of this complex, as shown in eq. (5)

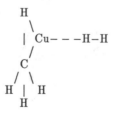

$$\qquad (5)$$

This abstraction reaction pathway presents a barrier of about 7 kcal/mol. Thus we would expect the rate constants to be k(insertion) $\gg k$(abstraction). This of course would be disregarding any tunneling effects of an H moiety into the abstraction pathway barrier. However, such tunneling would not be present if D_2 is used in the reaction. At this stage in conclusion we would necessarily infer from our theoretical results (M.E. Ruiz; G.A. Ozin and O. Novaro, work in progress) that this reaction would indeed lead to no products of the type CuD or D, in complete agreement with the observed facts.

Nitrogen Matrix Hindrance of H_2 Activation by Copper

An amazing discovery (Gracie, C., M.Sc. Thesis, University of Toronto, 1985) was made when the photoexcitation reaction

$$Cu(^2P) + H_2 + D_2 \xrightarrow{Kr} CuH + CuD + H + D \qquad (6)$$

previously mentioned and which had a reaction rate ratio of about $k_H/k_D \sim 1.5$ [66] was carried out replacing the rare gas matrix by a N_2 molecular matrix, with all other conditions (temperature, reactants, etc.) kept equal. The new results showed the reaction products

$$Cu(^2P) + H_2 + D_2 \xrightarrow{N_2} CuH + H + D_2 \qquad (7)$$

with the CuD and D subproducts being much scarcer, by a factor of about one thousand (in effect, $k_H/k_D \approx 10^3$). The suggestion was made based on infrared, optical and epr spectroscopical observations (Ozin, G.A., Gracie, C. and Mattar, S.M., Toronto University, unpublished data, 1984) that a CuN_2 complex existed of the type

$$Cu - (N \equiv N)_n \qquad (8)$$

When a Copper atomic resonance line $^2P \leftarrow {}^2S$ is photoexcited for Cu deposited in solid N_2 the existence of such complexes conceivably might hinder the activation of hydrogen by $Cu(^2P)$. Again a suggestion was made for us to study the $Cu+N_2$ system and test the stability and competition of such Cu-N≡N complexes towards hydrogen activation. Our results (Sánchez–Zamora, M., Novaro, O. and Ruiz, M.E., submitted for publication) are condensed in Figs. 8–9. From them we conclude the following: that the photoexcited Cu atom forms a stable complex with the N_2 molecule in a head-on (linear) complex with an interaction energy of almost 30 kcal/mol. Such a complex is certainly capable of hindering the Cu* capacity to react with H_2 (or D_2), particularly considering that the relevant Cu orbitals to capture and activate the hydrogen molecule are precisely those already involved in the Cu*-N-N bonding situation. The strong attractions of N_2 come from the excited $Cu(^2P)$ and (^2D) states, (^2S) only attracts a relaxed N_2. Still it might seem odd that Cu binds N_2 strongly when reportedly this is not so for ethylene [54]. We however must remember that CuN_2 is a *linear* complex derived by interaction with the nitrogen lone pair electrons, which are not present in C_2H_4. Furthermore we have already stated our belief that d-electrons should be treated explicitly, so their inclusion in the pseudopotential, as was done in [54], is *not* and adequate approximation. Perhaps more accurate calculations eventually will show a much more stable CuC_2H_4 complex. In any case, the activation barrier for the $Cu^*+H_2+D_2$ reactions on a N_2 matrix ($\sim$ 60 kcal/mol) is much higher than the 20 kcal/mol barriers [44] for the same reaction in rare gas matrices. The activation of the H_2 molecule by Cu* is not really affected by the use of the N_2 matrix because of proton tunneling that the lighter protons can perform. The deuteron cannot do likewise. We therefore may conclude that while for the case of rare gas matrix supported Cu*, the Cu^*+H_2 and Cu^*+D_2 reactions naturally show the same order of magnitude of their rate constants: $k_H \sim k_D$, this is no longer true for N_2 matrices. In effect a high activation barrier ($\sim$ 60 kcal/mol) towards hydrogen capture is introduced by the formation of the D_2Cu^*-N-N complex and it is only by proton tunneling (Sánchez-Zamora, M.; Novaro, O. and Ruiz, M.E., to be published) that the k_H is still large. On the other hand Deuterium cannot do likewise and necessarily $k_D \ll k_H$ thus explaining the apparent paradox posed by Ozin.

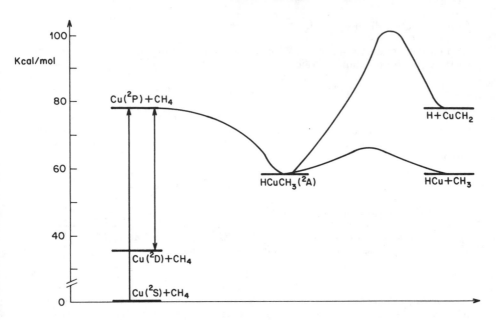

Figure 7. Potential energy curves relating the reactants (methane and photoexcited copper), the intermediates and the products of the photoinduced activation of methane by Copper.

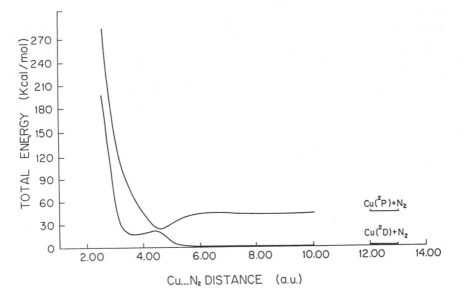

Figure 8. Side-on reaction of photoexcited Copper with a N_2 molecule. The curves have $^2B_2$ configurations in a C_{2v} symmetry.

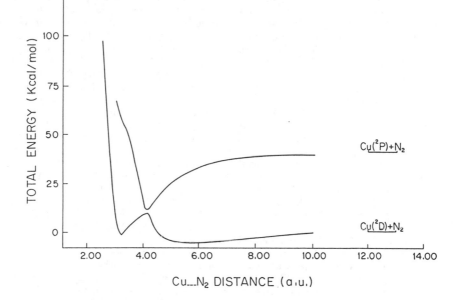

Figure 9. Head-on reaction of Cu with N_2. The curves have $^2\Pi$ configurations. Notice that these head-on configurations show greater stabilization than those of Fig. 8.

Outlook and Perspective

The theoretical prediction of different geometrical arrangements of the PdH_2 system [19] and its experimental verification [20] gives a hopeful picture of the role that quantum mechanical calculations on d- and f-electron systems may play in the future of the chemistry of transition-metal complexes and their interactions. Similar predictions for the PtH_2 system [36] (with the important differences with regard to PdH_2 that in Pt it is an excited state that captures H_2 and that it activates it in a manner that Pd cannot) have however not yet, to our knowledge, been tested experimentally. It is true that the $Pt+H_2$ reaction is much harder to control in low temperature experiments (Ozin, G.A. and García-Prieto, J., private communication). Also the quantum mechanical calculations of Pt, with its complex structure of low-lying excited states [46-48] are of a particularly difficult nature. Yet here is a system that perhaps merits particular attention from theoreticians and experimentalists.

The potential energy surfaces of the Cu^*+H_2 reaction, partially depicted in Fig. 6 have been obtained for the $Cu(^2S)$, (^2P) and (^2D) states which are involved in the photoactivation of H_2 [65]. The overall picture derived from theory is very clear and explains the reasons why the photoactivation of Cu is a prerequisite for H_2 capture [37], here the products are reached whether the reactants are moving on the ground state potential energy surface $(Cu(^2S)+H_2)$ or else on one of the excited $^2B_2$ surfaces of Fig. 6 [44]. The intermediate complexes CuH_2 however, capable of existing in two forms, linear and bent [43] have not been observed yet. Theoretically it is clear that due to the relatively small barriers that connect these intermediate complexes with the final products plus the large kinetic energy that the Cu^* and H_2 reactants gain *via* the radiationless transitions [43] one does not expect that it is easy to trap CuH_2. Still efforts are being made (Ozin, G.A., personal communication, 1988) to trap it by using different matrices. This, in fact, was the origin of the N_2 matrix experiments that lead to the discovery of the interesting isotope effects discussed in the previous subsections.

Speaking of these isotope effects, the amazing observations that the presence of methane practically eliminates any activation of D_2 by Cu^* and that the use of a molecular nitrogen matrix makes the CuD and D subproducts become almost negligible (using rare gas matrices these subproducts were almost as abundant as CuH and H) is one of the most attractive open questions in these experiments (Ozin, G.A.; Gracie, C. and Mattar, S.M., unpublished data, 1984).

We have mentioned above some attempts to explain these facts theoretically. Preliminary calculations show that Cu^* insertion of a C-H bond in methane is competitive with the abstraction reaction in H_2, so that the activation of hydrogen (specially when the deuterium isotope is used) is less effective in the presence of methane. Also CIPSI calculations on the Cu^*+N_2 interactions shows the formation of the $Cu(N_2)_n$ complexes which were first proposed by Gracie (Gracie, C., M.Sc. Thesis, University of Toronto, 1985). Such complexes make the interaction of Cu^* with D_2 much less effective, raising the activation barriers and explaining the scarcity of the CuD and D subproducts (Sánchez-Zamora, M.; Novaro, O. and Ruiz, M.E., submitted for publication). These modest attempts however only make it more imperative to carry out more theoretical studies both to obtain better potential energy surfaces and to do trajectory calculations for Cu^*+D_2 reactions.

Literature Cited

1. Ozin G.A., *Coord. Chem. Rev.* 1979, **28**, 117 and references therein.
2. Ozin G.A., *Cat. Rev. Sci. Eng.* 1977, **16**, 91 and references therein.
3. Van Zee, R.J.; DeVore, T.C.; Weltner, W., *J. Chem. Phys* 1977, **71**, 2051.
4. Breckenridge, W.H.; Relund, M.A., *J. Phys. Chem.* 1979, **83**, 1145.
5. Haratty, M.A.; Carter, E.A.; Beauchamp, J.L., Goddard, W.A.; Illies, A.J.; Bowers, M.T., *Chem. Phys. Lett.* 1986, **123**, 239.

6. Wolkow, R.A.; Moskovits, M., *J. Chem. Phys.* 1987, **87**, 5858.
7. Bach, S.B.II.; Taylor, C.A.; Van Zee, R.J.; Vala, M.T.; Weltner, W., *J. Am. Chem. Soc.* 1986, **108**, 7104.
8. Huber, H.; Ozin, G.A.; Power, W.J., *Inorg. Chem.* 1977, **16**, 779.
9. Breckenridge, W.H.; Umemoto, H. in *Dynamics of the Excited State* Lawley, K.P. Ed., Wiley NY 1982 and references therein.
10. Ozin, G.A., McCaffrey, J.G.; McIntosh, D.F. *Pure and Appl. Chem.* 1984, **56**, 111.
11. Roos, B.O.; Taylor, P.R.; Siegbahn, P.E. M., *Chem. Phys.* 1980, **48**, 157.
12. Siegbahn, P.E. M., *Int. J. Quant. Chem.* 1983, **23**, 1869.
13. Upton, T.H.; Goddard, W.A., *J. Am. Chem. Soc.* 1978, **100**, 321.
14. *Ab-Initio Methods in Quantum Chemistry*; Lawley, K.P. Ed; *Adv. Chem. Phys.* Vol. 69, Wiley, N.Y., 1987.
15. Huron, B.; Malrieu, J.P.; Rancurel, P. *J. Chem. Phys.* 1973, **58**, 5745.
16. Ozin, G.A.; Power, W.J.; Upton, T.H.; Goddard, W.A., *J. Am. Chem. Soc.* 1978, **100**, 4751.
17. Poirier, R.A.; Ozin, G.A.; McIntosh, D.F.; Csizmadia, I.G.; Daudel, R., *Chem. Phys. Lett.* 1983, **101**, 221.
18. Ruíz, M.E.; García-Prieto, J.; Poulain, E.; Ozin, G.A.; Poirier, R.A.; Mattar, S.M.; Csizmadia, I.G.; Gracie, C.; Novaro. O., *J. Phys. Chem.* 1986, **90**, 279.
19. Jarque, C.; Novaro, O.; Ruíz, M.E.; García-Prieto, J. *J. Am. Chem. Soc.* 1986, **108**, 3507.
20. Ozin, G.A.; García-Prieto, J., *J. Phys. Chem.* (1986), **108**, 3099.
21. Van Zee, R.J.; Garland, D.Λ.; Weltner, W., *J. Chem. Phys.* 1986, **85**, 3237.
22. Bagus, P.S.; Schaefer, H.F., *J. Chem. Phys.* 1973, **58**, 1894.
23. Carter, E.A.; Goddard, W.A., *J. Phys. Chem* 1984, **88**, 1485.
24. Bauschlicher, C.W.; Siegbahn, P.E.M., *J. Chem. Phys.* 1986, **85**, 2802.
25. Rappe, A.; Goddard, W.A., *J. Am. Chem. Soc.* 1982, **104**, 448.
26. Bauschlicher, C.W.; Petterson, L.G.M.; Siegbahn, P.E.M., *J. Chem. Phys.* 1987, **87**, 2129.
27. Bauschlicher, C.W.; Bagus, P.S.; Nelin, C.J.; Ross, B.O., *J. Chem. Phys.* 1986, **85**, 359.
28. Carsky, P.; Dedieu, A., *Chem. Phys.* 1986, **103**, 265.
29. Andzelm, J.; Russo, N.; Salahub, D.R., *J. Chem. Phys.* 1987, **87**, 6567.
30. Cotton, F.A.; Dunbar, K.R.; Price, A.C.; Schwotzer, W.; Walton, R.A., *J. Am. Chem. Soc.* 1986, **108**, 4843.
31. Thorn, D.L.; Hoffmann, R., *J. Am. Chem. Soc.* 1978, **100**, 2079.
32. Durand, Ph.; Barthelat, J.C., *Theor. Chim. Acta* 1975, **38**, 283.
33. Barthelat, J.C.; Durand, Ph.; Serafini, A., *Mol. Phys.* 1977, **33**, 159.
34. Serafini, A.; Barthelat, J.C.; Durand, Ph., *Mol. Phys.* 1978, **36**, 1341.
35. Pélissier, M., *J. Chem. Phys.* 1981, **75**, 775.
36. Poulain, E.; García-Prieto, J.; Ruíz, M.E.; Novaro, O., *Int. J. Quant. Chem.* 1986, **29**, 1181.
37. Ruíz, M.E.; García-Prieto, J.; Novaro, O., *J. Chem. Phys.* 1984, **80**, 1529.
38. Pacchioni, G.; Fantucci, P.; Giunchi, G.; Barthelat, J.C., *Gazz Chim. Ital.* 1984, **110**, 183.
39. Kołos, W., *Theor. Chim. Acta* 1979, **51**, 219.
40. Bauschlicher, C.W.; Walch, S.P.; Siegbahn, P.E.M., *J. Chem. Phys.* 1982 **76**, 6015.
41. Blomberg, M.R.A.; Siegbahn, P.E.M., *J. Chem. Phys.* 1983, **76**, 6015.
42. Malrieu, J.P., *J. Chem. Phys.* 1979, **70**, 1055.
43. García-Prieto, J.; Ruíz, M.E.; Poulain, E.; Ozin, G.A.; Novaro, O., *J. Chem. Phys.* 1984, **80**, 1529.
44. García-Prieto; J. Ruíz, M.E.; Novaro, O., *J. Am. Chem. Soc.* 1985, **107**, 5636.
45. Péllisier, M., *J. Chem. Phys.* 1983, **58**, 5745.
46. Balasubramanian, K., *J. Chem. Phys.* 1987, **87**, 2800.

47. Poulain, E.; Castillo, S.; Novaro, O.; Ruíz, M.E.; Submitted for publication.
48. Low, J.J.; Goddard, W.A., *Organomet.* 1986, **5**, 609.
49. Kasai, P.H.; McLeod, D.; Watanabe, T., *J. Am. Chem. Soc.* 1980, **102**, 179.
50. Windmark, P.O.; Roos, B.O.; Siegbahn, P.E.M., *J. Phys. Chem.* 1985, **89**, 2180.
51. Siegbahn, P.E.M., *Theor. Chim. Acta* 1986, **69**, 119.
52. Dewar, M.J.S., *Bull. Soc. Chim. Fran.* 1951, **18**, 79.
53. Chatt, J.; Duncanson, L.A.; *J. Chem. Soc.* 1953, 2939.
54. Nicolas, G.; Barthelat, J.C., *J. Phys. Chem.* 1986, **90**, 2870.
55. Typsoe, W.T.; Nyberg, G.L.; Lambert, R.M., *J. Phys. Chem.* 1984, **88**, 1960.
56. García-Prieto, J.; Novaro, O., *Mol. Phys.* 1980, **41**, 205.
57. Jarque, C.; Novaro, O.; Ruíz, M.E., *Mol. Phys.* 1987, **62**, 129.
58. Brandemark, U.B.; Blomberg, M.R.A.; Peterson, L.G.M.; Siegbahn, P.E.M., *J. Phys. Chem.* 1984, **88**, 4617.
59. Low, J.; Goddard, W.A., *J. Am. Chem. Soc.* 1984, **106**, 8321.
60. Yonezawa, T.; Nakatsuji, H.; Hada, M., *Croatica Chem. Acta* 1984, **57**, 1371.
61. Nakatsuji, H.; Hada, M., *J. Am. Chem. Soc.* 1985, **107**, 8264.
62. Pacchioni, G.; Koutecky, K., *Surf. Sci.* 1985, **154**, 126.
63. Low, J.J.; Goddard, W.A., *J. Am. Chem. Soc.* 1984, **106**, 6928, and references therein.
64. Ozin, G.A.; Gracie, C., *J. Phys. Chem.* 1984, **84**, 643.
65. Ozin, G.A.; Mitchell, S.A.; García-Prieto, J., *Angew. Chem. Suppl.* 1982, 785.
66. Ozin, G.A.; Mitchell, S.A.; García-Prieto, J., *J. Phys. Chem.* 1982, **86**, 473.
67. Ozin, G.A.; McIntosh, D.F.; Mitchell, S.A.; García-Prieto, J., *J. Am. Chem. Soc.* 1981, **103**, 1574.
68. Poulain, E.; Castillo, S., Bravo, G.; Cogordan, J.A.; Novaro, O., *Rev. Inst. Mex. Petr.*, 1987, **19** (3), 140.

RECEIVED December 28, 1988

Chapter 9

Mechanism for H_2 Dissociation on Transition Metal Clusters and Surfaces

Itai Panas, Per Siegbahn, and Ulf Wahlgren

Institute of Theoretical Physics, University of Stockholm, Vanadisvägen 9, S–11346 Stockholm, Sweden

Basic rules are set up for the requirements on a cluster for dissociatively chemisorbing H_2. These rules are built on previous experience from model calculations on transition metal complexes and surfaces. If these rules are combined with the knowledge of the atomic spectra of the transition metals, an understanding of both the low reactivity of the Co_6-Co_9 clusters compared to the corresponding nickel clusters and the similarly large reactivities of Co_{10} and Ni_{10} can be obtained. It is found that all the Co_6-Co_9 clusters have stable closed valence shell configurations. The present model for dissociating H_2 does not involve the electron donation from the cluster to H_2 that has been suggested previously. The correlation between ionization energies and reactivities of the clusters is shown to be a secondary consequence of the electronic structure of the clusters.

A few years ago Smalley and coworkers were able to obtain detailed experimental information about the reactivity of specific transition metal clusters with hydrogen molecules (1). The results for copper and nickel clusters were essentially as expected from the known results for surface and metal complex activities. For copper no clusters were able to dissociate H_2 whereas for nickel all clusters were active with a slow, steady increase of activity with cluster size. For the other transition metals studied, cobalt, iron and niobium, a completely different picture emerged. For these metals a dramatic sensitivity of the reactivity to cluster size was detected. No convincing explanation for these surprising results has hitherto been suggested. It should be added that there are no dramatic differences in the activity towards H_2 for the metal surfaces (or the metal complexes) of nickel on the one hand and iron, cobalt and niobium on the other.

In another set of interesting experiments Whetten et al

0097–6156/89/0394–0125$06.00/0

measured the ionization energies as a function of cluster size for
iron (2) and niobium (3). A strong correlation was observed between
these ionization energies and the reactivities of the clusters. The
simple explanation given for this correlation was that hydrogen
chemisorption requires charge transfer from the cluster. This
picture is in line with theoretical Extended Hückel results
obtained by Saillard and Hoffmann (4), who found that the critical
factor in surface activation is electron transfer from the metal.
The clusters with a higher Fermi level should thus have an enhanced
ability to give electron density to the antibonding σ^* orbital of
H_2. This picture is, however, quite different from the one we have
obtained for activation of H_2 on surfaces. In our calculations on
cluster models we have not found any correlation between the ability
to break the bond in H_2 and the cluster ionization potential.
Instead the critical factor is the overlap between σ and an
occupied cluster orbital of the same symmetry. In particular for
on-top dissociation, a requirement of the on-top atom is an easily
accessible atomic state with singly occupied 3d and 4s orbitals
(5-7). When this condition is fulfilled, the on-top dissociation
becomes the most favorable pathway. This is the reason why nickel
succeeds and copper fails to break the H_2 bond (7,8). The
correlation between the ionization energies and the reactivities of
the clusters is therefore, in our opinion, the most surprising of
all the above mentioned experimental results.

The simple model for H_2 activation proposed by Whetten et al
(2) was later questioned by Smalley and coworkers: ".. one wonders
how a simple electrostatic model can possibly explain the known
sensitivity of H_2 and N_2 chemisorption to particular surface
geometries, and to the presence of open d orbitals at the reactive
site" and they continue by raising the question "Is the
correspondence between I.P. and reactivity causal, or casual?" (9).
To answer this question Smalley et al studied the reactivity pattern
of ionized clusters. It turned out that the positive clusters had a
very similar reactivity pattern as the neutrals and sometimes
reacted with even greater cross sections even though these clusters
have a far less tendency to donate eletronic charge to the neutral
adducts. This finding is consequently in conflict with the simple
electrostatic model of Whetten et al.

In this paper we will first review our findings concerning
dissociative activation of H_2 by transition metal complexes and
clusters. Of major importance in the present context is our recent
understanding of the oscillations of the atomic hydrogen
chemisorption energies with cluster size (10). Also of importance is
our understanding of the dominant electronic mechanism for breaking
the H_2 bond (5-7). If these two factors are combined with the
knowledge of the atomic spectrum of the metal atoms (11) a large
step towards a complete understanding of the oscillations of
reactivity with cluster size can be obtained. It is also clear why
nickel does not show the same type of oscillations as iron and
cobalt. It is finally also possible to understand why the
reactivities correlate with the ionization energies of the clusters
without rationalizing this correlation by the simple electrostatic
model suggested by Whetten.

Computational details

The calculations in the present paper are very simple and
qualitative. The purpose is not to produce final definite numbers
but rather to illustrate the simple principles suggested here.
    Calculations are performed for both nickel and cobalt clusters.
When the metal atoms are in the atomic $d^{n+1}s$ state (see below) the
atoms are treated computationally as one-electron systems by the use
of Effective Core Potentials (ECP's). This is a crude but
qualitatively correct description and treats the core including the
3d shell by potentials and projection operators. Similarly for atoms
in the $d^{n}s^{2}$ state a two-electron ECP is used.
    The ECP's are constructed based on the frozen orbital ECP
technique (12). In this technique some of the core orbitals are
expanded in the valence basis set and frozen in atomic shapes. This
reduces the demand on the accuracy of the ECP potentials and the
projection operators. One-electron ECP's constructed by this
technique for nickel and copper have been shown to give results of
quantitative accuracy for surface problems, particularly for
hydrogen chemisorption which is treated here (13,14). In the
previous studies the one-electron ECP's included a frozen 3s
orbital. In the present case, states with a large occupancy of 4p
appeared for the s^{2} type configurations in a cluster surrounding.
In order to improve the description of the 4p orbitals a frozen 3p
orbital was included in both ECP's.
    The parameters of the ECP's were first determined by fitting
orbital energies and shapes for the corresponding atoms. The final
calibration was made in a comparison with calculations on $CoNi_{2}$
with the central cobalt atom both in a $d^{n+1}s$ state and in a
$d^{n}s^{2}$ state, using an all-electron description for Co and a
one-electron ECP description for Ni. The cluster binding energies at
the all-electron level and at the ECP level agreed to within a few
kcal/mol.
    Orbitals were generated in the single-configuration SCF
approximation and dynamical correlation was treated in one-reference
state Contracted CI (CCI) calculations (15). The Davidson correction
(16) was finally added to the CCI energies to account for unlinked
cluster contributions.

Atomic spectra

For transition metal atoms there are three different atomic
occupations which usually lead to states which are energetically low
lying. These chemically different states are in general terms
denoted $d^{n}s^{2}$, $d^{n+1}s$ and d^{n+2}. Bulk transition metals (and
surfaces) of iron, cobalt, nickel and copper are dominated by the
$d^{n+1}s$ atomic occupation so that there is essentially one s
electron contributing to the conduction band. The d levels will be
energetically deeper and will often be chemically rather inert. This
picture is particularly true for the metals to the right in the
periodic table such as nickel and copper. For the isolated atoms of
nickel and copper the $d^{n+1}s$ state is also the lowest (j-averaged)

state but this is often not the case for other atoms (11). For
cobalt the d^7s^2 state is 0.43 eV (10.0 kcal/mol) lower than the
d^8s state and for iron the d^6s^2 state is even lower, 0.86 eV
(19.8 kcal/mol) below than the d^7s state. It is immediately
obvious that this important difference between the atomic states of
iron and cobalt on the one hand and nickel and copper on the other
is one key to the understanding of the experimental finding that the
reactivity oscillates for the two former metals but not for the
latter two. The question that has puzzled us for the past years is
only how.

In passing it could be interesting to note that the importance
of accurately describing the above mentioned lowest states of the
transition metals has been recognized by quantum chemists for a long
time. But to obtain a correct splitting between these states has
turned out to be a formidable computational task. Since the number
of d-electrons differs between these states the correlation energy
of the d-shell will enter directly in the energy splitting. Only
recently has an acceptable error of 0.1 eV been reached
computationally for nickel (17). Large multi-reference CI
calculations with about 20 reference states and a basis set
including g-functions were required.

The H_2 Dissociation Mechanism

For H_2 to dissociate on a metal cluster two independent criteria
must be satisfied. First, the reaction must be exothermic. We will
discuss the conditions for exothermicity at the end of this section
and in the next section. Second, a dissociation mechanism must exist
such that the barriers for dissociation will not be too high. During
the past years we have studied the H_2 dissociation both for single
transition metal atoms and complexes and for clusters as models for
surfaces. Single iron, cobalt, nickel and copper atoms have been
studied (5,6) as well as models of nickel and copper surfaces (7,8).

For single metal atoms there are two requirements for a low
barrier: first the metal atom must be in a $d^{n+1}s$ state, including
open shell d-orbitals, and second one singly occuped d-orbital must
be low-spin coupled towards the 4s-electron. For nickel the only
low-lying state with a low barrier for dissociating H_2 is the
$^1D(d^9s)$ state. This state is only 9.8 kcal/mol above the ground
state. The lowest $^3F(d^8s^2)$ and $^3D(d^9s)$ states have very
high barriers. From the requirement that there must be open shell
d-orbitals, it is clear that the copper atom will not dissociate
H_2. The first state of copper that does dissociate H_2 is the d^9sp
state which is 112 kcal/mol above the ground state. Iron and
cobalt will have dissociative states very similar to the ^1D state
of nickel but these states will be fairly highly excited. For cobalt
the ^2F state is 21 kcal/mol above the ground state and for iron
the dissociative ^3F state is excited by as much as 34 kcal/mol.

The reason that the above mentioned states of nickel, cobalt
and iron are dissociative are two-fold. First, the low-spin coupling
between the d-shell and the s-electron allows for an efficient
sd-hybridization which will reduce the repulsion between the doubly
occupied σ-orbital of H_2 and the metal atom as H_2 approaches.

Second, the singly occupied d-orbital and the singly occupied s-orbital can start to form the two bonds to the hydrogen atoms as the H_2 bond is broken.

The above finding for single metal atoms is directly transferrable to the case of on-top dissociation of H_2 on metal clusters and on surfaces (7,8). The on-top dissociation of H_2 on a nickel surface consequently proceeds without any barrier (8), whereas the same dissociation pathway has a very large barrier for a copper surface. For copper other dissociation pathways will have lower barriers, such as dissociation over a hollow site. The barrier for this dissociation can be estimated to be of the order of 12-18 kcal/mol for a copper surface. The same barrier will be somewhat reduced for a nickel surface by 5-10 kcal/mol (it is possible that this reduction could be somewhat larger if the effect of the yet uninvestigated d-orbital participation turns out to be larger than expected). The presence of these relatively high barriers is the reason why no copper cluster was found to dissociate H_2. The exothermicity for H_2 dissociation would probably have been sufficient for dissociation on some clusters, since the exothermicity is sufficient for a copper surface (14).

For metal surfaces the $d^{n+1}s$ state will dominate the atomic description even if other states are dominant for the isolated atoms. This means that for both iron and cobalt surfaces, a barrierless dissociation over on-top sites is expected. For metal clusters with more than a single atom at least one atom is expected to be in the $d^{n+1}s$ state, which means that a dissociation mechanism with a low barrier should exist in principle for all these iron, cobalt and nickel clusters. For the single metal atoms the low spin coupling between the d- and the s-hole was found to be important to reduce the repulsion to H_2. The same type of spin-coupling is also important for clusters and surfaces. For clusters the spin-coupling will in general require a small excitation energy, which can cause the appearence of small barriers in the reaction with H_2 (8).

We have emphasized the importance of open d-orbitals and a proper atomic state if H_2 should dissociate with a low barrier on a transition metal surface. For clusters, however, the same type of dissociation puts up another requirement, which will turn out to be even more significant in the present context: There must be at least one open shell valence orbital (of s-character) on the cluster, otherwise the sd-hybridization will not take place (8). For an infinite surface, this requirement can always be satisfied since states with open valence orbitals must at least be reachable by a low energy excitation. For clusters, the same type of excitation may be much more expensive. Since all nickel clusters dissociate H_2 it seems clear that a dissociative state is reachable in all cases for nickel. The question that has worried us for the past years is why the same type of states do not always seem reachable for iron and cobalt clusters. The answer to this question is discussed in section VI.

In summary, the mechanism for H_2 dissociation (at an on-top site) requires that the cluster has a low lying state with at least

one atom in the $d^{n+1}s$ state with open d orbitals, and that this
cluster wavefunction has at least one open shell valence orbital.
(Possible dissociations at other sites, which in our opinion are
less likely, will have the same requirements on the valence
occupation). Exactly the same requirement holds also for formation
of a dihydride state of the same type as found on metal complexes
since such a state is on the same potential surface. Even the
recently found state of molecularly chemisorbed H_2 lies on this
potential surface (18,8). Both the formation of the dihydride and
the molecularly chemisorbed state will also normally be exothermic
if the dissociative state is present in the cluster. It should be
noted that atomically chemisorbed hydrogen, of the form observed on
infinite metal surfaces, is distinctly different from the dihydride
form on clusters (8), and consequently has other requirements for
its formation (see the next section).

Cluster Requirements for Atomic Chemisorption

During the last few years we have used the cluster model in several
studies of atomic chemisorption of mainly hydrogen and oxygen. In
the first of these studies (13,14), clusters centered around a
four-fold hollow site of a (100) surface were used. If the
chemisorption energy was calculated as the difference between ground
state energies at short and long distances between the adsorbate and
the cluster, a noncomplicated convergence towards experimental
surface results was obtained. This method (called D_2 in Tables
1-2) consequently allows both the total spatial and spin symmetry to
change during the chemisorption process. In Table 1 it is compared
to another method (D_1), where the wavefunctions were constrained
to have lowest possible spin. The reason for this restriction is
that dangling bonds are avoided and also that contributions to the
magnetism of the clusters are confined to the d orbitals as is
approximately the case for infinite surfaces. The comparison in
Table 1 for atomic hydrogen chemisorption in the four-fold hollow
site of Ni(100) shows that the D_2 method is clearly superior.
    In later studies, first in the case of chemisorption of two
hydrogen atoms on a Ni(100) surface (8) and second in the case of
adsorption of single hydrogen atoms on a Ni(111) surface, a much
more complicated convergence pattern with increasing cluster size
was obtained, see Table 2. Even clusters with as many as 40-50 atoms
sometimes gave very poor chemisorption energies using method D_2
described above. The solution to this problem was the realization
that also these large clusters need to be prepared for bonding (10).
The preparation for bonding consists of an electronic excitation to
a state with open shell valence orbitals (of s character) of the
same symmetries as the adsorbing hydrogen atomic 1s orbitals. The
energy required for such an excitation for an infinite surface
should be close to zero whereas for the clusters it was sometimes
found to be as large as 30-40 kcal/mol. If the chemisorption energy
is calculated relative to states which are prepared for bonding the
chemisorption energy is quite stable when the cluster size is varied
(method D_3 in Tables 1-2). In fact even clusters with only 10
atoms give chemisorption energies in quite reasonable agreement with

Table 1. Chemisorption of hydrogen in the four-fold hollow position of Ni(100)

D_1 is the chemisorption energy calculated as the energy difference between the energetically lowest low spin states at short and long distances.
D_2 is the energy difference between ground states irrespective of space and spin symmetry.
D_3 is the energy obtained when the clusters are prepared for bonding (see Section V).
$\bar{D}$ is the mean value and σ is the standard deviation.
The experimental value is 63 kcal/mol.
The electronic states at short and long distances respectively, are shown after each energy-value.

Cluster	D_1 kcal/mol	D_2 kcal/mol	D_3 kcal/mol
$Ni_5(4,1)$	42.0 $^1A_1-^1E$	53.1 $^3A_2-^3E$	54.1 $^3A_2-^3A_2$
$Ni_{21}(12,9)$	37.4 $^1A_1-^1E$	43.0 $^5E-^5E$	61.1 $^3A_2-^3A_2$
$Ni_{21}(12,5,4)$	65.0 $^1A_1-^1A_1$	63.0 $^3A_2-^3A_2$	63.0 $^3A_2-^3A_2$
$Ni_{25}(12,9,4)$	60.9 $^1A_1-^1A_1$	58.9 $^1A_2-^1A_2$	58.9 $^1A_2-^1A_2$
$Ni_{29}(16,9,4)$	58.5 $^1A_1-^1A_1$	53.5 $^1A_1-^1A_2$	58.5 $^1A_1-^1A_1$
$Ni_{41}(16,9,16)$	61.1 $^2A_1-^2E$	61.1 $^1A_1-^1E$	63.3 $^1A_1-^1A_1$
$Ni_{50}(16,9,16,9)$	76.6 $^2A_2-^2A_1$	61.6 $^4E-^4E$	61.6 $^4E-^4E$
$\bar{D}$	57.4	56.3	60.3
σ	12.5	6.5	3.0

Table 2. Chemisorption of hydrogen in the three-fold hollow position of Ni(111). The same symbols are used as in Table 1. The experimental value is 63 kcal/mol

Cluster	D_2 kcal/mol	D_3 kcal/mol
$Ni_4(3,1)$	55.9	55.9
$Ni_{10}(3,7)$	39.4	69.5
$Ni_{17}(3,7,7)$	63.0	63.0
$Ni_{20}(3,7,7,3)$	56.1	56.1
$Ni_{13}(12,1)$	63.7	63.7
$Ni_{19}(12,7)$	27.5	65.7
$Ni_{22}(12,7,3)$	37.8	58.3
$Ni_{25}(12,7,6)$	42.4	61.7
$Ni_{28}(12,7,6,3)$	50.8	66.2
$Ni_{40}(21,13,6)$	46.5	57.2
$\bar{D}$	48.3	61.7
σ	11.1	4.5

experimental surface results. It should be added that by using
method D_3 the electronic structure of the cluster at short and
long distances will be very similar. The resulting chemisorption
energies will thus be less sensitive to the accuracy of the
description of the cluster than they will be if method D_1 or D_2
are used.

In summary, if the two hydrogen atoms after an H_2
dissociation should be atomically chemisorbed on a cluster in a
similar way as they are on an infinite surface, the cluster should
have a low lying state with two open shell valence orbitals.

Reactivity Versus Cluster Size for Cobalt and Nickel Clusters

In the preceding sections we have outlined the requirements a
cluster has to fulfill in order to dissociatively chemisorb H_2. In
summary, the cluster first has to contain at least one atom with a
d^{n+1} occupation including at least one open d-orbital. Second,
there has to be at least one open shell valence (s-character)
orbital in the cluster wave-function. If there is only one open
shell orbital, a dihydride or possibly a molecularly chemisorbed
state will be formed. If there are at least two open shell orbitals,
atomically chemisorbed hydrogen atoms of the type found on surfaces
will be formed. The formation of the latter state is normally more
exothermic. Finally, if these requirements are not fulfilled by the
ground state wave-function of the cluster, excitation to a low lying
state which satisfies the requirements and which has an excitation
energy less than the exothermicity ($\sim$ 20 kcal/mol) will lead to
dissociation of H_2.

We will first examplify the above principles on a six atom and
a seven atom cluster. These two clusters are the most striking
examples of the differences beween nickel and cobalt clusters. These
cobalt clusters are the two least reactive of all the cobalt
clusters, whereas for nickel these two clusters are as reactive as
clusters of other sizes. In the final subsection we will discuss
also the four, five, eight, nine and ten atom clusters of which in
particular the eight and nine atom clusters have a markedly lower
reactivity for cobalt than for nickel. All calculations described in
this section used a one- or two- electron ECP level description of
the metal atoms. No all-electron atoms were included.

The Seven Atom Cluster. The most straightforward application
of the above principles occurs for the seven atom cluster. It is
clear that the Ni_7 cluster will be reactive: all the atoms will
have a d^{n+1} occupation and there is an odd number of valence
electrons with consequently at least one open shell. For this reason
all odd nickel clusters will be reactive. The Ni_7 cluster was
found to be a doublet, which means that a dihydride state will be
formed after addition of H_2. We have also investigated the most
preferred geometrical structure by essentially going through the few
most optimal structures for lithium clusters (19), but restricting
the nearest neighbour distance to be constant (equal to 4.71 a_0,
which is the bulk nickel, and nearly the bulk cobalt, bond
distance). About the same energy differences were found between

different nickel clusters as for the corresponding lithium clusters, so this should be a reasonably reliable procedure. (The same procedure was used for both the nickel- and the cobalt-clusters). The optimal structure for Ni_7 is then a D_{5h} structure with a five-membered ring in the plane and one atom above and one below this plane.

For Co_7 it was found that one atom, one which is not in the five-membered ring, prefers to have a $d^n s^2$ occupation. This was found by calculating the binding energy with respect to the isolated atoms, in this case 6 $d^{n+1}s$ atoms and one $d^n s^2$ atom. By using the known atomic splitting between the $d^n s^2$ and the $d^{n+1}s$ atoms of 10.0 kcal/mol (11) the dissociation energy with respect to 7 ground state $d^n s^2$ atoms can be calculated. The value obtained is 112.7 kcal/mol. In the same way the dissociation energy of the cluster formed from 7 $d^{n+1}s$ atoms was calculated to be 105.2 kcal/mol. The energy gain by having one $d^n s^2$ atom is consequently 7.5 kcal/mol. It should be noted that the energy gain is somewhat less than the pure atomic energy gain in going from $d^{n+1}s$ to $d^n s^2$ of 10.0 kcal/mol. Since there is no atomic energy gain for nickel, the Ni_7 cluster will prefer to have all atoms in the $d^{n+1}s$ state. This has also been checked by doing calculations on the corresponding nickel cluster. Calculations were also made to check that the Co_7 cluster will have only one $d^n s^2$ atom. If an additional change from $d^{n+1}s$ to $d^n s^2$ is made, the cluster binding energy drops by 16 kcal/mol.

The consequence of introducing the $d^n s^2$ cobalt atom in the Co_7 cluster is that there will now be 8 valence electrons which will form a closed shell configuration. Following the above principles, this cobalt cluster will thus not be reactive to H_2. The lowest vertical valence excitation energy for this cobalt cluster is 36.6 kcal/mol, which is a higher value than the exothermicity for dissociating H_2. Valence excited states will therefore not be available for dissociating H_2 either. The reason Co_7 is at all reactive towards H_2 is that the $d^n s^2$ atom can transform to a $d^{n+1}s$ atom at a cost of 7.5 kcal/mol (see above). This requires an inner shell (d-shell) rearrangement and should therefore be an unlikely process. The closed shell character of the ground state for the Co_7 cluster is the reason why there will be only one $d^n s^2$ atom in the cluster. Clearly, closing the already open shell of the cluster is more advantageous than creating new open shells.

The ionization energy of the cluster is an interesting value in light of the experimental correlation between reactivity and ionization energies (2,3). There are no experimental measurements of the ionization energies of cobalt, but calculations show that the same correlation exists also for cobalt clusters (20). It turns out from our calculations, that there is an increase in ionization energy for every $d^{n+1}s$ atom that is transferred to a $d^n s^2$ atom. This is not surprising since the introduction of the $d^n s^2$ state essentially means that an increased nuclear charge is seen by the valence electrons. The increase in ionization energy for the Co_7 cluster when one atom is changed from $d^{n+1}s$ to $d^n s^2$ is 0.3 eV. The use of the above principles consequently explains the correlation between reactivities and ionization energies without introducing an electrostatic model for dissociating H_2. The dissociation of H_2 depends instead on the number of open shells in the cluster, which in turn depends on the number of $d^n s^2$ atoms in the cluster, which in its turn will affect the ionization energy. The present conclusions are therefore in complete agreement with those of Smalley et al (9), who criticized the electrostatic model.

The Six Atom Cluster. For the six atom cluster two different structures are of interest. There is a nearly planar C_{5v} structure, with one atom above a five-membered ring, and there is the octahedral (O_h) structure. If all atoms are in the $d^{n+1} s$ state the C_{5v} structure is preferred and the wavefunction is a closed shell singlet. This does not mean that Ni_6 is non-reactive to H_2 since the energy difference to the O_h structure is only 10 kcal/mol, and the O_h structure has a triplet ground state with two open shells. Since the exothermicity is most certainly larger than 10 kcal/mol, this structural change is expected to occur during the interaction of H_2 with Ni_6.
    The triplet state of the O_h structure will be transferred to a closed shell singlet if two atoms are changed from $d^{n+1} s$ to $d^n s^2$. This is what happens for Co_6 which will have an O_h structure. The increase in the binding energy by changing two atoms is 15 kcal/mol. If a third atom is changed to $d^n s^2$ the cluster binding energy is lowered by 16 kcal/mol. Again, filling the open shells leads to an increased binding, while creating new open shells decreases the binding energy. As for Co_7 the gain in binding energy is somewhat less than the gain in energy for the two atoms of $2*10.0$ kcal/mol. Ni_6 will consequently only have $d^{n+1} s$ atoms since there is no corresponding gain in energy for the nickel atom.
    The lowest vertical valence excitation energy for Co_6 is 29 kcal/mol, which means that valence excited states will not be available for dissociating H_2 either. The alternative process for dissociating H_2 for Co_6 would be to change one of its $d^n s^2$ atoms to $d^{n+1} s^2$, which only costs 8.8 kcal/mol but is an inner shell rearrangement process.
    The ionization energy change when the $d^{n+1} s$ atoms are changed to $d^n s^2$ atoms is more pronounced for Co_6 than it is for Co_7. With only $d^{n+1} s$ atoms Co_6 has an ionization energy of 5.13 eV, and with two $d^n s^2$ atoms the ionization energy is increased to 5.98 eV.

Other Clusters. The cobalt clusters with six and seven atoms are the ones which differ most dramatically from the corresponding nickel clusters. The eight and nine atom clusters are also much less reactive for cobalt than for nickel, while the Co_{10} cluster is as strongly reactive as the Ni_{10} cluster. The experimental result for the four and five atom cobalt clusters is not completely clear to us. For the five atom cluster it is on the one hand claimed that it has reacted almost completely but on the other hand the relative reactivity is given in a figure as smaller than for the unreactive Co_9 cluster (1). For all these clusters we have obtained some preliminary results, which are described below.

The four atom cluster with only d^{n+1}s atoms has a closed shell singlet ground state. The geometry is a planar rhomboidal structure with the shortest diagonal equal to the side length. Ni_4 is still reactive towards H_2, however, since the square planar structure with a triplet ground state is only 10 kcal/mol higher in energy. Replacing two d^{n+1}s atoms with $d^n s^2$ atoms in the square planar structure, which is the most stable of these types of structures we have found, leads to a nonreactive closed shell singlet state with a high lowest excitation energy. Our calculations gave, however, a loss of binding energy of 1.2 kcal/mol for this process on the square planar cluster, and this form of Co_4 is thus 11 kcal/mol higher in energy than the rhomboidal structure with only d^{n+1}s atoms. Our calculations thus clearly indicate that the electronic structure of Co_4 and Ni_4 should be similar and that Co_4 thus should be reactive. As mentioned above, the experimental result for this cluster is not completely clear to us.

Ni_5 with five d^{n+1}s atoms has a planar parallel trapezoidal structure. Since there is an odd number of valence electrons there has to be at least one open shell and it is thus reactive towards H_2. Co_5 with one $d^n s^2$ atom prefers the same planar structure. The most optimal position for a $d^n s^2$ atom is as the central atom in the base of the trapezoid. This structure has a stable closed shell configuration with a high lowest excitation energy and it is therefore nonreactive towards H_2. The results of the calculations, however, actually gave a preference for the case with only d^{n+1}s atoms also for Co_5. This occupation was preferred by 6 kcal/mol. Even though our calculations are rather approximate errors of this size seems unlikely to us in view of the results on the seven atom cluster. We therefore suggest that also the Co_5 and Ni_5 clusters should have similar electronic structures and that Co_5 also should be reactive. Again, the experimental result is not clear to us.

Co_8, with all atoms in the d^{n+1}s state, has a very stable closed shell configuration with a T_d geometry (compare Ref.19). Co_8 should thus be nonreactive towards H_2 in agreement with experiments. The particular stability of the eight atom cluster is by now well established and actually follows from the simple ellipsoidal shell model for monovalent metals (21). Recently, Pettiette et al (22) have shown in the case of copper, that the HOMO-LUMO gap is much larger for the eight atom cluster than for any other cluster. It is therefore remarkable that the Ni_8 cluster is almost as reactive towards H_2 as all the other nickel clusters. This is clearly in contradiction with the present principles and we have no explanation for this finding at present.

For the nine atom clusters it is first clear from the rules above, that the Ni_9 cluster should be reactive, as found experimentally, since it will have an odd number of valence electrons. Of the structures we have investigated Ni_9 prefers the C_s-A structure from Ref.19. In the case of Co_9, apart from finding the optimal geometry, there is also the additional problem of where the $d^n s^2$ atom(s) should be placed. The approach we used to systematize the search of the ground state of Co_9 was to match together smaller clusters. We found that the optimal Co_9 cluster

can be described as a tetrahedron (with only $d^{n+1}s$ atoms) bound to
a pyramid with one d^ns^2 atom at the top. Both these constituents
will have two open shells and the binding between them will thus be
a double bond and lead to a closed shell singlet wavefunction. A
simple molecular analogue to the binding is thus the carbon monoxide
molecule. The singlet cluster is bound by 141.3 kcal/mol, while the
most stable Co_9 cluster with 9 $d^{n+1}s$ atoms is bound by 137.9
kcal/mol. The vertical excitation energy in the singlet cluster is
28 kcal/mol. Hence, we conclude that the Co_9 cluster should also
be unreactive towards H_2.

The Co_{10} cluster was investigated using the same
super-molecule model as was used for Co_9. With ten $d^{n+1}s$ atoms,
the optimal cluster can be described as two pyramids in their lowest
quartet states bound together by their square bases. Hence, a
triple-bonded species (one "σ_g" bond and two "π_u" bonds)
analogous to N_2, is obtained. Both the staggered and eclipsed
conformations were studied and the latter turned out to be the more
stable conformation by 17.3 kcal/mol. This cluster is similar to the
most stable Li_{10} structure given in Ref.19. The ground state is a
singlet but the vertical singlet-triplet excitation energy (a σ_g^2
to $\sigma_g^1 \pi_g^1$ excitation) is only 9.9 kcal/mol, compared to the
28-37 kcal/mol obtained for the inert clusters. The comparatively
small excitation energy is due to the small overlap between the
orbitals ("p_z"-orbitals) forming the σ_g bond. This small
overlap in turn is due to the location of the p_z orbital inside
the pyramid, with its node between the square base and the top of
the pyramid. The total binding energy of this cluster is 166.2
kcal/mol.

The next possibly non-reactive electronic state of Co_{10} has
two d^ns^2 and eight $d^{n+1}s$ atoms. The most stable cluster of
this type can be viewed as a double bond between two pyramids with
the two d^ns^2 atoms at the pyramidal tops. For this cluster,
however, the pyramids are associated via two triangular faces. An
alternative way to picture this cluster is as an octahedron bound to
two Co_2 $d^{n+1}s$ systems on opposite sides of the octahedron, thus
forming the two pyramids. This cluster state turned out to be 23.9
kcal/mol less bound than the above Co_{10} cluster state and is thus
excluded.

The Co_{10} cluster consisting of nine $d^{n+1}s$ and one d^ns^2
in the same conformation as Co_{10} with eight $d^{n+1}s$ and two
d^ns^2 atoms, was found to be 15.9 kcal/mol less bound than the
Co_{10} cluster state with ten $d^{n+1}s$ atoms.

We conclude that the Co_{10} (and the Ni_{10}) ground state is a
singlet and that the reactive state is reached via an excitation to
a triplet state. This excitation energy is one third of what is
needed to activate the "inert" Co_6-Co_9 clusters and sufficiently
low for dissociating H_2.

Conclusions

The basic requirement for a cluster to dissociate H_2 is that there
are open shell valence orbitals (with s-character) on the cluster.
If this is the case the presence of open d-orbitals of the atomic

$d^{n+1}s$ state will be sufficient for breaking the H_2 bond. If there is only one open shell orbital, a dihydride state of the type formed in transition metal complexes will be formed. If there are two open shells, atomically chemisorbed hydrogen atoms of the type formed on metal surfaces can be formed.

For cobalt and iron clusters a mechanism exists for changing the open valence shells to closed shells whereby the reactivity of the cluster is drastically decreased. This mechanism involves the change of an atom in the cluster from a $d^{n+1}s$ to a d^ns^2 state, which effectively increases the number of valence electrons. The reason this mechanism is effective on iron (in particular) and cobalt is that these atoms have a d^ns^2 ground state. For nickel, which has a $d^{n+1}s$ ground state, this mechanism will not be as effective.

Simple model calculations show that it is likely that all cobalt clusters in the range Co_6 to Co_{10} will have only closed valence shells. The large excitation energies needed to create open shells in the cases of Co_6–Co_9 effectively prevent reaction with H_2. In the case of Co_{10}, however, this excitation energy has been reduced to only one third of its value for the Co_6–Co_9 clusters, which makes the dissociation of H_2 on Co_{10} exothermic. The reason why the larger cobalt clusters are more reactive is thus that the relevant excitation energies decrease with increased cluster size. This does not mean that the possibility for some centres to attain the d^ns^2 electron configuration will become unimportant all together, but rather that its impact on the cluster reactivities will diminish.

The results of the present study indicate that the correlation found experimentally between cluster ionization energy and reactivity is not directly causal but merely appears as a secondary consequence of the electronic structure of the clusters. The correlation is not completely casual either since the change of atoms from the $d^{n+1}s$ to the d^ns^2 state which leads to the decreased reactivity also leads to an increase in the ionization potential of the cluster. This is due to the increased nuclear charge felt by the valence electrons. The best example of the effect on the ionization energy by the change of the atomic states is found for Co_6 where the ionization energy increases from 5.13 eV to 5.98 eV as two atoms change state and the valence wavefunction becomes closed shell. In our model of the H_2 dissociation the reaction is not governed by an electron jump from the cluster to H_2, and it is thus not difficult to understand why the ionized clusters show an equal reactivity as the neutral clusters (9). The tendency to change atomic states in order to close the valence orbitals must be equally important also for the ionized clusters.

To understand the high reactivity of the Ni_8 cluster is a remaining problem. All studies, theoretical and experimental, agree that the monovalent eight atom cluster is a particularly stable cluster, due to its closed shell character. Even though Ni_8 should be a monovalent cluster, the reactivity of this cluster does not seem to be lower than for the other clusters. Work to resolve this problem is currently in progress.

Literature Cited

1. M.D. Morse; M.E. Geusic; J.R. Heath and R.E Smalley, J. Chem.
   Phys. 1985, 83, 2293; M.E. Geusic; M.D. Morse and R.E.
   Smalley, J. Chem. Phys. 1985, 82, 590.

2. R.L. Whetten; D.M. Cox; D.J. Trevor and A. Kaldor, Phys. Rev.
   Lett. 1985, 54, 1494.

3. R.L. Whetten; M.R. Zakin; D.M. Cox; D.J. Trevor and A. Kaldor,
   J. Chem. Phys. 1986, 85, 1697.

4. J.-Y. Saillard and R. Hoffmann, J. Am. Chem. Soc. 1984, 106
   2006.

5. M.R.A. Blomberg and P.E.M. Siegbahn, J. Chem. Phys. 1983,78
   5682; ibid 1983, 78, 986.

6. P.E.M. Siegbahn; M.R.A. Blomberg and C.W. Bauschlicher,Jr, J.
   Chem. Phys. 1984, 81 , 1373.

7. P.E.M. Siegbahn; M.R.A. Blomberg and C.W. Bauschlicher,Jr, J.
   Chem. Phys. 1984, 81, 2103.

8. P. Siegbahn; M. Blomberg; I. Panas and U. Wahlgren, Theoret.
   Chim. Acta in press.

9. P.J. Brucat; C.L. Pettiette; S. Yang; L.-S. Zheng; M.J.
   Craycraft and R.E. Smalley, J. Chem. Phys. 1986, 85, 4747.

10. I. Panas; J. Schüle; P. Siegbahn and U. Wahlgren, Chem. Phys.
    Letters 1988, 149, 265.

11. C.E. Moore, Atomic Energy Levels, U.S. Department of Commerce,
    Nat. Bur. Stand., U.S. Government Print Office: Washington, D.C.,
    1952.

12. L.G.M. Pettersson; U. Wahlgren and O. Gropen, Chem. Phys. 1983
    80, 7.

13. I. Panas; P. Siegbahn and U. Wahlgren, Theoret. Chim. Acta in
    press.

14. A. Mattsson; I. Panas; P. Siegbahn; U. Wahlgren and H. Åkeby,
    Phys. Rev. 1987, B36, 7389.

15. P.E.M. Siegbahn, Intern. J. Quantum Chem. 1983, 23, 1869

16. E.R. Davidson, in The World of Quantum Chemistry, edited by R
    Daudel and B. Pullman, Reidel, Dordrecht, 1974.

17. C.W. Bauschlicher,Jr; P. Siegbahn and L.G.M. Pettersson,
    Theoret. Chim. Acta in press.

18. A.-S. Mårtensson; C. Nyberg and S. Andersson, Phys. Rev. Letters 1986, 57, 2045.

19. I. Boustani; W. Pewestorf; P. Fantucci; V. Bonacic-Koutecky and J. Koutecky, Phys. Rev. 1987, B35, 9437.

20. A. Rosen and T.T. Rantala, in Metal Clusters, edited by F. Träger and G. zu Pulitz, Springer-Verlag Berlin, Heidelberg, 1986.

21. W.D. Knight; K. Clemenger; W.A. de Heer; W.A. Saunders; M. Chou and M.L. Cohen, Phys. Rev. Lett. 1984, 52, 2141.

22. C.L. Pettiette; S.H. Yang; M.J. Craycraft; J. Conceicao; R.T. Laaksonen; O. Cheshnovsky and R.E. Smalley, J. Chem. Phys. 198 88, 5377.

RECEIVED October 24, 1988

Chapter 10

Reaction of Methane with Nickel [111] Surface

Hong Yang and Jerry L. Whitten

Department of Chemistry, State University of New York at Stony Brook, Stony Brook, NY 11794-3400

The dissociation of CH_4 on the [111] surface of Ni is treated by a many-electron embedding theory to describe the electronic bonding. The lattice is modelled as a 26-atom, three layer cluster which is extracted from a larger cluster by an orbital localization transformation. Ab initio valence orbital CI calculations carried out on a local surface region permit an accurate description of bonding at the surface; 3d orbitals are explicitly included on four nickel atoms on the surface. The dissociation of CH_4 to CH_3 + H is calculated to be 0.1 eV endothermic, and the most likely dissociation pathway is predicted to involve atop Ni adsorption followed by dissociation to separated 3-fold sites for CH_3 and H. The Ni 3d orbitals contribute to the bonding by directly mixing with CH_3 orbitals and through an interaction of $3d^9$ and $3d^{10}$ states for atop atom adsorption.

The chemisorption and reaction of hydrogen and hydrocarbon fragments on catalytically active transition metal surfaces has received a great deal of attention due to the commercial importance of hydrocarbon formation reactions (1-15). The methanation reaction on nickel has been studied by a variety of surface science techniques. Yates, et al. (1), and others, have concluded that the following steps must occur on Ni(111):

C(ads) + H(ads) = CH(ads)

CH(ads) + H(ads) = CH_2(ads)

CH_2(ads) + H(ads) = CH_3(ads)

CH_3(ads) + H(ads) = CH_4(g)

From kinetic studies using temperature programmed desorption, Yates, et al. concluded that the rate determining step must occur at an earlier stage than the CH_3(ads) + H(ads) = CH_4(g) process. Lee, Yang and Ceyer (2) have recently studied the dynamics of the

0097–6156/89/0394–0140$06.00/0

activated dissociative chemisorption of CH_4 on Ni(111) by molecular beam techniques coupled with high resolution electron energy loss spectroscopy. A barrier to dissociative chemisorption of CH_4 was found, and products of the dissociation were identified by EELS as adsorbed CH_3 and H. From estimates of CH_3 adsorption energies, and both calculated and experimental values for H adsorption, the dissociation of CH_4 would be expected to be nearly energetically neutral. It is therefore a challenging theoretical problem to calculate the energetics with sufficient accuracy to contribute to the understanding of the dissociation mechanism. High quality calculations are also needed to sort out the role of the Ni 3d electrons.

The present work treats the adsorption of CH_4, CH_3 and H on a Ni(111) surface in the context of a many-electron theory that permits the accurate computation of molecule-solid surface interactions at an ab initio configuration interaction level. The adsorbate and local surface region are treated as embedded in the remainder of the lattice electronic distribution which is modelled as a 26-atom, three layer cluster, extracted from a 62-atom cluster by an orbital localization transformation.

In the investigation of the reaction mechanism, our goal is to find the geometry, binding energy, vibrational frequencies and the bonding properties of hydrogen and the hydrocarbon species on the [111] surface of Ni. Our calculations predict that CH_4 will dissociate only if it is possible to produce CH_3 and H in separated 3-fold sites. Across bond dissociation to produce coadsorbed fragments in adjacent 3-fold sites is energetically very unfavorable. Even for the most favorable, widely separated sites, the dissociation of CH_4 is predicted to be slightly endothermic. The construction of the cluster, surface sites and the binding properties are discussed in the following sections.

GENERAL THEORY

The assumption underlying the present theoretical approach is that a description of molecule-surface interactions and dissociative processes on metal surfaces may necessitate a reasonably sophisticated treatment of the surface region to account for changes in polarization and electron correlation accompanying reactions. In the embedding theory, a local region is defined as an N-electron subspace extracted from the remainder of the system by a localization transformation (16,17). The adsorbate and local region are then treated at high accuracy as embedded in the Coulomb and exchange field of the remainder of the electronic system. The use of electron exchange as the basis of a localization transformation is discussed in Refs. 16-20. In the present work, we model the boundary atoms of a small cluster of 26 atoms to represent the s-band attributes of the larger cluster on which the initial s-band calculations are carried out.

There are two clusters relevant to the present argument, a large cluster of M atoms (in the present case M = 62) that provides the initial model for the metal lattice valence s electrons, and a smaller cluster of M' atoms (M' = 26 in the present case) derived from the larger lattice, see Figure 1. The objective is to modify the boundary atoms of the small cluster to take into account the fact that these atoms should be bonded to the now missing remainder of the lattice. In the present case, the Ni_{26} cluster has three layers of 14, 7 and 5 atoms, respectively. The larger 62 atom cluster, representing the Ni(111) surface, consists of Ni_{26} and all of its nearest neighbor atoms in the first three layers at unperturbed fcc lattice sites. The s-band of the larger cluster is

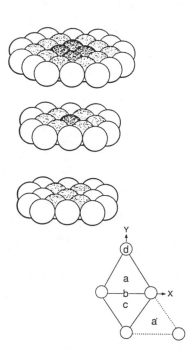

Figure 1. Cluster geometry and local region of the nickel cluster used to model the (111) crystal face of nickel. The three layer, 62-atom cluster, consists of a surface layer of 28 atoms, a second layer of 17 atoms and a third layer of 17 atoms. Embedding theory is used to reduce the Ni_{62} cluster to a 26 atom model depicted as shaded atoms. Atoms surrounding the four central atoms in the surface layer and those surrounding the one central atom in the second layer are described by effective potentials for (1s-3p core)$(3d)^9(4s)^{1/2}$ and (1s-3p core)$(3d)^9$ $(4s)^{1/3}$ configurations, respectively. Effective potentials for the shaded atoms in the third layer describe the (1s-3p core)$(3d)^9(4s)^{3/5}$ configuration. Unshaded atoms have neutral atom (1s-3p core)$(3d)^9(4s)^1$ potentials. All atoms have Phillips-Kleinman projectors $\Sigma |Q_m \rangle\langle Q_m| (-\epsilon_m)$ for the fixed electronic distribution.

described simply using a single 4s orbital per atom and a pseudopotential for all core and d-type electrons. A SCF calculation on
this cluster, followed by a localization of the cluster(molecular)
orbitals is used to define the number of electrons of the small
cluster strongly involved in bonding with atoms outside the
cluster; the latter atoms are referred to as the bulk. Orbitals
are localized by exchange maximization with the bulk atoms, and for
each resulting orbital ϕ_k' the electron density is analyzed to
determine the degree of localization (21,22). A set of occupied localized orbitals large enough to neutralize the nuclear charge on
the boundary atoms is then selected; in the present problem this
corresponds to the accommodation of 62 - 26 = 36 electrons.
Analysis of the electron density of these orbitals shows that, in
addition to covering the boundary atoms, the density spreads partly
onto the neighboring atoms of the 26 atom cluster. This spread in
density can be thought of as the fraction of the 26 atom cluster
strongly bound to the remainder of the lattice. If the electron
density associated with the boundary atoms were sufficiently distant
from the adsorption sites of interest to be taken as invariant, the
remaining electronic subspace could be treated by representing the
Coulomb and exchange fields of the static distribution and by
including a Phillips-Kleinman potential, $\Sigma \; |\phi_k><\phi_k|(-\lambda_k)$ to repre-
sent the occupied space. To provide an adequate response of the
lattice electron distribution to a large nearby perturbation such as
a bond forming or breaking reaction requires, however, that basis
functions actually be present in the "static" region (23). In the
present work, we assign that fraction of the boundary atom density
that is tightly bound to the remainder of the lattice to a spherical
density f s(1)s(1) where s is a normalized linear combination of
gaussians and f is the fraction of assigned charge (21). As would
be expected, the charge assignments determined from the localization
analysis qualitatively agree with a calculation based simply on the
number of nearest neighbors removed from bonding with a given atom
when the cluster is reduced in size. Figure 1. gives the fraction
of charge for each boundary atom assigned to the effective
potential.

Since there is uncertainty in the assignment of the boundary
electron density and by its representation by a spherical effective
potential, it is important to examine the characteristics of the SCF
solution of the final 26 atom cluster model. The calculated 4s-band
width is 9.0 eV which is reasonably close to the width measured by
photoemission of 10 eV (24). The ionization energy or work function
of the cluster is 5.29 eV, calculated as $E(Ni^+)$ - $E(Ni)$; and 5.39 eV
calculated by Koopmans' theorem. Both values are close to the experimental values of 5.15-5.35 eV for Ni(111) (6-8). In addition,
the calculated eigenvalue spectrum for the 4s-band is moderately
dense, particularly near the Fermi level (see Figure 2), which is
important to assure charge flow on adsorption. It would appear,
therefore, that the present model gives a good account of the essential features of the nickel 4s band.

Figure 2 also shows a d-band, arising from the four nickel
atoms with d electrons explicitly included, extending downward from
about -0.5 a.u. for the clean surface, adsorbed CH_4 and coadsorbed
CH_3 and H cases. In a Ni atom, for this basis, the average d orbital energy is -0.44 a.u., a value close to the Hartree-Fock result.
Photoemission measurements position the d ionization peaks of nickel
near the Fermi level, a result also obtained by most density functional treatments of nickel clusters. Application of Koopmans'
theorem would therefore suggest that the present d-ionization

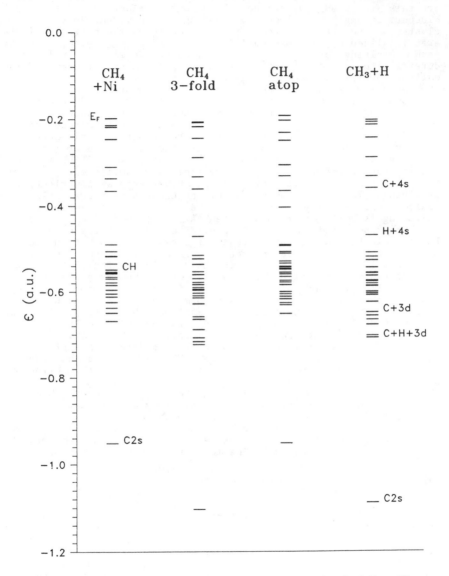

Figure 2. SCF eigenvalues for CH$_4$ and coadsorbed H + CH$_3$ on Ni(111). Principal C and H levels, and Ni 3d + C levels are indicated. The left most eignevalue spectrum is for the clean surface plus CH$_4$; C and H levels are indicated.

energies are unrealistically large. However, in <u>ab initio</u> treatments of d-electron systems, it is known that the explicit inclusion of d-d Coulomb and exchange interactions leads to substantial orbital relaxation on ionization of d-electrons (<u>25,26</u>). For example, unrestricted SCF calculations on the ionization of nickel surface atoms show a reduction in d-electron ionization energy from a Koopmans' value of 13.2 eV to 7.2 eV when calculated as $E\{Ni^+\}-E\{Ni\}$ (<u>22</u>). In contrast, ionization energies of s-electrons are much more accurately given by Koopmans' theorem.

The same set of effective potentials is used in all subsequent calculations on the Ni(111) surface and for all adsorption calculations.

CALCULATIONS

The cluster model of the [111] surface of Ni described in the previous section is shown in Figure 1; it consists of three layers: the surface layer of 14 atoms, a second layer of 7 atoms, and a third layer of 5 atoms. For the local surface region of four nickel atoms, see Figure 1, a [1s-3p] core potential is used and 3d orbitals are explicitly included. Other Ni atoms are described by 4s basis functions only and an effective core potential for [1s-3d] electrons. For all boundary atoms, and those in the third layer, the core potential is further modified to account for bonding to the bulk, as described above and in Ref. 21. The Ni core basis and core density and exchange expansions are the same as used in a previous study of nickel surface states (<u>19,22</u>). Polarization functions (4p) perpendicular to the surface are included. A double zeta basis is used for C and H. The sites considered for CH_4, CH_3 and H adsorption are as follows: a hollow three-fold site where there is no second layer nickel atom underneath (fcc extension of the lattice), a filled three-fold site with a second layer nickel atom underneath (hcp extension of the lattice), a bridge site and an atop atom site, as shown in Figure 1. Distances from the surface were optimized. Calculations are performed by first obtaining self-consistent-field (SCF) solutions for the nickel cluster plus adsorbed species. The occupied and virtual orbitals of the SCF solution are then transformed separately to obtain orbitals spatially localized about the four-atom surface region shown in Figure 1 and the adsorbate(s). This unitary transformation of orbitals is based on exchange maximization with the valence orbitals of atoms belonging to the surface region and is designed to enhance convergence of the configuration interaction (CI) expansion (<u>16-18</u>). The CI calculations involve excitation within a 26-electron subspace to 30 possible virtual localized orbitals. All configurations arising from single and double excitations with an interaction energy greater than $1x10^{-4}$ hartree with the parent SCF configuration are explicitly retained in the expansion; contributions of excluded configurations are estimated using second order perturbation theory. For all sites calculated, the SCF solution is the only dominant configuration. Details of the procedure are given in Ref. 20.

RESULTS

In Table I, adsorption energies, equilibrium distances and vibrational frequencies are reported for CH_3 and H on Ni(111). Both CH_3 and H are found to be most strongly adsorbed at 3-fold sites. The CH_3 adsorption energy is 1.8 eV for adsorption at the 3-fold (hcp) site at an equilibrium C-Ni surface (perpendicular) distance of 1.74

Table I. CH_3 and H adsorption on Ni(111)

Adsorption energies, equilibrium distances and vibrational frequencies from CI calculations on CH_3 and H adsorbed at different sites on a cluster model of the Ni(111) surface. Embedding theory is used to reduce a Ni_{62} cluster to a 26 atom model shown in Figure 1. The calculated equilibrium geometry of CH_3 is close to tetrahedral for all sites. Energies reported are for an HCH angle of $109.5°$ and a CH distance of 1.08 Å

Site[a]	R (Å)[a]	Ads. Energy (eV)[b]	$\omega_e(cm^{-1})$
H			
3-fold (fcc, c)	1.20	2.50	1043
bridge (b)	1.32	2.57	1183
3-fold (hcp, a)	1.18	2.66	1176
atop (d)	1.61	1.94	2332
Expt (3-fold)	1.15[c]	2.70[d]	1122[e]
	1.17[f]		1137[f]
CH_3			
3-fold (fcc, c)	1.71	1.26 (1.73)[g]	395
bridge (b)	1.78	1.27	391
3-fold (hcp, a)	1.74	1.78[g]	383
atop (d)	1.99	0.61	488
Expt			370[h]

[a] Sites are labeled as indicated in Figure 1. R is the perpendicular distance form the surface plane of nuclei. The 3-fold (hcp) site for CH_3 is depicted in Figure 3a.

[b] Energies are relative to H and CH_3 at infinite distance from the surface.

[c] Ref. 4. [d] Ref. 3. [e] Ref. 9. [f] Ref. 5.

[g] Modified third layer potential, see Sect. 4. [h] Ref. 2.

Å. In this work, only symmetric distortions of CH_3 are considered, and the minimum energy occurs for a near tetrahedral geometry with the hydrogens in a plane parallel to the surface, as depicted in Figure 3a. The calculated H atom adsorption energy is 2.7 eV for the most stable 3-fold (hcp) site which agrees with an experimental value of 2.70 eV (3). For hydrogen, the hollow and filled 3-fold sites and the bridge site are found to be of comparable stability in agreement with calculations by Siegbahn, et al. (15). The atop site is considerably less favorable. The nickel-hydrogen bond distance is 1.86 Å for the three-fold site, corresponding to a 1.18 Å perpendicular distance from the surface. This H-Ni distance is in excellent agreement with both LEED and neutron inelastic scattering values (4,5). Calculated vibrational frequencies, reported in Table I, for H in 3-fold and bridge sites are in good agreement with electron energy loss measurements of 1122 cm^{-1} (9) and neutron inelastic scattering measurements of 1137 cm^{-1}(5). Our data for CH_3 are insufficient to permit calculation of an asymmetric stretching frequency. For the CH_3-Ni surface perpendicular stretch, the calculated frequency is 383 cm^{-1}. A value of 370 cm^{-1} is attributed to the CH_3-Ni stretching mode in EELS measurements by Lee, et al. (2).

Figure 2 compares SCF eigenvalues for the clean surface and chemisorbed systems. For CH_3 + H coadsorbed in separated 3-fold sites, the figure shows the principal H level at 7.0 eV below the Fermi level, slightly deeper than the H feature seen in photoemission at 5.8 eV below E_F for H on Ni(111) (8). The charge distribution data show that on bonding of both CH_3 and H to the surface, electron transfer from the lattice to the adsorbate occurs accompanied by image charge formation in the first layer and a deepening of the Ni 3d levels. The calculations also provide considerable information about orbital interactions on CH_3 and H adsorption. Analysis of the SCF results in Figure 2 shows that CH_3 and H interact strongly with the Ni 4s-band. In addition, however, CH_3 and H orbitals mix significantly with the Ni 3d orbitals. For CH_3 and H coadsorbed in separated 3-fold sites, the main d-CH_3 mixing occurs in three low lying doubly occupied orbitals with populations (0.6d, 0.2C, 0.2H), (0.6d, 0.3C, 0.1H) and (0.7d, 0.1C, 0.2H), respectively. For atop Ni adsorption of CH_3, the mixing occurs almost entirely with the $3d_z^2$ orbital normal to the Ni surface, in one doubly occupied molecular orbital with populations (0.3d, 0.5C, 0.2H).

As noted above, the use of effective potentials to link the active electronic subspace with the bulk is at an early stage of development. There is evidence that increasing the number of active electrons in the second layer of the cluster, for example by increasing the polarizability of the third layer, favors adsorption in the hollow (fcc) site for both H and CH_3. It is only for CH_3 that the effect is large, however, leading to an increase in the adsorption energy for the hollow site by 0.4 eV as shown in Table I. Basis superposition corrections can also influence the relative stability of the two types of 3-fold sites and these corrections are not yet available for the CH_3 and CH_4 adsorption cases. From the evidence available, it appears that the hollow and filled 3-fold sites may be of comparable stability for both H and CH_3 adsorption.

Next we consider the methane molecule and its approach to the nickel surface. The main factor determining the mechanism for dissociation of CH_4 is the difficulty of stretching the CH bond. The total energy of a gas phase CH_4 molecule is calculated to increase by 0.8 eV on stretching one of the CH bonds from its equilibrium

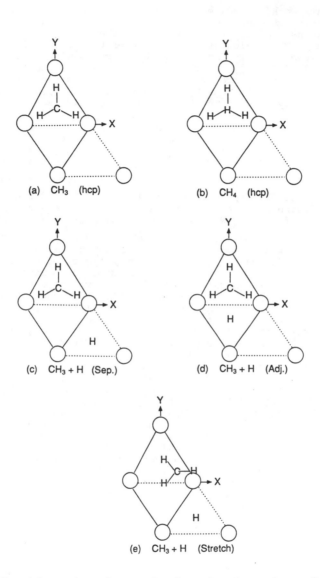

Figure 3. Adsorption sites and selected orientations of CH_4 and CH_3 plus H on Ni(111). Atop atom and bridge sites for CH_3 correspond to a vertical translation of CH_3 in 3(a); the fcc site is a reflection of the CH_3 group through the xz plane. The atop atom site for CH_4 corresponds to a vertical translation of the CH_4 group in 3(b). CH_3 has a tetrahedral geometry in all cases with H's in a plane parallel to the surface. In 3(e), the CH_3 adsorption site is midway between a Ni atom and the center of the 3-fold site.

distance of 1.08 Å to 1.46 Å, i.e., the increase in bond length necessary to reach the center of a three-fold site for C adsorbed atop Ni. To distort gas phase CH_4 to a transition state corresponding to pyramidal structure with a <u>planar</u> CH_3 base and H at the apex requires 1.1 eV. Thus, it is possible to dissociate CH_4 with a low energy barrier only if the distortion enables C or H bonding with Ni to compensate for the energy of distortion and other repulsions with the surface.

Table II shows that the approach of undistorted methane toward the Ni surface is repulsive for all orientations of methane. For a C-Ni surface (perpendicular) distance of 2.4 Å, and a 3-fold adsorption site, the three symmetric CH_4 orientations are (in order of increasing energy): 3 H's down, 1 H down, and 2 H's down. The latter is unfavorable due to interactions of hydrogens with the Ni-Ni bridge site. Thus, these data show the expected repulsive interaction between the saturated methane molecule and the surface. Figure 3b shows the most favorable orientation of the hydrogens for tetrahedral CH_4.

Attempts to stretch a single CH bond starting with either C or H above a 3-fold site always produced high energy intermediates. The reason such stretches do not result in sufficiently strong CH_3/Ni and H/Ni bonding to offset the breaking of the CH bond can be understood by examining the energies of CH_3 and H coadsorbed in 3-fold sites. Results in Table II show that coadsorption of CH_3 and H in adjacent 3-fold sites sharing a Ni-Ni bond (sites <u>a</u> and <u>c</u> in Figure 1 and Figure 3d) is very high in energy, 3.6 eV unbound with respect to desorbed CH_4. Although the energy would be reduced by further optimization of the geometry, it it very unlikely that a large reduction in repulsion will occur. In contrast, coadsorption of CH_3 and H in separated 3-fold sites (hcp sites <u>a</u> and <u>a</u>' in Figure 1 and Figure 3c) is much lower in energy, 1.1 eV unbound with respect to desorbed CH_4. The data in Table II also show that the energy of CH_4 in both 3-fold and atop sites increases on formation of the flattened, pyramidal structure (planar CH_3 base), but the increase is only 0.4 eV compared to 1.1 eV for gas phase CH_4. We conclude therefore that the only low energy distortion of CH_4 that could lead to dissociation is one that places CH_3 and H in separated 3-fold sites.

With H and CH_3 coadsorbed in separated 3-fold sites, rotation of CH_3 (keeping the H's in a plane parallel to the surface) leads to a 0.1 eV barrier when H's are above Ni-Ni bridge sites. Stretching one of the CH bonds in CH_4 by moving CH_3 and H halfway toward the centers of separated 3-fold sites <u>a</u> and <u>a</u>' produced high energy intermediates, but the degree of optimization of the fragment geometries is insufficient to establish a minimum energy pathway for dissociation. Figure 3e depicts the orientation of the tetrahedral CH_3 in this case, and in case where CH_3 and H are further separated by moving H to the center of the 3-fold site a'.

It is clear that Ni is playing an important role in developing a bond with the pyramidal distortion of CH_4 since the energy required for the distortion is less than that for gas phase methane. As noted earlier, direct mixing of the Ni 3d orbitals with CH_3 orbitals occurs. In addition, the CI calculations reveal a mixing of $3d^9$ and $3d^{10}$ configurations in the atop atom case. Without this interaction, the ability of a single Ni to induce dissociation would be diminished.

Table II. CH_4 adsorption on Ni(111)

Adsorption energies relative to desorbed CH_4 from CI calculations on CH_4 adsorbed at different sites on an embedded cluster model of the Ni(111) surface. A minus sign indicates a repulsive interaction (endothermic relative to gas phase CH_4). All CH distances are the same as in CH_4 unless indicated otherwise

	R (Å)[a]	Ads. Energy (eV)
CH_4 above surface		
3-fold site[b]		
1 H down	2.4	-1.0
2 H's down	2.4	-1.2
3 H's down	2.4	-0.9
CH_4 on surface		
3-fold site, 3 H's down[b]		
tetrahedral	2.0	-1.5
distorted	2.1	-1.4
flattened	2.0	-1.9
atop atom, 3 H's down[b]		
distorted	2.1	-1.8
flattened	2.0	-2.2
Coadsorbed CH_3 + H		
3-fold (adjacent, Fig. 3d)	H 1.45, C 2.2	-3.6
3-fold (separated, Fig. 3c)	H 1.19, C 1.72	-1.1
3-fold (∞ separation)	H 1.19, C 1.72	-0.1
CH_3- H Stretch		
3-fold (Fig. 3e)	H 1.19, C 1.98	-1.9
	R_{H_3C-H} = 2.05 Å	

[a] R is the perpendicular distance from the surface plane of nuclei
[b] Flattened geometries correspond to a pyramidal geometry with a planar CH_3 base and H at the apex. Distorted geometries are intermediate between the tetrahedral and flattened structures corresponding to a H(apex)C H(base) angle of 95.6°. The atop atom geometry corresponds to a translation of CH_4 shown in Figure 3b to the upper most atom in the figure.

CONCLUSIONS

The conclusions of the present study can be summarized as follows:

(1) CH_3 and H bind to Ni(111) in 3-fold sites with binding energies of 1.8 eV and 2.7 eV, respectively. The dissociation of CH_4 to CH_3 + H is endothermic by only 0.1 eV for widely separated CH_3 and H on the surface.

(2) Interactions between CH_4 and the Ni surface are replusive for all distances and orientations considered.

(3) CH_4 can dissociate only if it is possible to produce CH_3 and H in separated 3-fold sites. Across bond dissociation to produce coadsorbed fragments in adjacent 3-fold sites is energetically very unfavorable. Atop Ni adsorption of CH_4 is therefore the site most likely to produce dissociation. No low energy pathways to dissociation have been identified.

(4) The principal interactions of CH_3 and H orbitals with the surface involve the Ni 4s-band, however, significant mixing of Ni 3d orbitals with orbitals of C and H occurs. In addition, the d shell of Ni plays an important role in atop Ni adsorption of CH_3 by mixing the Ni $3d^9$ and $3d^{10}$ configurations.

ACKNOWLEDGMENTS

Support of the work by the U.S. Department of Energy is gratefully acknowledged. The authors are grateful to Professor J. Lauher for making available his graphics programs.

LITERATURE CITED

1. Yates, J. T., Jr.; Gates, S. M.; Russell, J. N., Jr. Surf. Sci. 1985, 164, L839. Beebe, T. P., Jr.; Goodman, D. W.; Kay, B. D.;Yates, J. T., Jr. J. Chem. Phys. 1987, 87, 2305.
2. Lee, M. B.; Yang, Q. Y.; Ceyer, S. T. J. Chem. Phys. 1987, 87, 2724.
3. Christmann, K.; Schober, O.; Ertl, G.; Neumann, M. J. Chem. Phys. 1974, 60, 4528.
4. Christmann, K.; Behm, R. J.; Ertl, G.; Van Hove, M.A.; Weinberg, W. H. J. Chem. Phys. 1979, 70, 4168.
5. Cavanagh, R. R.; Kelley, R. D.; Rush, J. J. J. Chem. Phys. 1982, 77, 1540.
6. Baker, B. G.; Johnson, E. B.; Marie, G. I. C. Surf. Sci. 1971, 24, 572.
7. Himpsel, F. J.; Knap, J. A.; Eastman, D. E. Phy. Rev. 1979, B19, 2872.
8. Demuth, J. E. Surf. Sci. 1977, 65, 369.
9. Ho, W.; DiNardo, N. J.; Plummer, D. E. J. Vac. Sci. Technol. 1980, 17, 314.
10. Somorjai, G. A. Chemistry In Two Dimensions: Surfaces; Cornell University Press: Ithaca, N.Y., 1981.
11. Rhodin, T. H.; Ertl, G. The Nature of The Surface Chemical Bond; North-Holland: Amsterdam, 1979.
12. Van Hove, M. A.; Tong, S. Y. The Structure of Surfaces; Springer-Verlag: Berlin, Heidelberg, Germany, 1985, p 18.
13. Conrad, H.; Ertl, G.;Kuppers, J.; Latta, E. E. Surf. Sci. 1976, 58, 578.

14. Goodman, D. W. J. Vac. Sci. Technol., 1984, A2, 873.
15. Siegbahn, P. E. M.; Blomberg, M. R. A.; Bauschlicher, C. W.,
    Jr. J. Chem. Phys. 1984, 81, 2103.
16. Whitten, J. L.; Pakkanen, T. A. Phys. Rev. 1980, B21, 4357.
17. Whitten, J. L. Phy. Rev. 1981, B24, 1810.
18. Cremaschi, P.; Whitten, J. L. Surf. Sci. 1985, 149, 273, and
    references contained therein.
19. Madhavan, P. V.; Whitten, J. L. Chem. Phys. Lett. 1986, 127,
    354.
20. Madhavan, P. V.; Whitten, J. L. J. Chem. Phys. 1982, 77, 2673.
21. Cremaschi, P.; Whitten, J. L. Theor. Chim. Acta. 1987, 72, 485.
22. Trentini, F. v.; Madhavan, P.; Whitten, J. L. Prog. in Surface
    Science 1987, 26, 201.
23. Chattopadhyay, A.; Whitten, J. L. Localization studies and
    boundary atom potentials, to be published.
24. Herman, F.; Dalton, N. W.; Koehler, T. R. Computational Solid
    State Physics; Plenum Press: New York-London, 1972, p 39.
25. Newton, M. D. Chem. Phys. Lett. 1982, 90, 291.
26. Miyoshi, E., Takewaki, H.; Nakamura, T. J. Chem. Phys. 1983,
    78, 815.

RECEIVED November 8, 1988

Chapter 11

Calculations on Transition Metal Complexes

Ernest R. Davidson

Chemistry Department, Indiana University, Bloomington, IN 47401

Calculations on transition metal complexes are reviewed.
The limitations of ab initio calculations are shown to
make predictions of spectra, bond lengths, bond
energies, and spin state very difficult. Calculations
on bisammineporphinatoiron, hydridocobaltcarbonyls, and
scandium carbonyl are reviewed.

Accurate ab initio calculations on complexes between organic ligands
and the first transition metal series remain an elusive goal. The
major difficulty is that, in spite of the widespread use of semi-
empirical molecular orbital schemes to describe these molecules, the
ab initio Hartree-Fock approximation is in fact hopelessly in error.
The extra-molecular correlation energy is of similar magnitude as the
total ligand binding energy, so it has a large effect on bond lengths
(1-4). The metal's atomic correlation energy also varies from state-
to-state by amounts comparable to the promotion energy. This
strongly affects conclusions about the relative energy of the various
conceivable molecular states.

   Other difficulties with transition metals present only practical
rather than conceptual problems. Relativistic corrections in the
first transition series have a small, but noticeable, effect on bond
lengths and relative energies (5). Rather large basis sets are
required to prevent basis set superposition errors (6) and to
describe the large relaxation effects associated with s→d promotion
on the metal (7). The correlation energy of a many electron system
must be described in a size-consistent manner for many different
states, which is difficult to do with configuration interaction (8).
Because the ligand field strength is usually no larger than the
atomic state separations, many of the states require several Slater
determinants in order to write down a zero'th order description
(9,10). This renders the usual perturbation expansions, which assume
one dominant Slater determinant, inappropriate (11).

   In order to progress at all, it is customary in this field to
ignore these problems and carry out calculations which are marginally
justified, but perhaps are an insightful complement to experiment and
semi-empirical theory.

0097–6156/89/0394–0153$06.00/0

Atomic Correlation Problems

As can be seen in Table I, systematic errors in relative SCF energies favor the sp excited state. The SCF energy is also biased against the d^{n+2} and $d^{n+1}s$ configurations for the last half of the transition series. To make the situation even more difficult, most atomic L,S states require several Slater determinants to express the Hartree-Fock wavefunction.

For example, the lowest nickel atomic states computed in the reduced symmetry of a linear molecule with a single determinant SCF wavefunction must have errors that are enormous compared with the expected ligand binding energy. The lowest $^1\Sigma^+$ state should be the $M_L=0$ component of the d^9s^1 ^1D state of nickel, which has two open shells. A single closed shell Slater determinant description would instead give either the d^{10} ^1S or a mixture of the d^8s^2 ^1S, ^1D, and ^1G states. The single determinant ROHF description of the $M_S=1$ component of the $^3\Sigma^+$ state would similarly produce some mixture of d^8s^2 ^3P and ^3F or d^9s ^3D. Because of the bias in the atomic SCF energies, the d^8s^2 configurations will be too low. This will be partially compensated by the inability of a single Slater determinant to represent the $M_S=1$, $M_L=0$ component of the ^3F ground state.

Table I. Numerical Hartree-Fock Excitation Energies[a]

$d^ns^2 \rightarrow$	$d^{n+1}s$		d^{n+2}		d^nsp		n
	calc	expt	calc	expt	calc	expt	
Sc	8105	11520	36053	33764	7704	15672	1
Ti	4357	6556	34321	28773	7478	15877	2
V	1001	2112	26366	20202	7251	16361	3
Cr	-10219	-7751	46421	27647	7528	17220	4
Mn	26840	17052	73836	44979	8601	18402	5
Fe	14494	6929	60178	32874	8788	19351	6
Co	12328	3483	56846	27497	12800	23612	7
Ni	10289	205	44147	14729	15591	25754	8
Cu	-2998	-11203			16700	27816	9
Zn					21315	32311	10

a. ΔSCF energies (cm^{-1}) between the lowest LS multiplets

Thus the order of the states is generally wrong and the use of a single configuration in a lower symmetry point group is inadequate. When CI is included in the description of the states there is still a tendency to overestimate the importance of configurations involving 4p or higher Rydberg states compared to configurations which put more electrons into the 3d shell. CI also has a difficult time compensating for the broken symmetry introduced by carrying out calculations on non-spherical atoms ($L\neq0$) in reduced symmetry, as must be done when considering the atom as part of a dissociating complex.

Molecular Correlation Problems

Molecular SCF energies are seriously in error because of the atomic errors at the single determinant level as discussed above. Perturbation of the atom by the ligand environment is generally comparable to the mixing matrix elements between Slater determinants which differ in the arrangement of electrons within the open shells of the free metal atom (12). So the high-field limit, which is assumed in most MO calculations, is usually not accurate (13). In addition, inter-molecular correlation effects between the metal atom and the ligand are a large fraction of the total binding energy. Also, large differential intra-atomic correlation errors between different atomic states cause a quite incorrect zero'th order picture of the spectrum.

These facts cause a break-down in the state-of-the-art methods for obtaining wavefunctions. All of these methods count on an SCF or small MC-SCF calculation to produce reasonable estimates of the molecular equilibrium geometry and the zero'th order wavefunction. Calculations have shown, however, that SCF potential curves systematically underestimate ligand binding and systematically produce excessively long estimates of the bond lengths. When inter-molecular correlation is included, the bond lengths become much shorter. In molecules where there is a high-low spin reversal as the molecule is formed, this change in bond length is often essential to predicting the correct spin state.

For excited states it is necessary to predict the relative order of four quite different types of excitation. Open shell atom states are split by the ligand field as predicted by typical Tanabe-Sugano diagrams. In addition, there are excited states of the atom with different $d^n s^m p^k$ configurations which are also ligand field split. Further, there are ligand-field-split metal to ligand and ligand to metal charge transfer states (and their excited atomic configurations) corresponding to the other common oxidation states of the atom. Finally, there are excited states of the ligand itself. All of these excited states are subject to differential relaxation and correlation effects. Some ligand excited states, such as the $^1(\pi\pi^*)$ Soret bands of porphine, are subject to actual and artifactual Rydberg-valence mixing. Even in the absence of the metal, these can only be unraveled by calculations which include extensive sigma-pi correlation.

A Study of Iron Porphyrins

Because of the importance of iron porphine type compounds as a model for the heme group, they have been the subject of a number of ab initio calculations (14-16). In this chapter we will focus on our own results for some of these molecules (17-19).

Traditionally, (20) the free iron porphine ($Fe \cdot N_4 C_{20} H_{12}$) is regarded as $Fe^{2+} \cdot Por^{2-}$ with the Fe(II) in a $d^6 s^0$ configuration. Such an iron is treated in the Tanabe-Sugano diagram (12) as having a "B" parameter of about 1000 cm^{-1} a "C" parameter of about 4000 cm^{-1} and a spin-orbit coupling parameter of about 500 cm^{-1}. This causes the high spin ^5D state to lie about 20,000 cm^{-1} below the intermediate spin ^3P state and 30,000 cm^{-1} below the low spin ^1I state in the free Fe(II) ion (21).

The symmetry of the ground state of Fe•Por is not known with certainty, but is generally believed (16,17) to be ^3E. The Tanabe-Sugano diagram for a square planar d^6 complex allows an intermediate spin ground state for Dq/B in the range 1.3-1.8. On the other hand, the SCF calculation corresponds more nearly to Dq/B of 1.0 and gives a high spin ground state. Our limited CI corrected this somewhat, but still was unable to overcome the SCF bias toward singly-occupied d orbitals.

The adequacy of the ligand field model can be questioned because the actual Mulliken charge on the iron is only +1.2 due to the dative bonds with the nitrogens. Also, the CI calculations for the lowest states tended to find 20-30% Fe(III) character in the wavefunction. Nevertheless, this model does predict correctly that there should be a large manifold of states of ligand-field-split d^6 character within 3 eV of the ground state.

In the traditional interpretation of iron porphine spectra, Gouterman (20) looked only for "charge-transfer" and $\pi\pi^*$ excitations of the porphyrin ring. The ab initio calculation suggests that each of these categories of states is really a manifold of states split by the ligand field. The $\pi\pi^*$ states cover the range 3-6 eV, but are poorly represented unless one uses at least a double-zeta representation of the π and π^* orbitals and correctly describes the σ-π electron correlation. The charge transfer states to form d^5 Fe(III) and Por^{3-} were predicted in the CI calculation to lie in the 1-4 eV range. These energies are probably underestimated. Many additional states corresponding to Rydberg excited states of the iron or $\pi\pi^*$ excited charge transfer bands were also predicted at low energy. Because systematic errors of ± 1 eV are expected between states of different type in these crude calculations, the exact order of the 200 or so states within 6 eV of the ground state clearly cannot be predicted. At best the calculation produced a biased catalog of possibilities including some which had not previously been considered.

Another technical problem in doing this calculation was the use of one set of d orbitals for all states. It is well-known (22) that d orbitals change size slightly between d^n and d^{n+1} states. This relaxation introduces a few eV shift in the excitation energy which is difficult to describe with limited CI. Consequently CI energies tend to be biased toward the d configuration for which the SCF equations were solved. When a molecule has a very large number of closely spaced states, it is not possible to solve the SCF equations separately for each state, so the same orbitals must be used for all states even though the resultant accuracy will be low.

We also considered the effect of axial ligands, modeled by NH_3 groups. This octahedral complex followed the d^6 Tanabe-Sugano diagram, in which the intermediate spin is never the ground state, and had intermediate spin states shifted above both the high and low spin states. The SCF results actually gave high spin slightly below the singlet state in contradiction to experiment. From the Tanabe-Sugano diagram this indicates that the actual value of Dq/B is not very different from 2. The other noticeable effect from the axial ligands is the shift of the ionization potential, and the charge transfer bands, to lower energy. Conceptually, this comes from orthogonalizing the d shell to the ammonia lone pair orbitals. In fact at this level of calculation the ground state in the CI has a large fraction of Fe(III) character. Rydberg states are also shifted down to the 3 eV region.

In agreement with these Fe(II) results, calculations on Fe•Por•$(NH_3)_2^+$ clearly have the iron in the Fe(III) oxidation state. Charge transfer bands for this molecule are predicted to lie in the 2-5 eV range. As before, there is a manifold of ligand-field-split d^5 states below 3 eV. The Tanabe-Sugano diagram requires a Dq/B of almost 3 for the switch from high (S=5/2) to low (S=1/2) spin. Experimental examples with each spin type as the ground state are known. Our calculations gave Dq/B in the range of 2 and clearly predicted a high spin ground state. Since this calculation was done at an assumed geometry, it is likely that a smaller cavity for the iron would give a different spin state.

Colbalt Complexes

We have carried out calculations on cobalt complexes in order to study pieces of the hydroformylation reaction (23-25). Optimized SCF geometries for the following complexes were obtained:

$Co(CO\ )(H)(CO)_3$ (S=0,   Co(I)d^8)
$Co(C_2H_4)(H)(CO)_3$ (S=0,   Co(I)d^8)
$Co(H\ \ )\ \ (CO)_3$ (S=0 and 1, Co(I)d^8 and d^7p)
$Co(HCO\ )\ \ (CO)_3$ (S=0 and 1, Co(I)d^8 and d^7p)
$Co(C_2H_5)\ \ (CO)_3$ (S=0,   Co(I)d^8)

The effects of CI and basis sets on the results were explored.

In the split-valence basis set, RHF SCF approximation for the five coordinated species the bond lengths are systematically too long (26). Consequently, the ligand field is too small and the complex is predicted to be high spin. Improvement in the basis set has little effect. Inclusion of CI, however, shortens the bond and reverses the spin state. This comes primarily from the fact that there is a substantial dependence of the correlation energy on bond length in all states due to inter-molecular electron correlation (otherwise known as London dispersion forces). This has a very sharp bond length dependence, so that it substantially shortens the equilibrium bond-length in all states by 0.1-0.2 Å. In our calculations, the singlet state was the lowest state at the experimental geometry primarily because that bond length is fairly high on the repulsive side of the triplet SCF potential curve. Thus merely changing the bond length from the SCF singlet optimum value to the experimental value reversed the situation of having the triplet 0.8 eV below the singlet to having it 0.6 eV above without any significant differential correlation effects. Differential correlation effects were expected because, at large metal-ligand separations, SCF calculations systematically place states with two singly occupied orbitals too low compared to closed shell singlet states. The singlet atomic states also require several Slater determinants to express the L,S wavefunction, so the single determinant approximation for the molecule is also expected to be inadequate at large bond lengths.

It is interesting to note that, at the experimental bond lengths, the London dispersion type electron correlation contributes about 10 kcal/mol to each metal-carbonyl bond energy. This is approximately

half of the total bond energy. In spite of the fact that the singlet
SCF energy of HCo(CO)$_4$ is actually 20 kcal/mol higher at the
experimental bond lengths than at the SCF minimum, the computed total
CI energy is 13 kcal/mol lower due to the 33 kcal/mol increase in
inter-molecular correlation energy in this short distance.

With these reservations about the significance of SCF geometries,
it is still interesting to examine the shape of the molecule at the
calculated geometry. For HCo(CO)$_4$, the singlet state is trigonal
bipyramidal. In this form the hydrogen could occupy either an apical
or equatorial position. Minima of each shape are found. The SCF
energy slightly favors the equatorial position. While the computed
energy difference is too small to be significant, it does illustrate
the fact that hand-waving arguments, which show that either the axial
or equatorial position "must" be favored, are certainly vacuous.

Replacement of one carbonyl by a π complexed ethylene molecule
presents many more possibilities. The lowest energy form is found to
be one with the ethylene and hydrogen both in the equatorial plane.
When both carbons of the ethylene are in this plane, the ethylene
binding approaches a bidentate situation and the complex can be
regarded as octahedral Co(II). Turning the ethylene by 90°, or
placing it in the axial position, leads to a trigonal bipyramid Co(I)
complex about 10-20 kcal/mol higher in energy. Because these
geometry optimizations were done with symmetry constraints, they may
not all be true minima on the potential surface.

Four coordinated complexes are more difficult to describe. The
singlet spin state seems to have several true minima in the potential
surface which correspond to all ways to distribute a vacancy, a
ligand •R (•R = •H, •CH(O), or •CH$_2$CH$_3$), and three carbonyls on the
vertices of a trigonal bipyramid. The minimum SCF singlet energy
occurs for the configuration with a carbonyl and the vacancy in axial
positions. The triplet spin state of RCo(CO)$_3$ also has many possible
shapes. The lowest of these was the C$_{3v}$ tetrahedral configuration
with a 115° R-Co-C bond angle as predicted by Elian and Hoffmann
(27).

Scandium Carbonyl

The work on cobalt compounds raised two questions which could best be
answered by studying a smaller model compound. One was the role of
dispersion forces in metal ligand binding. The other was the elusive
role of pi back bonding which is supposed to account for a large
fraction of the metal carbonyl bond strength.

In Figure 1, we (28) show the calculated RHF potential curves,
for various electron configurations, as a function of Sc-C distance
for a fixed CO distance. These curves are very different from the
true potential curves because of the large error at large R. The
$4s3d\pi4p\pi$ configuration, which lies too low because of the quite low
electron correlation energy compared to sd^2 configurations, shows up
as an intruder state at all distances. Figure 2 shows the calculated
UHF/MP2 correlation energy for the $^2\Pi$ (s^{2}dπ) and $^4\Sigma^-$(s^{1}d$_x$,dπ_y)
states. The correlation energy for large R is the correction for Sc
atom and CO. The extramolecular correlation energy is due mainly to
double excitations of the type used in describing London dispersion
forces. Conventionally, this effect is thought to be small, as
indeed it is if evaluated at the van der Waals' minimum for the

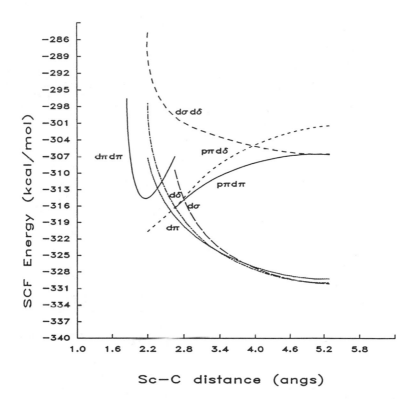

Figure 1. RHF potential curves for various configurations of ScCO at a fixed CO distance.

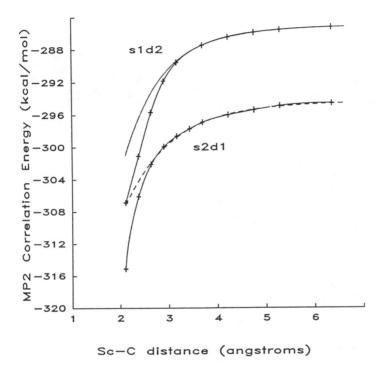

Sc—C distance (angstroms)

Figure 2. The MP2 correlation energy of ScCO. The calculation was done with the Wachters' "contraction 3" s,p Sc basis with Hay's augmented 5d basis and an additional "4p" function. The 6-31G* basis set was used for C and O. The C-O bond length was constrained to be the MP2 optimized bond length, 2.175 bohr, of free CO.

repulsive $^2\Pi$ state. The bond-length dependence of this correlation energy, however, is not very state dependent. For states with Sc-C equilibrium distances near 4 bohr, this would contribute about 20 kcal/mol to the binding energy. Thus this effect is large compared to the bond energy in scandium carbonyl complexes. When this MP2 energy is added to the UHF potential curves, as shown in Figure 3, the complex is still not quite predicted to be bound. However, if the whole s^1d^2 curve is shifted downward by the error in the MP2 promotion energy of the Sc atom, a reasonable potential curve results.

In the modeling of potential surfaces by force fields, non-bonded interactions are often modeled by a c/R^6 term for each atom pair. The constant "c" is roughly proportional to the product of the atomic polarizabilities. Following this model for ScCO gives a very poor fit to the inter-molecular correlation energy. A modified term such as $c/(R^2+A^2)^3$ provides a much better fit to the MP2 correlation energy. Of course, in the empirical potential a repulsive term is included which can model both the short range overlap repulsion in the SCF energy, and the inadequacy of the R^{-6} term.

The pi back-bonding model of Dewar (29), Chatt, and Duncanson (30) has been widely invoked as an explanation of a variety of features of transition metal complexes. While extended Huckel theory clearly shows such mixing of orbitals, ab initio calculations have found them more elusive (31,32).

One can look for pi back-bonding either in population or energy effects. Orbital population in the free carbon monoxide π^* orbital could arise from mixing with the CO π orbital as well as from mixing with a metal $d\pi$ orbital. Population can also arise from purely basis set superposition errors if an inadequate basis is used for either the metal or carbonyl. An additional problem arises because Mulliken populations are basis set dependent and become less meaningful as large diffuse basis sets are used. Populations in London orthogonalized basis sets are even more problematical because these will change if completely unused basis functions are introduced on neighboring atoms. One alternative procedure is to project the density matrix onto a fixed atomic Hartree-Fock or split-valence plus polarization (SVP) basis (33). The Mulliken population analysis of this projected density will lose any electron density outside of this basis, but will give a definite result as more diffuse basis functions are included.

By way of example, we have examined the populations for the ScCO linear molecule in various states. For the repulsive s^2d^1 linear states at 2.2 Å using an extended basis set projected onto a SVP basis, there is a distinct difference between the $d\sigma$, $d\delta$, and $d\pi$ states. The $d\delta$ state shows a charge transfer in the dative bond with the carbonyl which gives the scandium 0.16 e$^-$ in $3d\sigma$, 0.04 e$^-$ in $4p\sigma$ and a loss of 0.03 e$^-$ from 4s. The $d\pi$ state similarly shows a sigma charge transfer of 0.18 e$^-$ into $3d\sigma$, 0.07 e$^-$ into $4p\sigma$, and 0.06 e$^-$ out of 4s. But the $d\pi$ state also shows a transfer from the nominal $d\pi^1$ orbital of 0.16 e$^-$ into CO π^* and 0.02 e$^-$ into $4p\pi$. The $d\sigma$ state shows a sigma charge transfer of only 0.05 e$^-$ into $3d\sigma$, 0.09 e$^-$ into $4p\sigma$, and 0.05 e$^-$ out of 4s.

The $s^1d\pi^2$ $^4\Sigma^-$ state is interesting because it is probably bound relative to ground state Sc and CO (34). This has sigma charge transfer of 0.06 e$^-$ into $3d\sigma$, 0.12 e$^-$ into $4p\sigma$, and 0.02 e$^-$ out of

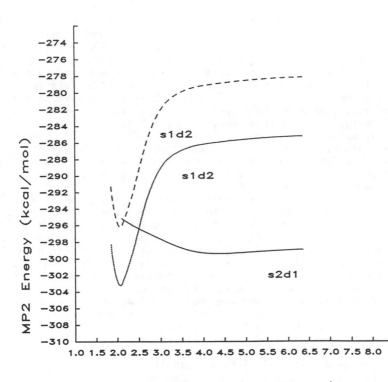

Figure 3. UHF/MP2 potential curves for ScCO. The dotted curve
results from shifting the computed s^1d^2 curve to give the correct
excitation energy at large R.

4s. The pi back bond is quite evident in this state with 0.39 e⁻
going into CO π^* and 0.09 e⁻ into 4pπ. Thus, the populations show
the expected pi back-bond with about 20% of the nominal dπ population
going into CO π^*.

The energetic consequences of pi back-bonding are also evident in
this case. Among the s^2d^1 states at 2.2 Å, the dπ is 9.8 kcal/mol
below the dδ, presumably entirely due to pi back-bonding. The dσ is
15.5 kcal/mol above dδ because of the repulsion between the dσ
electron and the CO lone pair.

The energetics of the $s^1d\pi^2$ $^4\Sigma^-$ state are an interesting
illustration of the inadequacy of any simple model. The state is
barely bound. The pi back-bonding is worth about 20 kcal/mol for
this state. The inter-molecular electron correlation contributes
another 20 kcal/mol. But the atomic promotion energy from s^2d^1 ^2D to
s^1d^2 ^4F is 33 kcal/mol. The sigma dative bond is very weak and may
be slightly anti-bonding. Hence the net bond energy is very small.

Conclusion

The ab initio theory of bonding in transition metal complexes is
considerably more complicated than the usual semi-empirical MO model
(35). Pi back-bonding and sigma overlap effects do make a
significant contribution to the energy. But many other effects are
equally important. Several Slater determinants are usually required
to describe the wavefunction for an open-shell atom and these
continue to be important for the molecule. Inter-molecular electron
correlation, analogous to London dispersion forces, plays a large
role in stabilizing the complex. For metals in the zero'th oxidation
state, the promotion energy to remove an (anti-bonding) 4s electron
is a major barrier to bond formation.

Literature Cited

1. Demuynck, J.; Strich, A.; Veillard, A. Nouv. J. Chim. 1977, 1,
    217.
2. Almlof, J.; Faegri, K.; Schilling, B. E.; Luthi, H. P. Chem.
    Phys. Lett. 1984, 106, 266.
3. Luthi, H. P.; Siegbahn, P. E.; Almlof, J. J. Phys. Chem. 1985,
    89, 2156.
4. Luthi, H. P.; Ammeter, J. H.; Almlof, J.; Faegri, K. J. Chem.
    Phys. 1982, 77, 2002.
5. Martin, R. L. J. Chem. Phys. 1983, 78, 5840.
6. Martin, R. L.; Hay, P. J. J. Chem. Phys. 1981, 75, 4539.
7. Hay, P. J. J. Chem. Phys. 1967, 66, 4377.
8. Davidson, E. R.; Silver, D. W. Chem. Phys. Lett. 1977, 53, 403.
9. Tanabe, Y.; Sugano, S. J. Phys. Soc. (Japan) 1956, 11, 864.
10. Condon, E. U.; Shortly, G. H. The Theory of Atomic Spectra;
    Cambridge University Press, 1967.
11. Møller, C.; Plesset, M. S. Phys. Rev. 1934, 46, 618.
12. Figgis, B. N. Introduction to Ligand Fields; Wiley Interscience:
    New York, 1966.
13. Ballhausen, C. J.; Gray, H. B. Molecular Orbital Theory;
    Benjamin: New York, 1964.
14. Dedieu, A.; Rohmer, M.; Veillard, H.; Veillard, A. Nouv. J.
    Chim. 1979, 3, 653.

15. Dedieu, A.; Rohmer, M.; Veillard, A. Adv. Quantum Chem. 1982,
     16, 43.
16. Kashiwagi, H.; Obara, S. Int. J. Quantum Chem. 1981, 20, 843.
17. Rawlings, D. C.; Gouterman, M.; Davidson, E. R.; Feller, D. Int.
     J. Quantum Chem. 1985, 28, 773.
18. Rawlings, D. C.; Gouterman, M.; Davidson, E. R.; Feller, D. Int.
     J. Quantum Chem. 1985, 28, 797.
19. Rawlings, D. C.; Gouterman, M.; Davidson, E. R.; Feller, D. Int.
     J. Quantum Chem. 1985, 28, 823.
20. Gouterman, M. In The Porphyrins; Dolphin, D., Ed.; Academic:
     New York, 1978; Vol. 3.
21. Konig, E.; Kremer, S. Ligand Field Energy Diagrams; Plenum: New
     York, 1977.
22. Pelissier, M.; Davidson, E. R. Int. J. Quantum Chem. 1984, 25,
     483.
23. Antolovic, D.; Davidson, E. R. J. Am. Chem. Soc. 1987, 109, 977.
24. Antolovic, D.; Davidson, E. R. J. Am. Chem. Soc. 1987, 109,
     5828.
25. Antolovic, D.; Davidson, E. R. J. Chem. Phys. 1988, 88, 4967.
26. Daniel, C.; Hyla-Kryspin, J.; Demuynck, J.; Veillard, A. Nouv.
     J. Chim. 1985, 9, 581.
27. Elian, M.; Hoffman, R. Inorg. Chem. 1975, 14, 1058.
28. Frey, R. F.; Davidson, E. R. (unpublished)
29. Dewar, M. J. S. Bull. Soc. Chim. Fr. 1951, 18, C71.
30. Chatt, J.; Duncanson, L. A. J. Chem. Soc. 1953, 2939.
31. Bagus, P. S.; Roos, B. O. J. Chem. Phys. 1981, 75, 5961.
32. Bauschlicher, C. W.; Bagus, P. S. J. Chem. Phys. 1984, 81, 5889.
33. Davidson, E. R. J. Chem. Phys. 1967, 46, 3313.
34. Jeung, G. H.; Koutecky, J. Chem. Phys. Lett. 1986, 129, 569.
35. Blomberg, M. R. A.; Brandemark, U. B.; Siegbahn, P. E. M.;
     Mathisen, K. B.; Karlstrom, G. J. Phys. Chem. 1985, 89, 2171.

RECEIVED October 24, 1988

Chapter 12

Density Functional Theories in Quantum Chemistry

Beyond the Local Density Approximation

Axel D. Becke

Department of Chemistry, Queen's University, Kingston, Ontario K7L 3N6, Canada

A review of the "local density" exchange-correlation approximation (LDA) in quantum chemistry is provided, with particular emphasis on the calculation of molecular bond energies, bond lengths, and vibrational frequencies. The LDA, surprisingly successful in itself, is improved even further by addition of so-called "gradient correction" terms, and following a very brief discussion of the history of gradient-corrected exchange-correlation approximations, recent work in this area is presented. Experience to date, and new results of the present work, indicate that beyond-LDA density functional theories yield molecular spectroscopic properties of near chemical accuracy, even in transition-metal systems, with minimal computational effort.

Due to the complexity of *ab initio* quantum chemical methods in transition-metal chemistry, approximate molecular orbital schemes such as the well-known "Xα" method have enjoyed considerable popularity in this area (1). Though the Xα method has been very useful generally, its performance in predicting molecular binding energy curves is somewhat erratic, and it suffers basic theoretical deficiencies as well. Fortunately, the seminal work of Hohenberg, Kohn, and Sham (2) has inspired the development of a new approach to many-electron problems known as "density functional theory" (DFT), of which the Xα approximation may be viewed as the simplest possible variation. Contemporary DFT is a physically and mathematically rigorous theory, and provides remarkably accurate molecular properties such as bond lengths, vibrational frequencies, and even dissociation energies.

Here we will not discuss the density functional formalism in depth, but we refer the reader to Ref. (3) for access to the extensive and growing literature on its fundamental theoretical

0097–6156/89/0394–0165$06.00/0
© 1989 American Chemical Society

aspects.   We present only a minimal complement of concepts and
formulas required to appreciate current efforts in this field.

Exchange-Correlation Hole Functions

It can be shown (2) that the total electronic energy of a
many-electron system is expressible in the form

$$E = \frac{1}{2} \sum_i \int |\nabla \psi_i|^2 \, d^3\underline{r} + \int \rho V_{nuc} d^3\underline{r}$$

$$+ \frac{1}{2} \iint \frac{\rho(1)\rho(2)}{r_{12}} \, d^3\underline{r}_1 d^3\underline{r}_2 + E_{XC}$$

(1)

where the first term is the kinetic energy of a *reference* system
of *non-interacting* electrons having the same total density ρ as
the actual system, the second and third terms are the usual
potential energy of interaction with the nuclei and the classical
electrostatic interaction energy respectively, and the last term
is the so-called "exchange-correlation" energy.   The
non-interacting reference orbitals ψ_i satisfy the following
single-particle equation:

$$-\frac{1}{2} \nabla^2 \psi_i + V\psi_i = \varepsilon_i \psi_i$$

(2)

with a local, orbital-independent potential V given by

$$V = V_{nuc} + V_{el} + V_{XC}$$

(3a)

$$\text{where} \qquad V_{el}(1) = \int \frac{\rho(2)}{r_{12}} \, d^3\underline{r}_2$$

(3b)

$$\text{and} \qquad V_{XC} = \frac{\delta E_{XC}}{\delta \rho}$$

(3c)

where Equation 3c represents the functional derivative, in a
variational sense, of the exchange-correlation energy with respect
to the density ρ on which it depends uniquely.   This formalism,
known as the "Kohn-Sham" formalism, is easily extended to
spin-polarized systems as well.
     Let us emphasize that the Kohn-Sham orbitals of DFT and the
Hartree-Fock orbitals of *ab initio* theory, though normally quite
similar numerically, are distinctly different from a theoretical
point of view.   Hartree-Fock orbitals satisfy a single-particle
equation with a mathematically non-local (i.e. orbital-dependent)

exchange potential, whereas the Kohn-Sham potential is *local* and
orbital-independent. Moreover, the sum of the Hartree-Fock
orbital densities in *ab initio* theory does *not* equal the total
interacting electron density, whereas the sum of the Kohn-Sham
orbital densities in DFT equals the total interacting density *by
definition*. These properties of the Kohn-Sham orbitals are
physically and computationally appealing, and are largely
responsible for the great popularity of the density functional
formalism in solid-state physics.

The exchange-correlation energy itself is given by an
expression of the form

$$E_{XC} = \frac{1}{2} \sum_{\sigma\sigma'} \iint \frac{\rho_\sigma(1)h_{XC}^{\sigma\sigma'}(1,2)}{r_{12}} d^3r_1 d^3r_2 \qquad (4)$$

where we now include an explicit spin index σ, and where the
four-component two-electron function $h_{XC}(1,2)$ is called the
exchange-correlation "hole" function. All information underlying
electronic "exchange" and "dynamical" correlation energies,
including the *kinetic* component of the latter, is contained
therein (see, for example, Ref.(4)). The exchange-correlation
problem reduces, therefore, to determination of simple (at least
in principle) two-electron functions only. Though a formal
expression for their derivation exists, determination of
exchange-correlation hole functions from first principles is, in
fact, an intractable task, akin to a full CI calculation. The
advantage of the density functional approach lies in the fact that
our hole functions rigorously satisfy certain *global constraints*
which allow the construction of simple but realistic hole-function
models. For example, denoting coordinate 1 in all subsequent work
as the "reference" coordinate, we have the following total charge
conditions:

$$\int h_{XC}^{\alpha\alpha}(1,2) \, d^3r_2 = -1 \qquad (5a)$$

$$\int h_{XC}^{\beta\beta}(1,2) \, d^3r_2 = -1 \qquad (5b)$$

$$\int h_{XC}^{\alpha\beta}(1,2) \, d^3r_2 = 0 \qquad (5c)$$

$$\int h_{XC}^{\beta\alpha}(1,2) \, d^3r_2 = 0 \qquad (5d)$$

valid at *any* reference point. Also, the Pauli exclusion principle
is reflected by the following constraints on the $\alpha\alpha$ and $\beta\beta$ holes
as coordinate 2 and the reference coordinate coalesce:

$$h_{XC}^{\alpha\alpha}(1,1) \;\; = \;\; - \rho_\alpha(1) \qquad\qquad\qquad (6a)$$

$$h_{XC}^{\beta\beta}(1,1) \;\; = \;\; - \rho_\beta(1) \qquad\qquad\qquad (6b)$$

Furthermore, since the 2-integration in Equation 4 extracts only the *spherical average* of the hole function with respect to the reference point 1, details of its *angular* dependence are unimportant. Spherically symmetric hole-function models are therefore perfectly justified. These constraints, among others discussed elsewhere (4), are satisfied in *any* many-electron system, whether a helium atom, a transition-metal cluster, a uniform electron gas, etc.. To the extent that properties such as these contain the essential physics of exchange and correlation phenomena, hole-function models provide a simple and convenient alternative to traditional *ab initio* technology.

The Local Density Approximation (LDA)

A system of fundamental theoretical importance in many-body theory is the uniform-density electron gas. After decades of effort, exchange-correlation effects in this special though certainly not trivial system are by now well understood. In particular, sophisticated Monte Carlo simulations have provided very useful information (5) and have been conveniently parametrized by several authors (6). If the exchange-correlation hole function at a given reference point r in an *atomic* or *molecular* system is approximated by the hole function of a *uniform electron gas* with spin-densities given by the *local* values of $\rho_\alpha(r)$ and $\rho_\beta(r)$, we obtain an exchange-correlation energy approximation of the form

$$E_{XC}^{LDA} \;\; = \;\; \int \rho(r) \; \varepsilon_{XC}(\rho_\alpha(r),\rho_\beta(r)) \; d^3r \qquad\qquad (7)$$

where $\varepsilon_{XC}(\rho_\alpha,\rho_\beta)$ is the exchange-correlation energy per particle of a uniform electron gas with spin-densities ρ_α and ρ_β obtained from the parametrization of Vosko, Wilk, and Nusair (6) in the present work. This approximation is known as the "local density" approximation (LDA), or, more precisely, the local *spin*-density approximation (LSD or LSDA).

The LDA has been widely applied to problems in atomic, molecular, surface, and solid-state physics with surprisingly good results overall (7). Apparently, the uniform gas model describes pair correlations even in inhomogeneous systems fairly well. In this communication, we shall focus our attention on molecular binding energies and related properties such as equilibrium bond lengths and vibrational frequencies. In Tables I, II, and III, we

Table I: Bond Lengths r_e of Diatomic Molecules (Å)

	Expt.[a]	LDA	PW1[b]	PW2[c]
H_2	0.74	0.76	0.74	0.75
Li_2	2.67	2.71	2.71	2.73
B_2	1.59	1.60	1.62	1.62
C_2	1.24	1.24	1.25	1.25
N_2	1.10	1.09	1.10	1.10
O_2	1.21	1.20	1.23	1.22
F_2	1.41	1.38	1.44	1.41
Na_2	3.08	3.00	3.11	3.09
Al_2	2.47	2.46	2.52	2.49
Si_2	2.25	2.27	2.30	2.29
P_2	1.89	1.89	1.91	1.90
S_2	1.89	1.89	1.92	1.91
Cl_2	1.99	1.97	2.04	2.01
Δ[d]		0.03	0.03	0.02
Cu_2	2.22	2.17	2.29	2.24

a) Ref.(20).

b) Present Work: Equation 15 plus correlation LDA.

c) Present Work: Equation 15 plus gradient-corrected
    correlation approximation of Perdew, Ref.(10).

d) Δ: rms deviation from experiment for H_2 through Cl_2.

Table II: Dissociation Energies D_e (eV)

	Expt.[a]	LDA	PW1[b]	PW2[c]
H_2	4.8	4.9	5.1	4.9
Li_2	1.1	1.0	0.9	0.9
B_2	3.1	3.9	2.8	3.2
C_2	6.3	7.3	5.6	6.0
N_2	9.9	11.6	10.1	10.3
O_2	5.2	7.6	5.6	6.1
F_2	1.7	3.4	1.9	2.2
Na_2	0.7	0.9	0.7	0.7
Al_2	1.6	2.0	1.3	1.6
Si_2	3.2	4.0	3.1	3.4
P_2	5.1	6.2	5.0	5.2
S_2	4.4	5.9	4.5	5.0
Cl_2	2.5	3.6	2.3	2.7
Δ[d]		1.2	0.3	0.4
Cr_2	2.0[e]	3.0	1.4	1.6
Cu_2	2.1	2.7	1.9	2.2

a)   Ref.(20).

b)   Present Work: Equation 15 plus correlation LDA.

c)   Present Work: Equation 15 plus gradient-corrected
     correlation approximation of Perdew, Ref.(10).

d) Δ: rms deviation from experiment for H_2 through Cl_2.

e)   Ref.(18) and references therein.

Table III: Vibrational Frequencies ω_e (cm^{-1})

	Expt.[a]	LDA	PW1[b]	PW2[c]
H_2	4400	4190	4450	4330
Li_2	350	330	340	330
B_2	1050	1030	990	1000
C_2	1860	1880	1810	1830
N_2	2360	2380	2320	2330
O_2	1580	1620	1490	1550
F_2	920	1060	930	990
Na_2	160	160	150	150
Al_2	350	350	310	330
Si_2	510	490	460	470
P_2	780	790	760	780
S_2	730	730	670	700
Cl_2	560	570	500	540
Δ[d]		70	50	40
Cu_2	270	290	240	260

a) Ref.(20).

b) Present Work: Equation 15 plus correlation LDA.

c) Present Work: Equation 15 plus gradient-corrected
correlation approximation of Perdew, Ref.(10).

d) Δ: rms deviation from experiment for H_2 through Cl_2.

therefore present LDA bond lengths, bond energies, and vibrational
frequencies respectively for a variety of homonuclear diatomic
molecules calculated with the non-basis-set fully-numerical
diatomic code of Ref.(8). The numerical precision of these data
is of the order 0.01 Å, 0.1 eV, and 20 cm^{-1} for bond lengths, bond
energies, and frequencies respectively. Observe that the LDA bond
lengths and vibrational frequencies agree remarkably well with
experiment despite the great simplicity of the theory. It is
evident in Table II, however, that the LDA tends to overestimate
molecular dissociation energies by as much as several eV in the
worst cases. This order of discrepancy is unacceptable for
chemical applications and improvement is certainly required in
this regard.

Gradient Corrections to the LDA

The integrand of the LDA exchange-correlation energy of Equation 7
depends explicitly on local spin-densities only. An atomic or
molecular system is not, however, homogeneous, even locally, and
we therefore seek improvements to the LDA by somehow incorporating
information from the inhomogeneity of the density distribution. A
formal systematic procedure called the "density gradient
expansion" has been proposed for this purpose (2). One assumes
that improved exchange-correlation functionals exist in which the
integrand of Equation 7 depends not only on the local value of the
density, but also on the local values of its various *derivatives*.

It can be shown, essentially by dimensional analysis (9),
that the *lowest order* gradient-corrected exchange-correlation
functional has the form

$$E_{XC} = E_{XC}^{LDA} - b \int \left[\frac{(\nabla\rho_\alpha)^2}{\rho_\alpha^{4/3}} + \frac{(\nabla\rho_\beta)^2}{\rho_\beta^{4/3}} \right] d^3\underline{r} \qquad (8)$$

where the first term is the usual LDA functional and b is a
constant. Actually, this lowest-order gradient term is
dimensionally consistent with "exchange" energy only, defined in
density functional theory as the exchange energy of the Slater
determinant constructed from the Kohn-Sham orbitals.
"Correlation" corrections are significantly more complicated, due
to their non-trivial dependence on the interelectronic coupling
strength e$^-$ as well as their non-trivial spin dependence, and
will therefore not be discussed here (see, however,
Refs.(4) and (10) for recent developments in this area).
Nevertheless, since exchange-correlation energy is dominated by
exchange, we make the reasonable first approximation that
corrections to the LDA *correlation* energy are of secondary
importance, and we shall concentrate our efforts on *exchange*
corrections only.

The gradient correction term of Equation 8 suffers the

well-known deficiency that its corresponding *potential* of Equation 3c *diverges* asymptotically in finite systems. Moreover, ensuing molecular spectroscopic properties are in poor agreement with experiment, displaying severe *under*binding as opposed to the previous *over*binding of the LDA (11). Consequently, various modifications of the gradient term have been proposed in order to damp its behaviour in the limit $x_\sigma \to \infty$, where x_σ is the dimensionless parameter

$$x_\sigma = \frac{|\nabla\rho_\sigma|}{\rho_\sigma^{4/3}} \tag{9}$$

corresponding to the asymptotic limit $r \to \infty$. Among several modifications in the literature, we have proposed the following (12):

$$E_X = E_X^{LDA} - b \sum_\sigma \int \rho_\sigma^{4/3} \frac{x_\sigma^2}{(1 + cx_\sigma^2)} d^3r \tag{10}$$

where the exchange-only LDA is given by the simple expression

$$E_X^{LDA} = -\frac{3}{2} \left(\frac{3}{4\pi}\right)^{1/3} \int (\rho_\alpha^{4/3} + \rho_\beta^{4/3}) d^3r \tag{11}$$

and where b and c are constants. Such modified gradient corrections may be viewed as infinite summations of a certain class of gradient terms in the presumed infinite-order gradient expansion and are thus formally (if not empirically) justified.

The particular functional form of Equation 10 is certainly not the only conceivable dimensionally-consistent possibility, but was chosen on the basis of algebraic simplicity (12). Viewed in a semi-empirical spirit, it is found that Equation 10, with suitable values of the constants b and c, fits exact atomic exchange energies throughout the periodic table with extremely high accuracy (i.e. roughly 0.2% error compared with 5-10% error for the exchange-only LDA). It should be noted in this connection that constants b and c are "universal" constants, and *not adjustable* from atom to atom as is the α parameter in so-called "muffin-tin" $X\alpha$ schemes.

Even more interesting, however, is the observation (12) that molecular *dissociation energies* determined by Equation 10, in combination with a variant of the LDA for correlation (13), are in excellent agreement with experiment for homonuclear diatomic molecules from H_2 through Cl_2 and even Cr_2, a well-known and stringent test of quantum chemical methods. Inspired by this discovery, further applications have been made by Ziegler et al (14) to a wide variety of transition-metal problems with very gratifying results, some of which are reproduced in Table IV (see also T. Ziegler, J.G. Snijders, and E.J. Baerends, this volume). It appears that the theory is capable of bond energy predictions

Table IV: Bond Energies[a] in Selected Transition-Metal Systems
(kJ mol^{-1})

	Expt.	GCX[b]	LDA
Average Bond Energies:			
$Cr(CO)_6$	110	107	–
$Mo(CO)_6$	151	126	–
$W(CO)_6$	179	156	–
M–H or M–CH$_3$ Dissociation Energies:			
$HMn(CO)_5$	213	225	–
$CH_3Mn(CO)_5$	153	153	–
$HCo(CO)_4$	238	230	–
First Metal–Carbonyl Dissociation Energies:			
$Cr(CO)_6$	162	147	276
$Mo(CO)_6$	126	119	226
$W(CO)_6$	166	142	247
$Fe(CO)_5$	176	185	263
$Ni(CO)_4$	104	106	192

a) Ziegler et al, Ref.(14).

b) GCX: Gradient-Corrected Exchange, Equation 10,
    plus LDA for "antiparallel-spins-only" Correlation (13).

of remarkable accuracy (i.e. 10-20 kJ mol^{-1}, a few kcal mol^{-1}, a few tenths of an eV).

Despite this evidence of its utility, the semi-empirical nature of our theory leaves much to be desired. Indeed, several *other* modified gradient-corrected functionals have been proposed on the basis of various empirical and theoretical considerations (15). Unpublished work by the author indicates that all such functionals give consistent bond energy results, within a few tenths of an eV of each other, as long as "reasonable" non-divergent asymptotic behaviour is enforced. This relative insensitivity to precise functional form is reassuring. Nevertheless, ambiguity regarding the optimum large-x_σ dependence is unsettling.

Gradient Corrections: Present Work

Very recently (16), we have found a particularly satisfying gradient-corrected exchange energy functional with significant advantages over previous versions. The exchange-only hole function $h_X^{\sigma\sigma}(1,2)$ generates a coulomb exchange *potential* $U_X^{\sigma\sigma}$ at reference point 1 given by

$$U_X^{\sigma\sigma}(1) = \int \frac{h_X^{\sigma\sigma}(1,2)}{r_{12}} d^3r_2 \tag{12}$$

in terms of which the total exchange energy can be written in the form (see Equation 4),

$$E_X = \frac{1}{2} \sum_\sigma \int \rho_\sigma U_X^{\sigma\sigma} d^3r \tag{13}$$

This exchange potential has the following *exact* asymptotic behaviour:

$$\lim_{r \to \infty} U_X^{\sigma\sigma} = -\frac{1}{r} \tag{14}$$

which is a simple consequence of the normalization condition of Equation 5. Though it may seem unlikely that a *gradient*-type, purely *density-dependent* exchange energy functional could ever reproduce such asymptotic behaviour, we have, in fact, found a suitable such functional:

$$E_X = E_X^{LDA} - b \sum_\sigma \int \rho_\sigma^{4/3} \frac{x_\sigma^2}{(1+6bx_\sigma \sinh^{-1} x_\sigma)} d^3r \tag{15}$$

the success of which relies on the fact that the density itself

has *exponential* asymptotic dependence. As a consequence of
enforcing the behaviour of Equations 13 and 14, we are left with
only *one* free constant b. Since previous functionals have
involved *two* constants or more (see, for example, Equation 10),
this is a further advantage of the present work. With the value
$b=0.0042$ a.u., the exact exchange energies of the six noble gas
atoms He through Rn may be fit with impressively low r.m.s.
deviation of only 0.11% (see Table V). Moreover, theoretical
arguments based on real-space exchange hole models have been
published (17) supporting a b value of this order. Our functional
is therefore very satisfying from both empirical and theoretical
points of view, and previous ambiguities concerning large-gradient
behaviour are now removed.

The new gradient-corrected exchange energy functional of
Equation 15, with addition of the uncorrected LDA for correlation,
has been applied to the calculation of molecular spectroscopic
properties. Results are presented in column PW1 of Tables I-III.
Observe that the overall agreement with experiment, especially in
the case of dissociation energies, is remarkably good, even though
the calculations are of *independent particle* simplicity. The
dissociation energies typically deviate from experiment by *tenths*
of an eV, of similar quality as for the functional of Equation 10
(12,14), and a dramatic improvement over the LDA.

Also, we examine the effect on molecular properties of
gradient corrections to the LDA *correlation* energy, neglected
until now, by combining Equation 15 for exchange with the recent
gradient-corrected correlation approximation of Perdew (10). The
resulting data are listed in column PW2 of our tables. Notice
that, relative to PW1, the inclusion of beyond-LDA *correlation*
corrections slightly improves bond lengths and vibrational
frequencies, but does not improve dissociation energies. Typical
deviations from experiment are of the order 0.02 Å for bond
lengths, 40 cm^{-1} for frequencies, and, as for PW1, tenths of an eV
for dissociation energies.

Special mention should be made of the Cr_2 data, as this
particular system is a celebrated challenge to quantum chemists of
all persuasions (see, for example, the review in Ref.(18)). The
present data have been obtained using "anti-ferromagnetic"
broken-symmetry spin-orbitals as is customary in the recent
density functional literature on this molecule. Unfortunately,
the many-electron state thus obtained is *not* a pure singlet state,
and is contaminated by higher-spin admixtures. An analysis of
this problem by Dunlap (19) suggests that the theoretical bond
energy predictions in Table II should probably be *increased* by a
few tenths of an eV. The apparently too low binding energy of the
present work may therefore prove quite satisfactory in the final
analysis. Note that, due to the great expense of the numerical
Cr_2 spin-unrestricted calculations, precise bond length and
frequency results are not yet available and the Cr_2 dissociation
energies in Table II have been computed at the experimental
equilibrium internuclear separation.

Table V: Atomic Exchange Energies (a.u.)

	Exact	LDA[a]	PW[b]
He	-1.026	-0.884	-1.025
Ne	-12.11	-11.03	-12.14
Ar	-30.19	-27.86	-30.15
Kr	-93.89	-88.62	-93.87
Xe	-179.2	-170.6	-179.0
Rn	-387.5	-373.0	-387.5

a) Local Density Approximation, Equation 11.

b) Present Work: Equation 15 with $b=0.0042$ a.u.

Conclusions

Currently, research in our laboratory continues on real-space
models of exchange and correlation hole functions in inhomogeneous
systems. We anticipate that this work will ultimately generate
completely non-empirical parameter-free beyond-LDA density
functional theories. The quality of molecular dissociation
energies and related properties obtainable with *existing*
semi-empirical gradient-corrected DFTs approaches chemical
accuracy, and we hope these future theoretical developments will
continue this trend.

Literature Cited

1. J.C. Slater, **The Self-Consistent Field for Molecules and
    Solids**, McGraw-Hill, New York, 1974.
2. P. Hohenberg and W. Kohn, **Phys. Rev. B** 1964, **136**, 864.
    W. Kohn and L.J. Sham, **Phys. Rev. A** 1965, **140**, 1133.
3. **Local Density Approximations in Quantum Chemistry and
    Solid-State Physics**, edited by J.P. Dahl and J. Avery, Plenum,
    New York, 1984.
    Density Functional Methods in Physics, edited by R.M. Dreizler
    and J. da Providencia, Plenum, New York, 1985.
    Density Matrices and Density Functionals, edited by R. Erdahl
    and V.H. Smith, Jr., Reidel, Dordrecht, 1987.
4. A.D. Becke, **J. Chem. Phys.** 1988, **88**, 1053.
5. D.M. Ceperley and B.J. Alder, **Phys. Rev. Lett.** 1980, **45**, 566.
6. S.H. Vosko, L. Wilk and M. Nusair, **Can. J. Phys.** 1980, **58**,
    1200.
    J.P. Perdew and A. Zunger, **Phys. Rev. B** 1981, **23**, 5048.
7. A.K. Rajagopal, **Adv. Chem. Phys.** 1980, **41**, 59.
    Theory of the Inhomogeneous Electron Gas, edited by
    S. Lundqvist and N.H. March, Plenum, New York, 1983.
8. A.D. Becke, **J. Chem. Phys.** 1982, **76**, 6037 ; 1983, **78**, 4787.
9. F. Herman, J.P. Van Dyke and I.B. Ortenburger, **Phys. Rev.
    Lett.** 1969, **22**, 807.
    F. Herman, I.B. Ortenburger and J.P. Van Dyke, **Int. J.
    Quantum Chem. Quantum Chem. Symp.** 1970, **3**, 827.
10. J.P. Perdew, **Phys. Rev. B** 1986, **33**, 8822.
11. A.D. Becke, **Int. J. Quantum Chem.** 1985, **27**, 585.
12. A.D. Becke, **J. Chem. Phys.** 1986, **84**, 4524.
13. H. Stoll, C.M.E. Pavlidou and H. Preuss, **Theor. Chim. Acta**
    1978, **49**, 143.
    H. Stoll, E. Golka and H. Preuss, **Theor. Chim. Acta** 1980, **55**,
    29.
14. T. Ziegler, V. Tschinke and A. Becke, **Polyhedron** 1987, **6**, 685.
    T. Ziegler, V. Tschinke and A. Becke, **J. Am. Chem. Soc.** 1987,
    109, 1351.
    T. Ziegler, V. Tschinke and C. Ursenbach, **J. Am. Chem. Soc.**
    1987, **109**, 4825.

15. J.P. Perdew and Wang Yue, **Phys. Rev. B** 1986, 33, 8800.
    A.D. Becke, **J. Chem. Phys.** 1986, 85, 7184.
    A.E. DePristo and J.D. Kress, **J. Chem. Phys.** 1987, 86, 1425.
    S.H. Vosko and L.D. MacDonald, in Proceedings of the 1986
    International Workshop on Condensed Matter Theories, Argonne
    National Laboratory, edited by P. Vashishta et al, Plenum,
    New York, 1987.
16. A.D. Becke, **Phys. Rev. A** 1988, 38, 3098.
17. A.D. Becke, **Int. J. Quantum Chem.** 1983, 23, 1915.
18. D.R. Salahub, **Adv. Chem. Phys.** 1987, 69, 447.
19. B.I. Dunlap, **Adv. Chem. Phys.** 1987, 69, 287.
20. K.P. Huber and G. Herzberg, **Molecular Spectra and Molecular
    Structure IV: Constants of Diatomic Molecules**, Van Nostrand
    Reinhold, New York, 1979.

RECEIVED December 9, 1988

Chapter 13

The LCGTO–Xα Method for Transition Metal Model Clusters

Recent Developments and Applications

N. Rösch[1], P. Knappe[1], P. Sandl[1], A. Görling[1], and B. I. Dunlap[2]

[1]Lehrstuhl für Theoretische Chemie, Technische Universität München,
8046 Garching, Federal Republic of Germany
[2]Theoretical Chemistry Section, Code 6119, Naval Research Laboratory,
Washington, DC 20375

Transition metal clusters as models for surface phenomena are discussed and contrasted with gas phase clusters. The accuracy of *ab initio* and local density methods for transition metal systems is briefly compared. The linear combination of Gaussian–type orbitals (LCGTO) Xα approach is presented as an efficient method to calculate the electronic structure of such systems within the local density functional theory. Gaussian broadening of the one–electron levels is used to both mimic the embedding of a cluster in a surface and to reduce the number of different SCF calculations which otherwise would be required due to the high density of states around the highest occupied (Fermi) level. The method is applied to nickel clusters of up to 17 metal atoms. In a comparative study of compact clusters it is found that an icosahedral structure is energetically favored for a Ni_{13} cluster over a cuboctahedron, a small piece of crystalline nickel. Spin–polarization is calculated to be 0.62 μ_B per metal atom for both structures investigated, a value that is rather close to bulk magnetization. Na and K chemisorbed on rather planar Ni_n clusters model the low coverage limit of alkali chemisorption at various sites of different crystalline surfaces. Calculated values for binding energy and induced 'surface' dipole moment show satisfactory agreement with experimental data. From the calculated adsorbate–substrate bond length a substantial covalent character of alkali bonding to transition metal surfaces is deduced in the low coverage limit. A spin–polarized calculation shows that the unpaired spin of the alkali atom is almost completely quenched upon chemisorption.

Clusters of transition metal atoms form a fascinating subject with direct relations to such diverse fields as inorganic (1) and metallorganic chemistry (2), physical chemistry (3,4), surface science (5) and catalysis (6). Their electronic structure, both the intermetallic bonds and the bonding between metal atoms and 'ligands' provides one of the strongest challenges for experimentalists and theoreticians alike (7,8).

NOTE: LCGTO is an abbreviation for linear combination of Gaussian-type orbitals.

Chemisorption cluster models

Cluster models have been quite popular for some time now as a basis for the discussion of chemisorption systems ($\underline{9}$ – $\underline{11}$), especially among quantum chemists who were able to contribute with their methods and tools to surface science via these constructs. (The references of this paragraph are intended to provide examples only since an exhaustive list would be too lengthy to be appropriate here.) Transition metal clusters have been the most intensively studied systems from the beginning due to the interesting chemisorptive and catalytic properties of such surfaces. At first one–electron aspects dominated cluster model applications ($\underline{12},\underline{13}$), photoelectron spectra providing the bridge between theory and experiment ($\underline{14}$). The simpler quantum chemical methods (Extended Hückel ($\underline{9}$), Xα Scattered Wave ($\underline{15}$), semiempirical SCF methods ($\underline{16}$)) gave way to more sophisticated treatments, based both on *ab initio* (Hartree–Fock SCF ($\underline{17}$), CI ($\underline{18}$ – $\underline{20}$)) and on local density methods (DVM–Xα ($\underline{21},\underline{22}$), LCGTO–Xα ($\underline{23}$), LCGTO–LSD–MP ($\underline{24}$)). These more elaborate studies furnish structural and energetic information on chemisorption bonds – within the restriction of a local cluster model ($\underline{9}$), of course.

However, from calculations on transition metal complexes whose structural and electronic properties are known with higher accuracy it became evident that *ab initio* treatments have to be carried to the level of configuration interaction ($\underline{25},\underline{26}$), at least for the late transition elements (iron group and beyond). A useful computational method for such systems must be able to deal with the quite diffuse valence s orbitals and the rather localized valence d orbitals with their characteristic directional properties in a balanced manner in order to achieve a proper description of transition metal ligand bonds ($\underline{25}$).

Taking transition metal carbonyl complexes, like $Ni(CO)_4$ and $Fe(CO)_5$, as benchmark systems ($\underline{27},\underline{28}$) for 'surface molecules' because much has been determined about them, one observes that the local density method underestimates the metal carbon bond distance and overestimates the corresponding binding energy whereas *ab initio* treatments exhibit the opposite trends for both observables. The results obtained from the Hartree–Fock method are particularly disconcerting ($\underline{29}$) since with increasing quality of the basis used, they show unusually large and, especially for the case of $Fe(CO)_5$, quite uneven deviations from the experimental bond lengths. More troublesome is the fact that a negative binding energy for the carbonyl metal bond was found. From a detailed comparison of various computational methods for $Ni(CO)_4$ and for the Ni atom it has been argued ($\underline{27}$) that the errors of the various methods correlate with their accuracy of reproducing the relative energies of relevant configurations of the transition metal atom, $d^9 s^1$, $d^8 s^2$ or $d^{10} s^0$ in the case of nickel. These findings severely limit the value of chemisorption model calculations for transition metal systems at the Hartree–Fock level and caution against conclusions drawn from them. On the other hand, we must mention that large overbinding ($\underline{23},\underline{27}$, $\underline{28},\underline{30}$) is often observed in local density functional calculations. A very promising extension of the LDF method that uses the gradient of the density reduced the deviation of the dissociation energies from experiment for fifteen homonuclear diatomic molecules from 1.2 eV to 0.3 eV ($\underline{31}$).

Quite a few studies of transition metal systems have been carried out with rather small clusters of five or less atoms ($\underline{17},\underline{20},\underline{32}$) modeling the chemisorption site, CI studies often being restricted to one or two transition metal atoms representing the surface ($\underline{19}$). The shortcomings of such small cluster models are apparent. Application of two–dimensional periodic boundary conditions ($\underline{33},\underline{34}$) provides one way to improve the realism of the computational model, another one would be to increase the size of the cluster to include several shells of neighbors of the adsorption site (Rösch, N.; Sandl, P.; Görling, A.; Knappe, P. *Int. J.*

Quantum Chem. 1988, S22, in press (RSGK),(35)), embedding the cluster would be yet another possibility (36).   All these improvements entail significantly larger computational efforts. For *ab initio* methods large cluster models have been feasible only by introducing some effective core approximation.

The two–dimensional supercell method and large cluster model ultimately complement each other since they aim at opposite experimental situations (especially, if the computing requirements are to stay within manageable bounds). The high coverage limit of chemisorption giving rise to small unit cells is least demanding to model when using the slab approach (which then also offers a chance to study lateral interactions within the adsorbed layer). The low coverage limit, on the other hand, remains the proper domain of the cluster approach to chemisorption. At any given level of methodological sophistication the latter procedure is by far less challenging computationally and at present still seems to be the method of choice if information on bond distances and energies is desired. Both these extensions of the small cluster approach to chemisorption have been implemented in the framework of local density theory taking advantage of the computational economy of this method.

Gas phase clusters

Despite all the merits and successes of the chemisorption cluster approach its fundamental approximation – modeling an infinite system by a finite one – is self–evident and a major source of concern. Nevertheless the demands of surface science have stimulated the development of electronic structure methods capable of handling reasonably large transition metal systems with good accuracy. A new field of application for these computational techniques has recently opened up due to a tremendous progress, now allowing almost routine generation and mass selection of beams of clusters of almost any element, including transition metals (37,38).   Combining the two applications in a sense, chemisorption of atoms and small molecules on such clusters of transition metal atoms has been studied experimentally (39,40). It is fair to say, however, that the most challenging problems in the study of gas phase clusters, the 'magic numbers' observed in mass spectra of ionized cluster, predicting reaction paths, and the question of cluster isomers calls for even further improvement of existing quantum chemical methods.   Undoubtedly these challenges will spur increasing sophistication of the local density implementations, such as the use of energy gradient techniques to facilitate the determination of the ground–state geometric structure of a cluster (41,42).   Such methodological improvements would be highly desirable for tackling additional questions even within the cluster approach to chemisorption, like the structural response of a transition metal surface to the interaction with an adsorbate. The structure problem presents a formidable task (4) even for smaller gas phase clusters (i.e. 5 to 8 atoms). It may be simplified by additional assumptions concerning the symmetry of the cluster under study, a common procedure which while often plausible is fundamentally incomplete without the computation of second order gradients.

In the following we shall illustrate the present status of the local density method as implemented in the LCGTO – Xα approach by applying it to transition metal clusters in both fields mentioned above. The examples will deal with nickel clusters of up to 17 atoms, but larger clusters seem to be within the reach of today's computational possibilities.

The LCGTO–Xα method

The linear combination of Gaussian–type orbitals (LCGTO) Xα method (43) is based on the local density functional (LDF) method using the Xα variant for the exchange potential:

$$v_{xc}\,(\mathbf{r}) = -\,\alpha\,\frac{3}{2}\,(\frac{3}{\pi})^{1/3}\,\rho(\mathbf{r})^{1/3}.$$

where $\rho(\mathbf{r})$ designates the electronic density of the system. This approach can be generalized in a straightforward manner to allow for spin–polarization in systems where one–electron levels are not completely filled. In the work presented here α has been set to 0.7, a value close to the one for the exchange only which is 2/3 (44). Other approximations for the exchange–correlation functional (45,46) may be treated in a similar fashion. (For a recent review of problems related to the different choices for this functional see (47)). A fundamental characteristic of the LCGTO–Xα approach is the use of two basis sets in addition to the one used to construct the usual one–electron wavefunctions (molecular orbitals). These auxiliary basis sets (43,48) are used to expand the total charge density

$$\rho(\mathbf{r}) \approx \overline{\rho}(\mathbf{r}) = \Sigma\ a_i\ f_i(\mathbf{r}),$$

and the exchange–correlation potential

$$v_{xc}(\mathbf{r}) \approx \overline{v}_{xc}(\mathbf{r}) = \Sigma\ b_i\ g_i(\mathbf{r}),$$

in single–center expansions, taking advantage of the smooth and positive definite nature of ρ and $\rho^{1/3}$. The method also takes advantage of the relatively easy way integrals involving Gaussian–type functions may be now evaluated given the vast amount of *ab initio* work on this problem. The resulting method features considerable computational economy due to the smaller number of required integrals (approximately N^3) in contrast to *ab initio* methods (N^4 or higher). All three basis sets are chosen as linear combinations of solid harmonics augmented by a Gaussian type radial dependence (49, Dunlap, B.I., unpublished) and are symmetry adapted.

Four types of functions are used to generate the fitting bases (50). s–type functions are derived from the the s–exponents of the orbital basis by proper scaling (2 in the case of the charge and 2/3 in the case of the exchange). Similarly d(r^2)–type functions (also radially symmetric about a nucleus) are constructed from the p–exponents of the orbital basis. To allow for larger flexibility these sets are augmented by polarization functions (p– , d– , f–type) or by bond centered functions (normally of s–type). Bond–centered s–type fitting functions may provide an efficient alternative to polarization–type fitting functions (50), especially for fitting the nodeless exchange potential. However, they are most practical for homonuclear bonds. Using contraction schemes for the orbital basis and for the two fit bases it is possible to reduce the computational effort while still maintaining reasonable accuracy. In this way it is possible to treat all electrons explicitly.

The total energy is calculated in a method that is stationary apart from the need to numerically fit the exchange potential (43).

The expansion coefficients a_i for the charge density ρ are determined variationally by minimizing the coulomb self–interaction energy of the difference between the fitted and the exact charge density (43). (One can also fit the coulomb potential itself variationally (51)). Because the electron–electron repulsion which provides the largest part of the fitted potential is treated variationally, one has, as a rule, a lower total energy the worse the fitting basis. Although a variational principle has been suggested (52) to evaluate the expansion coefficients b_i of the exchange potential, a least square fit procedure is used over a moderate size sample grid (typically about 1000 points per inequivalent transition metal atom). The molecular grid is a superposition of 'atomic' parts confined to a 'Wigner–Seitz' cell about each nucleus. Each atomic grid has a logarithmic radial dependence, the 26 node angular mesh consisting of points alternately placed on the cartesian axes, the corners of a cube and the corners of an icosahedron (53). A possible geometric preference for certain directions may be efficaciously countered by independently randomizing the orientation of the angular mesh at each radial distance (Jones, R.S.; Mintmire, J.W.; Dunlap, B.I. Int. J. Quantum Chem. 1988, S22, in press).

While the point group symmetry of the molecule can be efficiently used in the evaluation of the integrals, the handling of the exchange potential takes up about half of the computational effort in a LCGTO–Xα calculation. To reduce this effort an analytic fitting procedure is currently being implemented requiring an additional number of integrals (order N^3), however (52). The convergence of the SCF process is sped up by a perturbational scheme (54) in conjunction with dynamical damping (Knappe, P.; Rösch, N., unpublished).

The LCGTO–Xα approach described so far has been successfully applied to a large variety of systems, including main group molecules (50,52,53), transition metal compounds, e.g. carbonyl complexes (27,28,55,56) and ferrocene (57), and a number of transition metal dimers (47). Besides these investigations on ground state properties useful information has also been obtained for selected problems involving excited states (52), such as the photolysis of $Ni(CO)_4$ (58,59) and localized excitons in alkali halides (60) and in other ionic crystals (61).

A crucial problem of the cluster approach to chemisorption is connected to the fact that transition metal clusters (especially those of late transition elements) show a high density of states around the cluster 'Fermi' level, i.e. near the highest occupied molecular orbital. The resulting small gap in the energy spectrum of the cluster may lead to problems in the determination of the ground state energy surface. This finite size effect may lead to artifacts in chemisorption models where a variation of the adsorbate geometry is to be investigated. This quasi–degeneracy of states also impedes the SCF process due to occupation number changes. The interaction of a chemisorption cluster with the energy bands of the substrate leads to a broadening of the discrete cluster levels and, as a consequence thereof, to effective partial occupation numbers. (A further consequence of this 'embedding' would be a possible charge transfer between the chemisorption cluster and the remaining solid.)

These considerations lead one to suggest a modified cluster model that takes advantage of the fact that local density models admit partial occupation numbers (RSGK). We formally broaden in energy each level by σ and apply 'Fermi statistics' to this continuous system. We add infinitesimal fractions of an electron to the broadened levels in order, until all the electrons are used up, yielding a precise Fermi energy, ϵ_f, and the various occupation numbers

$$n_i(\epsilon_f) = m_i \int_{-\infty}^{\epsilon_f} (1/\sqrt{2\pi}\,\sigma)\exp[-(\epsilon - \epsilon_i)^2 / 2\sigma^2]\,d\epsilon.$$

Here, m_i designates the degeneracy of the level and σ the standard deviation of the Gaussian. The cluster Fermi energy ϵ_f is determined self–consistently equating the sum of the occupations to the total number of electrons in the cluster. This procedure essentially defines an average over low lying one–electron configurations and results in a well–behaved SCF process. Any results determined through this scheme will depend on the level broadening parameter σ. For transition metal clusters this dependence is found to be weak (see below) for a reasonable range of parameters, a result not unexpected in view of the rather similar spatial character of the orbitals around the cluster Fermi level, especially when symmetry is used. This Gaussian form of broadening leads to an averaging over much fewer configurations than would a Fermi distribution of the same width, the tails of which extend much further.

Basis sets

The use of three types of basis sets is a unique feature of the LCGTO–Xα method. It therefore seems warranted to elaborate on the arguments involved in their selection. Due to their intricate interdependence these basis sets have to be chosen judiciously and well balanced among each other. This has to be done separately for the chemisorption clusters and for the gas phase clusters in order to account for the two rather different physical situations without sacrificing basis set flexibility and computational efficiency.

Since the geometry of each nickel cluster was not to be varied during the investigation of alkali adsorption stronger contracted basis sets should be acceptable than in a geometry optimization of a transition metal complex. This should be especially true for nickel atoms not in the immediate neighborhood of the adsorption site. Properties such as bond distance, binding energy and frequency of the totally symmetric stretching vibration for the tetrahedral cluster Ni_4, as well as similar observables for a prototypical chemisorption cluster Ni_4Na, were computed. Comparisons were made between naked Ni, Ni_2, Ni_3 and Ni_4 clusters. We focus our discussion on the tetrahedral cluster Ni_4 since it reflects the consequences of various basis set modifications in a typical manner. Also some effects of orbital basis set contractions on the one–electron eigenvalues of tetrahedral Ni_4 have been studied and compared to effects of technical variations within the scattered–wave Xα approach ([62]).

The interatomic distance, the binding energy per atom and the frequency of the totally symmetric motion for a tetrahedral Ni_4 cluster are summarized in Table I for different choices of the molecular orbital basis and the two auxiliary basis sets. The starting point was basis set no. 1, a (15s,11p,6d) molecular orbital basis ([63]). It was derived from an atomic basis ([64]) by adding one s–expo-

Table I. Comparison of various contraction schemes for a tetrahedral Ni_4 cluster

No.	MO (s/p/d)	Charge (s/$d_{\Gamma 2}$/p)	Exchange (s/$d_{\Gamma 2}$/p)	d^a [Å]	$E_B{}^b$ [eV]	$\omega_0{}^c$ [cm^{-1}]
1	(15/11/6)	(15,11,5)	(15,11,5)	2.25	2.99	1445
2	(11/ 8/4)	(15,11,5)	(15,11,5)	2.25	2.99	1445
3	(9/ 5/3)	(15,11,5)	(15,11,5)	2.26	2.93	1445
4	(6/ 3/2)	(15,11,5)	(15,11,5)	2.26	2.75	1350
5	(11/ 8/4)	(10, 5,5)	(10,11,5)	2.25	2.99	1445
6	(9/ 5/3)	(6, 3,5)	(10, 6,5)	2.23	2.64	1230
7	(9/ 5/3)	(6, 3,0)	(10, 6,0)	2.35	2.99	1200

a) Ni–Ni distance.   b) Binding energy per atom.
c) Frequency of the totally symmetric vibration.

nent (0.346 atomic units) to smooth this series of exponents, two diffuse p–exponents (0.364, 0.104) to allow 4p participation in the bonding and one d–exponent (0.1316). The s–type and d(r^2)–type charge and exchange fitting functions were constructed as described above by proper scaling of the exponents of the s– and p–type orbital basis functions. This set was supplemented by five polarization functions of p–type (50) leading to fitting basis sets that will be denoted by (15,11,5). This basis set combination is the most extensive one used in the present study and will therefore be used as reference.

Basis sets no. 2 to 4 in Table I were derived by segmented contractions while performing an atomic calculation. A very modest contraction (11,8,4) which essentially fixes the 1s and 2p parts of the orbitals has essentially no effect on all displayed properties. Slight deviations are observed when the stronger contraction (9,5,3) is used where the binding energy is clearly a more sensitive indicator of the quality of a basis than the bond distance. The deviation from the reference are quite large for basis no. 4, not unexpected considering its near minimal quality (6,3,2). But even in this case the bond energy deviates less than 10 percent from the reference result.

Basis sets no. 5 to 7 in Table I are those actually used in the present chemisorption cluster study where the first one was intended for atoms directly at the adsorption site, the second one for first nearest neighbors and the last one at the edges of large clusters. They differ from the basis sets no. 2 to 4 by an additional contraction of the fitting bases, basis set no. 7 being without any polarization functions in the auxiliary bases. While slight contractions of the fitting bases (no. 5) produce no significant loss of accuracy reducing the number of fitting functions by a factor of two entails noticeable deviations, with the bond distance being again relatively less sensitive than the binding energy and the vibrational frequency. The intermetallic bond distance is affected by the very small basis set no. 7, however, which uses only spherical atom–centered fitting functions.

Since the charge density is variationally fit in our method, one finds surprisingly little effect on the total energy if no polarization functions are included to directly fit the bonding charge (43). Even in nonvariational methods, the fact that such functions are not critical has lead to the self–consistent charge method (65). Properly fitting this bonding charge density is crucial, however, if accurate potential surfaces are desired (66).

For sodium, the same (12,7,1) basis was employed as used previously (RSGK,60). It was derived from a (12,6,0) basis (67) by adding one p–type exponent (0.143647) and one d–type exponent (0.122). Only the orbital basis was contracted, to (9,7,1).

Since potassium belongs to the same period as the first row of transition metals the basis sets may be constructed similarly to the approach taken for nickel. Starting with a (14,9,0) basis (64) the s–type exponent series was smoothed by replacing one exponent (0.035668) by two others (0.120408, 0.045301). Again two diffuse p–type exponents were added (0.084744, 0.031408) and the resulting (15,11,0) basis was augmented with one polarization function of d–type (exponent 0.078).

In all chemisorption cluster investigations the Gaussian level broadening parameter σ was fixed at 0.3 eV. At least for our nickel clusters this seems a reasonable choice as orbitals near the Fermi levels are of similar spatial character. The results shown in Table II below support this view.

A slightly different strategy lead to the basis sets for the study of gas phase clusters, Ni_{13}. Here, a (14,9,5) basis (64) was augmented with diffuse p and d orbital functions of exponent 0.1. Since in *ab initio* work it was sometimes found necessary (19,26) to add f–type polarization functions to the orbital basis this possibility was investigated here, too. However, when a significantly larger than minimal basis of s–, p–, and d–type functions is used it appears impossible to add an f–type polarization function that is diffuse enough to allow f–type contributions to bonding. In any such attempt a numerically singular set of combined functions resulted in at least one irreducible representation. This result also holds for the larger 4d transition metal niobium, leading one to suspect that f–type polarization function cannot be used effectively in any compact metal cluster, unless special numerical provisions are taken. The orbital set finally used consisted of the minimal basis plus the four most diffuse s–type, the two most diffuse p–type and three most diffuse d–type Gaussians, leading to a (6,4,3) orbital basis.

The polarization part of the fitting basis sets was tested in calculations where a minimal orbital basis was used. Sets of five p– and five d–type fitting functions with exponents 0.04, 0.2, 0.6, 2.0 and 10.0 were used. The cluster symmetries are such that d exponents produce 5 I_h and and 10 O_h fitting functions on the surface atoms only. The effect of fitting functions beyond p–type polarization functions was negligible judged both from the cohesive energy and the radial bond length. This statement holds true also for the case were the p–type set was augmented by two sets of bond–centered functions (exponents 0.125 and 0.5). Nevertheless, auxiliary basis sets containing the full set of p–type and d–type polarization functions plus one set of s–type bond–centered functions (exponent 0.3) were used in the following calculations.

The Ni_{13} cluster exhibit virtually no energy gap between the highest occupied and the lowest unoccupied one–electron level (less than 0.1 eV for I_h). The density of states in this range is so high that conventional SCF procedures cannot easily be used to give the energy of a single electronic configuration. Instead, the formal level broadening described above to model the embedding of a surface cluster had to be used here as a device to generate convergence, turning the present set of gas phase cluster calculations into a model study, too. The effect of reducing the Gaussian level broadening parameter σ will be discussed below in further detail (see Table II).

The gas phase clusters Ni_{13}

The structure of crystalline nickel is face–centered cubic (fcc). The most compact 13–atom cluster that results from the constraint that each atom occupy an fcc lattice site is the cuboctahedron formed by the central atom and its 12 equivalent nearest neighbors (see Figure 1a). For this O_h symmetric cluster the radial bond distance, from the central atom to any surface atom, is identical to the tangential bond distance between any two nearest neighbors on the cluster surface. One could easily expect this cluster to be the ground state of the 13–atom nickel cluster. Experimentally the shape of the 13–atom cluster of rare gas atoms is an icosahedron (68) (see Figure 1b), however. Icosahedral symmetry is also found as the ground state structure of 13 atoms interacting via many different kinds of pair potentials (69). For such a cluster of I_h symmetry the tangential bond distance is five percent larger than the radial bond distance. Because these two clusters represent reasonable candidates for the ground state of the gas phase cluster, Ni_{13}, their Xα geometries have been optimized on the basis of spinpolarized calculations. The results are presented in Table II.

Of particular interest are the predicted magnetic moments and the question of whether or not isomers of transition metal clusters can be separated using inhomogeneous magnetic fields. To date, cluster isomers have only been detected via their different chemical reactivity (70 – 73). One would expect abnormally large magnetic moments for the I_h clusters if they had unusually high density of states at the Fermi level, as has been postulated for aluminum (74).

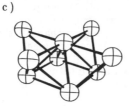
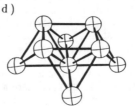
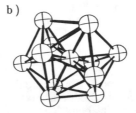
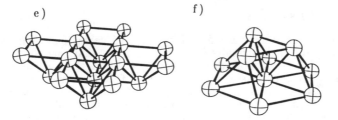

Figure 1. Nickel clusters used in the present study. Gas phase clusters: Ni_{13} O_h (a) and I_h (b); surface model clusters representing various adsorption sites on Ni(100): on–top position on Ni_9 (c), fourfold–hollow position on Ni_9 (d) and on Ni_{17} (e); and on Ni(111): threefold–hollow position on Ni_{10} (f) and on–top position on Ni_{10} (f, upside down).

Table II. Comparison of the icosahedral and octrahedral Ni_{13} isomers
and study of the Gaussian level broadening

σ^a [eV]	Spin–polarized Calculation	d^b [Å]		$E_B{}^c$ [eV]		s^d	
		I_h	O_h	I_h	O_h	I_h	O_h
0.5	no	2.26	2.33	3.22	3.08	–	–
0.5	yes	2.29	2.35	3.28	3.15	8.48	8.77
0.25	yes	2.28	2.34	3.46	3.29	8.18	8.28
0.125	yes	2.29	2.34	3.50	3.33	8.00	7.87
0.0625	yes	2.29	2.34	3.51	3.34	8.00	7.99

a) Gaussian broadening parameter. b) Bond distance to the central atom.
c) Binding energy per atom. d) Spin–polarization.

A spin–polarization of s = 6 electrons has been found in an Xα–SW study of a Ni_{13} cuboctahedron (75) due to a comparable high density of one–electron levels near the cluster Fermi energy. But in another study using this computational approach s = 8 was calculated (76).

It has been argued (47) that the Xα approximation for the local exchange potential overestimates the tendency towards spin–polarization. The results for another local density functional will be presented later, elsewhere, but for the moment these Xα calculations (with α = 0.7) seem appropriate because we would like to determine an upper limit on the effect of isomerization on the magnetic moment of this hopefully representative cluster of magnetic atoms.

Reducing the Gaussian level broadening parameter σ eventually leads to an integral difference in the number of electrons for each spin (see Table II). Remarkably that difference, s = 8, is the same for both isomers – and it is identical to the value that had been chosen previously (77) in a semi–empirical study of the relative stability of four Ni_{13} clusters. In that case, the argument had been based on the fact that a spin–polarization of s = 8 leads to a net magnetic moment of 0.62 μ_B per atom which is as close as possible to the bulk value 0.56 μ_B. It was also predicted that the icosahedral cluster lies lowest in energy (77). Judging from a figure (77), however, it would appear that the radial I_h bond distance was the same or slightly larger than the bulk nearest–neighbor distance 2.49 Å. Ours is eight percent smaller. While s = 8 is the result of the limiting process σ → 0 for a wide variation of the I_h bond distances, reducing the O_h bond distance by 0.3 Bohr yields s = 6. Still at the smallest value of σ considered, 0.0625 eV, which is much smaller than the overall accuracy of the calculation within the Xα model, does not come close to yielding a pure state. For this value of σ and the icosahedron there are eight levels above the highest fully occupied majority–spin t_{2g} orbital, and all of them are partially occupied. For the cuboctahedron, only three of the eight levels above the highest fully occupied majority spin level, a_{1u}, are partially occupied. Due to the dense level structure and the partial occupation of levels near the Fermi level of the icosahedral cluster a distortion away from perfect symmetry may be expected for this cluster.

A cluster density of states of Ni_{13}, literally (not just formally) broadened by 0.2 eV, is presented in Figure 2a for the icosahedral geometry near the equilibrium structure. The general features are roughly similar to that derived from a spin–polarized band structure calculation of bulk nickel. Near the Fermi level a very high density of states of the minority spin is found, the d 'band' of the

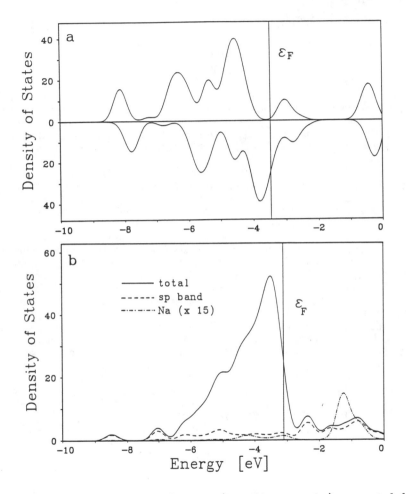

Figure 2.   Cluster density of states (in arbitrary units) generated by
Gaussian broadening of the one–electron energies and cluster Fermi energy
ϵ_f: a) icosahedral Ni$_{13}$ (spin–polarized calculation, majority spin above the
axis; $\sigma = 0.2$ eV).   b) Ni$_{17}$Na (solid line; $\sigma = 0.3$ eV).   Also shown are the
sum of the contributions from the s and p populations of the nickel atoms
(dashed line) and the population of the sodium atom (dotted line).

majority spin being completely filled. The exchange splitting of the one–electron levels near the Fermi energy is about 0.75 eV, a value 0.1 eV larger than found in a bulk band structure (78,79). The cohesive energy of 5.70 eV obtained in a bulk calculation (79) is significantly larger than that found in the present cluster study. A direct comparison has to be postponed until the present work has been extended to the same local density functional.

If the octahedral and icosahedral 13–atom nickel clusters are representative, an inhomogeneous magnetic field is not a good tool for separating isomers of transition metal clusters. This result is reminiscent of the fact that geometry plays no strong role in the ionization potential variation of alkali clusters (80). However, many calculations must be made before we can be definitive here.

Alkali Chemisorption on Nickel Surfaces: Energy and Geometry

Alkali adsorption on other metals has been the subject of many experimental and theoretical studies (18,81–84), the first investigation dating back to the early thirties (85). The interest in this field is motivated by possible consequences for technological applications such as heterogeneous catalysis or electrochemistry. In recent years many investigations have focused on the 'promoting' of catalysts by alkali coadsorption. Coadsorption of alkali atoms and carbon monoxide on various transition metal surfaces serves as a prototype system to understand these phenomena (for a recent review, see (86)). One of the most conspicuous effects of alkali coadsorption is the lowering of the CO stretching frequency by up to 700 cm⁻¹ (86). Several differing explanations have been suggested for the latter observations involving electrostatic effects, indirect charge transfer and direct orbital interaction (86). In order to build a realistic, but economical cluster model suitable for the study of alkali coadsorption some information on various properties of alkali chemisorption would be useful. However, despite many investigations in this field little seems to be known about the length of the alkali surface bond. The results for alkali chemisorption on nickel surfaces presented in the following will address this question, thus serving as an important first step towards a more complete modeling of alkali/carbon monoxide coadsorption.

The nickel clusters that have been used to model various sites on (100) and (111) surfaces are listed in the first column of Table III. The larger ones are shown in Figure 1c to 1f. All these clusters preserve the local symmetry, C_{4v} or C_{3v}, of the corresponding alkali chemisorption problem. Figure 1e shows the cluster that is used to model a fourfold–hollow position for an alkali atom on Ni(100) with 17 nickel atoms. The corresponding cluster with nine nickel atoms (see Figure 1d) is obtained from the former by omitting eight symmetry equivalent nuclei in the top lattice plane, deleting four more nickel atoms in the second lattice plane leads to the corresponding Ni_5 cluster. Figure 1c shows the Ni_9 cluster (actually the Ni_9 cluster of Figure 1d upside down) which is used to model on–top adsorption on Ni(100). The Ni_{10} cluster of Figure 1f represents a threefold–hollow site on Ni(111) and, again upside down, an on–top site on the same surface. The (111) clusters with three and four nickel atoms are obtained by omitting the appropriate atoms in the second plane of the Ni_{10} cluster. The cluster of six nickel atoms models the other threefold–hollow site that exists on a (111) surface, i.e. the one with no substrate atom situated directly under the adsorption position. The cluster is an octahedron lying on one of its triangular faces. For all systems at least seven calculations have been performed, one for the bare substrate cluster, five with varying alkali surface distance and one at the interpolated optimum geometry. The results for various clusters are collected in Table III.

Table III. Comparison of calculated results for various Ni_nM (M = Na, K) surface clusters modeling adsorption of the alkali atoms on various surfaces

	n	d^a [Å]	r^b [Å]	$E_B{}^c$ [eV]	$\Delta\epsilon_f{}^d$ [eV]	$\omega_0{}^e$ [cm^{-1}]	$\mu_{ind}{}^f$ [Debye]
Ni(100)/Na hollow	5	2.11	1.50	1.68	0.32	180	3.94
Ni(100)/Na hollow	9	2.27	1.63	1.92	0.25	175	4.21
Ni(100)/Na hollow	17	2.22	1.59	1.87	0.06	175	4.74
Ni(100)/Na top	9	2.49	1.25	1.72	0.33	210	4.92
Ni(111)/Na hollow	3	2.23	1.41	1.38	0.56	225	4.37
Ni(111)/Na hollow	4	2.37	1.52	1.09	0.13	170	4.01
Ni(111)/Na hollow	6	2.46	1.60	1.50	0.31	175	4.62
Ni(111)/Na hollow	10	2.46	1.60	1.78	0.41	170	5.19
Ni(100)/K hollow	9	2.76	2.03	2.16	0.47	115	7.15
Ni(100)/K hollow	17	2.67	1.95	2.25	0.18	115	6.98

a) Distance of adsorbate above first lattice plane. b) Effective radius of alkali atom. c) Binding energy of alkali atom. d) Change of cluster Fermi energy with respect to the bare cluster. e) Vibrational frequency of the adsorbate. f) Change of cluster dipole moment with respect to bare metal cluster.

The calculated binding energy for sodium on Ni(100) clusters in fourfold-hollow position range from 1.68 eV to 1.87 eV. This is in fair agreement with the corresponding experimental value, 2.52 eV, for sodium on Ni(100) extrapolated to the limit of zero coverage (81). Not unexpectedly, one finds a somewhat lower binding energy for the corresponding on–top position (0.20 eV difference for Ni_9, see Figures 1c and 1d). This is still an order of magnitude larger than the estimated activation energy for the surface diffusion of sodium (86). However, the transition state for that motion is most probably situated in a twofold-bridging position. The binding energies for the two Ni_9 clusters are also compared in Figure 3. For the surface Ni(111) the calculated values for the binding energy range up to 1.78 eV, deviating somewhat more from the corresponding experimental value of 2.54 eV (86). As expected, the calculated binding energy for potassium is higher (by about 0.4 eV) than for sodium. Available experimental data (81) for Ni(110) show a difference of 0.38 eV between Na and K chemisorption energies.

It is worth noting that no overbinding seems to occur in the local density description of alkali chemisorption on nickel clusters, in contrast to findings for carbon transition metal bonds (23,27,28). At present, it would be premature to correlate this difference with the character of the various bonds (covalent vs. ionic). Clearly, density gradient corrections to the energy functional (31) would be highly useful in deciding this question.

The frequency for the vibrational motion of the sodium atom perpendicular to the substrate is calculated to about 175 cm^{-1} for the hollow positions both on the (100) and on the (111) surface. The corresponding value for the larger potassium atom lies clearly lower (115 cm^{-1} for Ni_{17}).

The calculated bond distances of 2.11 Å to 2.46 Å for sodium chemisorption on nickel are considerably shorter than the value of 2.7 Å found in an *ab initio* calculation (18). The resulting surface bond distance is best appreciated

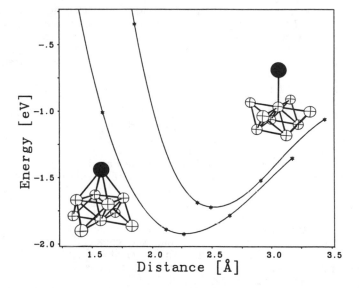

Figure 3. Binding energy (in eV) of Na in fourfold–hollow and on–top position on Ni$_9$ as function of the distance (in Å) of the adsorbate above the first lattice plane.

by converting it into an effective atomic radius of the alkali atom, i.e. by subtracting the metallic radius of nickel (1.245 Å) from the shortest alkali–nickel
bond distance. These radii are about 1.6 Å for sodium and about 2.0 Å for
potassium, lying between the corresponding ionic and metallic radii (Na: 0.98 /
1.82 Å; K: 1.33 / 2.26 Å (87)), as one would expect. But the values are surprisingly close to the metallic side. For all investigated adsorption sites geometric considerations thus suggest a rather strong covalent contribution to the
chemisorptive bond of alkali atoms to nickel surfaces. We hope to elaborate on
this topic by extending the present study of alkali adsorption to other metal
surfaces.

   In our model the dipole moments induced through adsorption are derived
by subtracting the value for the bare substrate cluster from that of the chemisorption model cluster. The calculated induced dipole moment for the largest
clusters representing the (100) and (111) surface are 4.74 and 5.19 Debye, respectively. Both values are considerably smaller than the corresponding experimental values (7.2 and 7.4 Debye (81)). However to appreciate this comparison one
has to keep in mind that the 'experimental' values are deduced from the change
of the workfunction at zero coverage limit via a rather simple model (88) based
on the classical image charge concept. From this point of view the results obtained from model clusters agree quite satisfactorily.

   More detailed information on the electronic structure of the model cluster
and on the corresponding chemisorptive bond may be obtained from the cluster
density of states (13,74,RSGK). In Figure 2b, as a representative example, the
density of states for the $Ni_{17}Na$ cluster is presented. Besides the total density of
states, the contributions of the nickel sp 'band' and of sodium are also shown in
Figure 2b. The density of states is dominated by the contribution from the
nickel d band (width about 4 eV) peaking near the Fermi energy and exhibiting
a behavior similar to that known from band structure calculations of bulk nickel
(78,79) and from photoemission experiments (89). The sp 'band' (width about
8 eV) extends to both sides of this peak, just like in the bulk band structure.
The Na density of states consists of two structures. The larger one above the
Fermi level at −1.2 eV has the form of a Na 3s resonance (82), the smaller and
broader one below the Fermi level represents the covalent contribution to the
sodium surface bond. Altogether, the sodium density of states resembles very
much the form found in density functional calculations with a semi–infinite
jellium half–space modeling the metal substrate (82,83).

   In Table IV the changes in the Mulliken populations due to chemisorption
of sodium (or potassium) are shown for the Ni_{17} cluster. Starting with the
sodium cluster we note that the alkali atom has given up 0.72 of its valence
electron charge which matches the increase of the p and d population of the four
nearest nickel atoms. Essentially the same conclusions have been drawn previously from a population analysis for the cluster Ni_9Na (RSGK). The jellium
model (82,83) also leads to a charge transfer from the alkali atom into the
bonding region. The LCGTO–Xα results for the analogous potassium cluster
$Ni_{17}K$ are actually rather similar to that for sodium, including the alkali valence
s population. Two significant exceptions are the changes in the charge of the
potassium atom and of its nearest neighbors. The increase in the potassium 4p
and 4d populations is partially balanced by a concomitant decrease in the 4s
population of the nearest neighbor Ni atoms. Therefore one is lead to conclude
that these population differences do not represent a true change in the spatial
charge distribution, but rather have to be taken as an artifact of the Mulliken
population analysis caused by the very diffuse nickel and potassium basis functions. This well–known shortcoming of the Mulliken procedure may also be

responsible for the apparent contradiction between the high ionicity and the large effective radius of the alkali atom determined in the geometry optimization (see Table III). It seems that the alkali s population underestimates the charge in the adsorbate region of space, since diffuse nickel basis functions are certainly able to contribute.

Table IV. Changes in the Mulliken populations of the valence orbitals (relative to the bare Ni_{17} cluster) and resulting atomic charges due to the formation of the chemisorption clusters $Ni_{17}Na$ and $Ni_{17}K$

Atom [a]		Change of Population			Charge/Atom
		s	p	d	
Na		−0.73	−0.01	0.02	0.72
K		−0.76	0.11	0.11	0.54
$Ni_4(1)$	(Na)	−0.02	0.14	0.06	−0.18
	(K)	−0.24	0.15	0.06	0.03
$Ni_8(1)$	(Na)	0.01	0.01	0.00	−0.02
	(K)	0.05	0.03	0.01	−0.09
$Ni_1(2)$	(Na)	0.02	−0.01	−0.03	0.02
	(K)	0.00	0.04	−0.00	−0.04
$Ni_4(2)$	(Na)	−0.03	−0.01	−0.01	0.05
	(K)	−0.01	0.00	0.00	0.01

[a] Numbers in parentheses indicate the lattice plane, subscripts indicate the number of equivalent atoms.

Table V. Mulliken populations n and spin polarization s of the valence orbitals of $Ni_{17}Na$ obtained from a spin polarized calculation

Atom [a]		Orbital type			Charge/Atom
		s	p	d	
Na	n	0.29	−0.02	0.00	0.73
	s	−0.01	0.00	0.00	
$Ni_4(1)$	n	0.37	0.99	8.93	−0.29
	s	−0.05	−0.03	0.69	
$Ni_8(1)$	n	0.70	0.24	8.92	0.14
	s	0.02	0.00	0.86	
$Ni_1(2)$	n	0.25	1.19	9.00	−0.44
	s	−0.05	−0.05	0.78	
$Ni_4(2)$	n	0.56	0.62	8.88	−0.06
	s	0.00	0.00	0.84	

[a] Numbers in parentheses indicate the lattice plane, subscripts indicate the number of equivalent atoms.

Inspection of Table IV also reveals that the chemisorption interaction for both adsorbates is confined essentially to the nearest neighbor substrate atoms, a finding that may be taken as an *a posteriori* justification of the cluster approach.

In conclusion, we briefly summarize our cluster calculations. For $Ni_{17}Na$ the binding energy (1.90 eV) and the induced dipole moment (4.43 Debye) agree with those for the non–spinpolarized case within the accuracy of the method. The Mulliken populations displayed in Table V are rather similar to those of the nonspinpolarized case, exhibiting an almost pure d^9 configuration for each type of nickel atom. The spin–polarization of the nickel atoms which averages to 0.79 μ_B per atom is due mainly to the d electrons. This value is somewhat larger than that found for the gas phase Ni_{13} clusters discussed above (0.62 μ_B). Of special interest are the changes in the magnetic moments of the substrate and the fate of the unpaired electron of the adsorbate, the latter being completely quenched upon chemisorption (see Table V). One finds again that charge and spin transfer are almost exclusively confined to the nearest neighbor atoms of the adsorbate whose magnetic moment is reduced by 0.08 μ_B each.

The chemisorption cluster studies presented here seem quite encouraging. Cluster investigations are currently under way which address the alkali/carbon monoxide coadsorption problem.

Acknowledgments

The work of the Munich group was partially supported by the Deutsche Forschungsgemeinschaft through Sonderforschungsbereich 128, by the Fonds der Chemischen Industrie and by the Bund der Freunde der Technischen Universität München. B.I.D. thanks the Naval Research Laboratory for a grant of CRAY X–MP computer time for the study of Ni_{13} clusters.

Literature Cited

1. Cotton, F.A.; Walton, R.A. Multiple Bonds between Metal Atoms; Wiley: New York, 1982.
2. Muetterties, E.L.; Rhodin, T.N.; Band, E.; Brucker, C.F.; Pretzer, W.R. Chem. Rev. 1979, 79, 91.
3. Bauman, C.A.; Van Zee, R.J.; Bhat, S.V.; Weltner, W. J. Chem. Phys. 1983, 78, 190.
4. Koutecky, J.; Fantucci, P. Chem. Rev. 1986, 86, 539.
5. Rhodin, T.N.; Ertl, G., Eds.; The Nature of the Surface Chemical Bond; North – Holland: Amsterdam, 1979.
6. Gates, B.C.; Guczi, L.; Knözinger, H., Eds.; Metal Clusters in Catalysis; Elsevier: Amsterdam, 1986.
7. Veillard, A., Ed.; Quantum Chemistry: The Challenge of Transition Metals and Coordination Chemistry; NATO ASI Series C, Vol. 176; Reidel: Dordrecht, 1986.
8. Benedek, G.; Martin, T.P.; Pacchioni, G., Eds.; Elemental and Molecular Clusters; Proceedings of the 13th International School, Erice, Italy, 1987; Springer: Berlin, 1988.
9. Messmer, R.P. In ref. 5; p 51.
10. Kunz, B. In Theory of Chemisorption; Smith, J.R., Ed.; Springer: Berlin, 1980; p 115.
11. Shustorovich, E. In ref. 7; p 445.
12. Waber, J.T.; Adachi, H.; Averill, F.W.; Ellis, D.E. Japan. J. Appl. Phys. Suppl. 1974, 2 Pt. 2, 695.

13. Rösch, N.; Menzel, D. Chem. Phys. 1976, 13, 243.
14. Rhodin, T.N.; Gadzuk, J.W. In ref. 5; p 113.
15. Rösch, N. In Electrons in Finite and Infinite Structures; Phariseau, P.; Scheire, L., Eds.; NATO ASI Series B, Vol. 24; Plenum: New York, 1977; p 1.
16. Blyholder, G. Surface Sci. 1974, 42, 249.
17. Bagus, P.S.; Hermann, K.; Bauschlicher, C.W. J. Chem. Phys. 1984, 81, 1966.
18. Upton, T.H.; Goddard III, W.A. Crit. Rev. Solid State Mater. Sci. 1981, 10, 261.
19. Blomberg, M.; Brandemark, U.; Panas, I.; Siegbahn, P.; Wahlgren, U. In ref. 7; p 1.
20. Pacchioni, G.; Koutecky, J. In ref. 7; p 465.
21. Post, D.; Baerends, E.J. J. Chem. Phys. 1983, 78, 5663.
22. Wästberg, B.; Rosén, A. Surface Sci. 1988, 193, L7.
23. Jörg, H.; Rösch, N. Surface Sci. 1985, 163, L627.
24. Andzelm, J.; Salahub, D.R. Int. J. Quantum Chem. 1986, 29, 1091.
25. Lüthi, H.P.; Ammeter, J.; Almlöf, J.; Fægri, K. J. Chem. Phys. 1982, 77, 2002.
26. Fægri, K.; Almlöf, J. Chem. Phys. Lett. 1984, 107, 121.
27. Jörg, H.; Rösch, N. Chem. Phys. Lett. 1985, 120, 359.
28. Rösch, N.; Jörg, H.; Dunlap, B.I. In ref. 7, p 179.
29. Bauschlicher, C.W.; Bagus, P.S. J. Chem. Phys. 1984, 81, 5889.
30. Gunnarson, O.; Jones, R.O. Phys. Rev. B 1985, 31, 7588.
31. Becke, A.D. J. Chem. Phys. 1986, 84, 4524; ibid. 1986, 85, 7184.
32. Bagus, P.S.; Nelin, C.J.; Bauschlicher, C.W. J. Vac. Sci. Technol. A 1984, 2, 905.
33. Jepsen, O.; Madsen, J.; Andersen, O.K. Phys. Rev. B 1982, 26, 2790.
34. Wimmer, E.; Fu, C.L.; Freeman, A.J. Phys. Rev. Lett. 1985, 55, 2618.
35. Bagus, P.S., Müller, W. Chem. Phys. Lett. 1985, 115, 540.
36. Grimley, T.C. In Electronic Structure and Reactivity of Metal Surfaces; Derouane, E.G.; Lucas, A.A., Eds.; Plenum: New York, 1976; p. 113.
37. Phillips, J.C. Chem. Rev. 1986, 86, 619.
38. Morse, M.D. Chem. Rev. 1986, 86, 1049.
39. Kaldor, A.; Cox, D.M.; Zakin, M.R. Adv. Chem. Phys. 1987, 70, 211.
40. Cox, D.M.; Reichmann, K.C.; Trevor, D.J.; Kaldor, A. J. Chem. Phys. 1988,88, 111.
41. Versluis, L.; Ziegler, T. J. Chem. Phys. 1988, 88, 322.
42. Satoko, C. Phys. Rev. B 1984, 30, 1754.
43. Dunlap, B.I.; Connolly, J.W.D.; Sabin, J.R. J. Chem. Phys. 1979, 71, 3396; 4993.
44. Slater, J.C. Adv. Quantum Chem. 1972, 6, 1.
45. von Barth, U.; Hedin, L. J. Phys. C: Solid State Phys. 1972, 5, 1629.
46. Janak, J.F.; Moruzzi, V.L.; Williams, A.R. Phys. Rev. B 1975, 12, 1257.
47. Salahub, D.R. Adv. Chem. Phys. 1987, 69, 447.
48. Sambe, H.; Felton, R.H. J. Chem. Phys. 1975, 62, 1122.
49. Fieck, G. Theor. Chim. Acta 1980, 54, 323.
50. Jörg, H.; Rösch, N.; Sabin, J.R.; Dunlap, B.I. Chem. Phys. Lett. 1985, 114, 529.
51. Mintmire, J.W.; Dunlap, B.I. Phys. Rev. A 1982, 25, 88.
52. Dunlap, B.I. J. Phys. Chem. 1986, 90, 5524.
53. Dunlap, B.I.; Cook, M. Int. J. Quantum Chem. 1986, 29, 767.
54. Dunlap, B.I.; Phys. Rev. A. 1982, 25, 2847.
55. Dunlap, B.I.; Yu, H.L. Chem. Phys. Lett. 1980, 73, 525.
56. Dunlap, B.I.; Yu, H.L.; Antoniewicz, P.R. Phys. Rev. A 1982, 25, 7.

57. Rösch, N.; Jörg, H. J. Chem. Phys. 1986, 84, 5967.
58. Rösch, N.; Kotzian, M.; Jörg, H.; Schröder, H.; Rager, B.; Metev, S. J. Am. Chem. Soc. 1986, 108, 4238.
59. Rösch, N.; Jörg, H., Kotzian, M. J. Chem. Phys. 1987, 86, 4038.
60. Rösch, N.; Knappe, P.; Dunlap, B.I.; Bertel, E.; Netzer, F.P. J. Phys. C: Solid State Phys. 1988, 21, 3423.
61. Bertel, E.; Memmel, N.; Jacob, W.; Dose, V.; Netzer, F.P.; Rosina, G.; Rangelov, G.; Astl, G.; Rösch, N.; Knappe, P.; Dunlap B.I. Appl. Phys. A 1988, 47, 87.
62. Messmer, R.P.; Lamson, S.H. Chem. Phys. Lett. 1982, 90, 31.
63. Spangler, D.; Wendoloski, I.J.; Dupuis, M.; Chen, M.M.L.; Schaefer III, H.F. J. Am. Chem. Soc. 1981, 103, 3985.
64. Wachters, A.J.H. J. Chem. Phys. 1970, 52, 1033.
65. Rosén, A.; Ellis, D.E.; Adachi, H.; Averill, F.W. J. Chem. Phys. 1976, 65, 3629.
66. Dunlap, B.I.; Mei, W.N. J. Chem. Phys. 1983, 78, 4997.
67. Veillard, A. Theor. Chim. Acta 1968, 12, 405.
68. Echt, O.; Sattler, K.; Recknagel, E. Phys. Rev. Lett. 1981, 47, 1121.
69. Haymet, A.D.J. Phys. Rev. B 1983, 27, 1725.
70. McElvaney, S.W.; Creasy, W.R.; O'Keefe, A. J. Chem. Phys. 1986, 85, 632.
71. Zakin, M.R.; Brickman, R.O.; Cox, D.M.; Kaldor, A. J. Chem. Phys. 1988, 88, 3555.
72. Hamrick, Y.; Taylor, S.; Lemire, G.W.; Fu, Z.–W.; Shui. J.–C.; Morse, M.D. J. Chem. Phys. 1988, 88, 4095.
73. Elkind, J.L.; Weiss, F.D.; Alford, J.M.; Laaksonen, R.T.; Smalley, R.E. J. Chem. Phys. 1988, 88, 5215.
74. McHenry, M.E.; Eberhart, M.E.; O'Handley, R.C.; Johnson, K.H. Phys. Rev. Lett. 1986, 56, 81.
75. Messmer, R.P.; Knudson, S.K.; Johnson, K.H.; Diamond, J.B.; Yang, C.Y. Phys. Rev. B 1976, 13, 1396.
76. Raatz, F.; Salahub, D.R. Surf. Science 1986, 176, 219.
77. Elsässer, C.; Fähnle, M.; Brandt, E.H.; Böhm, M.C. J. Phys. F. 1987, 17, L301.
78. Wang, C.S.; Callaway, J. Phys. Rev. B 1977, 15, 298..
79. Moruzzi, V.L.; Janak, J.F.; Williams, A.R. Calculated Electronic Properties of Metals; Pergamon: New York, 1978.
80. de Heer, W.A.; Knight, W.D.; Chou, M.Y.; Cohen, M.L. Solid State Phys. 1987, 40, 93.
81. Gerlach, R.L.; Rhodin, T.N. Surface Sci. 1970, 19, 403.
82. Muscat, J.P.; Newns, D.M. Surface Sci. 1978, 74, 355.
83. Lang, N.D.; Holloway, S.; Norskov, J.K. Surface Sci. 1985, 150, 24.
84. Lang, N.D. Phys. Rev. B 1971, 4, 4234.
85. Villars, D.S.; Langmuir, I. J. Am. Chem. Soc. 1931, 53, 486.
86. Bonzel, H.P.; Surface Sci. Rep. 1987, 8, 43.
87. Ashcroft, N.W.; Mermin, N.D. Solid State Physics; Holt–Saunders: New York, 1976.
88. Gadzuk, J.W. Surface Sci. 1967, 6, 133.
89. Eberhardt, W.; Plummer, E.W. Phys. Rev. B 1980, 21, 3245.

RECEIVED October 24, 1988

Chapter 14

Photoelectron Spectroscopy and Chemical Bonding

Valence Bond Model Viewpoint

R. P. Messmer[1,2], P. A. Schultz[1], S. H. Lamson[2], C. H. Patterson[1], and H. Wang[1]

[1]Department of Physics, University of Pennsylvania, Philadelphia, PA 19104

[2]General Electric Corporate Research and Development, Schenectady, NY 12301

Recent valence bond studies of multiple bonds in molecules with only s, p-orbitals indicate that "bent bonds" are preferred to the usual σ and π bonds. This has potentially important implications for the description of multiple metal-metal bonds. However, the description of Σ, Π and Δ ion states in photoemission from a ground state of "bent bonds" is not so obvious as in the σ, π, δ-molecular orbital model. We examine these issues in the present contribution.

In the course of investigating multiple bonds in molecules and complexes by the valence bond approach, we have recently found that such multiple bonds are more accurately described as "bent bonds" rather than as σ and π bonds (*1-5*). In order to understand the potential implications of these results for multiple metal-metal bonds, it is important to briefly review the basic assumptions of the valence bond model and compare them to those of the more familiar molecular orbital model of bonding.

There are many expansions of the total many-electron wave function which may be used in discussing the properties of an electronic system. For the present discussion, we assume a *resonating valence bond* expansion, which is represented by

$$\Psi = \sum_{\nu} d_{\nu} \Phi_{\nu} \qquad (1)$$

where $\Phi_{\nu} = \det \left[\prod_{i}^{N} \psi_{i\nu} \, \Theta_{\nu} \right]$ and $\Theta_{\nu} = \sum_{j} c_{j\nu} \, \theta_{j}$.

In this expansion, each of the Φ_{ν} are separate valence bond orbital (VBO) structures in which the orbitals are non-orthogonal. Each such VBO structure

0097–6156/89/0394–0199$06.00/0

can be thought of as being made up of valence bond spin (VBS) structures, which describe the various allowed spin pairings. Such a description will be useful if the wave function can be approximated by a small number of VBO structures and a few spin couplings. See Fig. 1 for a schematic representation of VBO structures of CO and VBS structures of CH_4.

If we assume a situation where one term (one VBO structure) dominates (*i.e.*, we can neglect resonance), then the wave function can be approximated as

$$\Psi \approx \Phi^{VB} = \det \left[\{core\} \prod_i^N \psi_i \, \Theta \right] \qquad (2)$$

where the electrons not being correlated by the valence bond description are denoted as {core}. If the ψ_i are determined self-consistently then the wave function is the basis of the generalized valence bond (GVB) model (*6*). If in addition, for the system of interest, there is one dominant bonding spin-pairing, then the perfect-pairing approximation is a good approximation to the full GVB model and only one term in the expansion of Θ need be considered. Thus, we arrive at the "perfect-pairing" (PP) approximation

$$\Psi \approx \Phi^{PP} = \det \left[\{core\} \prod_i^N \psi_i \theta \right], \text{ with } \theta = (\alpha\beta\text{-}\beta\alpha)(\alpha\beta\text{-}\beta\alpha) \, \cdots . \qquad (3)$$

The further assumption that the orbitals ψ_i are orthogonal except for those which are singlet paired (the strong orthogonality approximation), leads to the strongly orthogonal, perfect pairing (SOPP) wave function

$$\Psi \approx \Phi^{SOPP} = \det \left[\{core\} \prod_i^{N/2} \psi_{ia}\psi_{ib} \, \theta \right] \qquad (4)$$

which is the basis of the GVB-PP model of Hay *et al.* (*7*). Finally, if the orbitals of a pair, ψ_{ia} and ψ_{ib}, are assumed to be identical, ϕ_i, then we arrive at an approximate wave function which is the basis of the molecular orbital (MO) model, *i.e.*,

$$\Psi \approx \Phi^{MO} = \det \left[\{core\} \prod_i^{N/2} \phi_i \, \alpha \, \phi_i \, \beta \right]. \qquad (5)$$

GVB Model of Multiple Bonds

To begin, we discuss the results of calculations using the GVB-PP model, *i.e.*, results based on the use of the SOPP restrictions on the wave function, Eq. (4). GVB-PP calculations for CO_2 (*1*), C_2F_2 (*2*) and benzene (*3*) have given descriptions of the multiple bonds in these molecules as being made up of bent bonds (Ω-bonds) rather than the traditional σ and π bonds normally attributed to these molecules. Figure 2 shows a schematic representation of the calculated bonding in these molecules.

There has been some recent concern (8,9) however, that this bent bond description of multiple bonds derived from the GVB-PP model may be an artifact of the model. The concern takes two forms: first, that the SOPP restrictions on the GVB wave function are the source of the bent bonds and the full GVB model will produce the usual σ,π-bond description; and second, if an MCSCF or CI wave function which is more general than GVB is used, this will give back the σ,π description.

With regard to the second point, it is important to note that an approximate wave function which is more general than those of Eqs. (2)-(5) *cannot* be described in terms of *either* bent bonds or σ,π bonds. Such a wave function (general MCSCF, GVB-CI or CI) will be a complicated combination of these two descriptions (as well as others, *e.g.*, the atoms-in-molecule picture (10)) or in certain approximate wave functions the descriptions are related by a transformation and are thus in some sense equivalent (10). Hence the best one can do is decide on a criterion to measure the extent to which a particular picture is contained in the general wave function. One possible measure would be the overlap of a unique σ,π or unique bent bond description with the general wave function.

Thus the issue reduces to whether the SOPP restrictions on the GVB wave function are introducing bent bonds as artifacts. It should be noted that the earliest applications of the GVB-PP model (7) included ethylene and acetylene as examples and that the authors of the work concluded that the σ,π bond description of these molecules was lower in energy than the bent bond description. Since that time many additional examples of molecules with multiple bonds have been considered within the GVB-PP model, but none seem to have bent bonds as the lower energy description of the bonding. Furthermore, in our studies of multiple bonds (4,5) (involving more than 15 cases of double, triple and conjugated systems) a large majority is found to have the σ,π description as the lower energy wave function.

Thus, there is legitimate cause for concern that our few examples of bent bonds within the GVB-PP model may not reflect the situation for the full GVB model. In order to resolve this issue on the interpretation of multiple bonds within the valence bond framework it is necessary to perform full GVB calculations on a series of molecules to remove the strong orthogonality and perfect pairing restrictions of the GVB-PP model. This is essential in order to assess the validity of the bent bond concept for multiple bonds. Full GVB calculations have in fact been carried out and the details will be presented elsewhere (5). Here we provide a brief summary of the results.

The general picture which emerges from the large number of calculations performed is the following: (1) those cases (*e.g.*, CO_2 and C_2F_2) which were found to prefer bent bonds within the GVB-PP model are found to *retain the bent bond* description when the SO and PP restrictions are removed from the wave function, *i.e.*, full GVB model is used to evaluate the nature of the multiple bonds; (2) in those cases (*e.g.*, C_2H_2 and C_2H_4) for which the σ,π

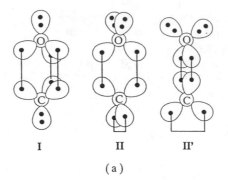

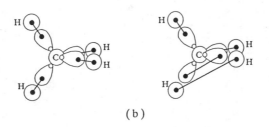

Figure 1. Schematic valence bond orbital (a) and valence bond spin (b) structures of CH_4.

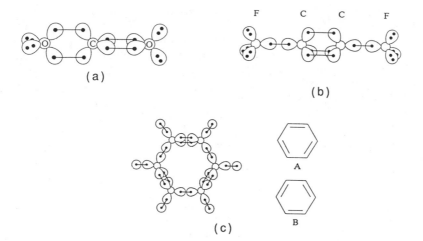

Figure 2. Bent bond representation of double and triple bonds to carbon in (a) CO_2, (b) C_2F_2 and (c) C_6H_6.

description was found to be lower in energy within the GVB-PP approximation, the situation is *reversed* using the full GVB model and bent bonds are found to be more stable than σ,π bonds. Hence, the conclusion based on extensive calculations for a group of typical multiple bonded molecules is that bent bonds are more stable at the full GVB level of approximation. Further, the SOPP restrictions are biased *against* the bent bond description rather than biased *toward* this description as previously suggested (8,9). Our results on multiple bonds using the full GVB model are the first of their kind. We believe they are significant because they represent the full implementation of the GVB model without the usual approximations and they give a nontraditional picture of multiple bonds.

Valence Bond Description of Photoelectron Spectroscopy

Certainly one of the first conceptual problems which arises if one describes the ground state of a molecule in terms of Ω-bonds is how to obtain a compatible description of the excited states and ion states which may have Σ, Π, or Δ symmetries. A discussion of how a compatible description is obtained has been given recently (11). However, since it is important for later discussions in this work, especially with respect to the connection between valence bond theory, localized molecular orbitals (LMOs) and canonical molecular orbitals (CMOs), a brief account is provided here.

In a time-dependent picture, one can imagine an electron being annihilated from a specific valence bond orbital ψ_i at time $t=0$. This is obviously not a stationary state of the ion and hence one must follow the time dependent evolution of the system from this initial condition. This requires solving the time-dependent Schrödinger equation which may be written as the following set of coupled equations for the time dependent probability amplitudes, $a_j(t)$:

$$i \sum_{j}^{N} \frac{\partial a_j(t)}{\partial t} S_{kj} = \sum_{j}^{N} a_j(t) H_{kj} \tag{6}$$

with $S_{kj} = \langle \Phi_k \mid \Phi_j \rangle$ and $H_{kj} = \langle \Phi_k \mid H \mid \Phi_j \rangle$, where Φ_k is the (N-1)-electron wave function with the electron missing from the VB orbital ψ_k. If we assume the Φ_k are not explicitly time dependent but may include electronic relaxation, then the time dependent VB description of the ion state is

$$\Phi_{ion}^{VB}(t) = \sum_{j}^{N} a_j(t) \Phi_j. \tag{7}$$

If we define the autocorrelation function as the overlap of Φ_{ion}^{VB} for the various initial states at time $t = 0$ with $\Phi_{ion}^{VB}(t)$, then the Fourier transform of this correlation function gives the photoelectron spectrum.

To make the connection with MO theory, consider the SOPP approximation of valence bond theory and write the ground state wave function as in Eq. (4). For the MO approximation ψ_{ia} and ψ_{ib} are the same and this leads to

Eq. (5) where the ϕ_i are localized molecular orbitals (LMOs) which are related to the familiar CMOs by a unitary transformation. Because of the orthogonality of molecular orbitals the autocorrelation function for the MO approximation is simply the sum of the time-dependent probability amplitudes for the hole to be in the various LMOs, ϕ_i:

$$C^{MO}(0,t) = \sum_l a_l(t). \tag{8}$$

Thus the spectrum which arises when Eq. (8) is Fourier transformed consists of a set of δ-functions at the energies corresponding to the stationary states of the ion (which *via* the theorem of Koopmans) are the one-electron eigenvalues of the Hartree-Fock equations). The valence bond description of photoelectron spectroscopy provides a novel perspective of the origin of the canonical molecular orbitals of a molecule. The CMOs are seen to arise as a linear combination of LMOs (which can be considered as uncorrelated VB pairs) and coefficients in this combination are the probability amplitudes for a hole to be found in the various LMOs of the molecule.

With this development in mind it is now quite straightforward to describe the Σ, Π and Δ ion states of a molecule. As an example we study the Σ and Π ion states of acetylene. Consider the description of C_2H_2 shown schematically in Fig. 3. The stationary states of the ion may be written in an approximate valence bond form as

$$\Phi_{ion}^{VB} = \sum_{i=1}^{5} (c_{ia}\Phi_{ia} + c_{ib}\Phi_{ib}). \tag{9}$$

But considering the large overlap within pairs that usually results from calculations, this expression can be further approximated by

$$\Phi_{ion}^{VB} = \sum_{i=1}^{5} c_i \Phi_i, \tag{10}$$

in order to illustrate the essential aspects of the argument, where the Φ_i are constructed from an in-phase combination of the orbitals ψ_{ia} and ψ_{ib}. It should be clear that basis states of the proper symmetry can be obtained by the following combinations of the Φ_i:

Σ_g: $c_{1\Omega}(\Phi_1 + \Phi_2 + \Phi_3)$; $c_{1CH}(\Phi_4 + \Phi_5)$

Σ_u: $c_{2CH}(\Phi_4 - \Phi_5)$

Π_{ux}: $c_{2\Omega}(2\Phi_1 - \Phi_2 - \Phi_3)$

Π_{uy}: $c_{3\Omega}(\Phi_2 - \Phi_3)$,

which, for the many-electron wave function, is exactly analogous to the formation of symmetry adapted molecular orbitals from atomic orbital basis functions in the one-electron case. Thus it is seen that the description of Σ and Π ion states is easily accommodated within a valence bond framework based on a ground state described in terms of bent multiple bonds (Fig. 3). This is also true for more general valence bond approximations to the wave function. As a consequence we realize that it is not at all necessary to have σ, π and δ orbitals in order to be able to describe Σ, Π and Δ ion states.

Transition Metals Molecules

To what extent is the insight gained from valence bond calculations on molecules which have only s, p-orbitals applicable to those which have d-orbitals as well? This is the topic we wish to explore in the present section. Although we intend to carry out full GVB calculations for multiple metal-metal bonds to learn if the experience of the s, p-atoms is confirmed for d-electron systems, no such calculations are available at present. Hence our discussion will focus on recent valence bond results for $Ni(CO)_4$, $Cr(CO)_6$, $PtCO$ and RcH_9^{2-}. We will explore briefly, however, the potential advantages of a conceptual framework based on multiple bent bonds and bent metal-metal bonds in general.

Consider the $Cr(CO)_6$ molecule (*12*). In the canonical molecular orbital (CMO) approach there are 33 doubly occupied delocalized valence orbitals. To obtain a bonding description from CMOs is rather tedious as there can be several CMOs contributing to a single covalent bond. By contrast, if the CMOs are transformed into localized molecular orbitals (LMOs) by the Foster-Boys (*13*) or Edminston-Ruedenberg (*14*) procedures only four unique orbitals are obtained. These are shown in Fig. 4 and can easily be identified as: (a) a carbon lone pair orbital representing the dative bond between a carbon and the Cr atom; (b) a lone pair orbital on the oxygen atom of CO; (c) a bent CO bond orbital (one of three equivalent orbitals for a CO molecule) and (d) a d-orbital which delocalizes slightly onto the bonding CO molecules. There are six orbitals of type (a), six of type (b), 18 of type (c) and three (d_{xy}, d_{xz}, d_{yz}) of type (d), accounting for the 33 occupied valence orbitals.

It will be recalled that the transformation to localized orbitals leaves the molecular orbital wave function, charge density and all other ground state properties completely unchanged. The relationship between LMOs, which are convenient for describing bonding and the CMOs, which give a zeroth order description of the one-electron ion states, was described in the last section.

If we allow for correlation effects within each pair using the GVB-PP method we obtain the SOPP orbitals also shown in Fig. 4. The close relationship between the SOPP orbitals and the LMOs is apparent. Each pair in an LMO is either "left-right" correlated in the case of a bond or "in-out" correlated in the case of lone pairs. The d_{xy} LMO or the corresponding pair from the GVB-PP calculation are the only orbitals to exhibit considerable delocalization. In the MO case, this behavior is referred to as back-bonding from the metal to the CO, but in the valence bond situation, our experience is that all

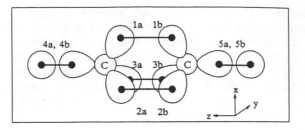

Figure 3. Schematic valence bond orbital representation of the CH and CC bonds in acetylene.

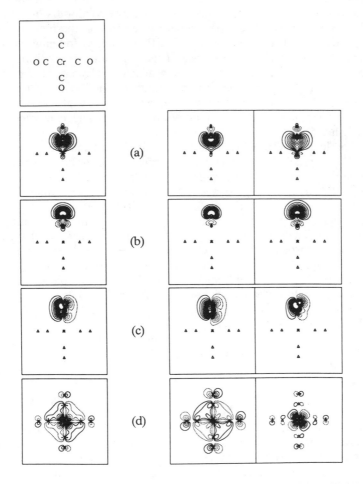

Figure 4. Unique orbitals of $Cr(CO)_6$ obtained by localization of Hartree-Fock molecular orbitals (left) or introduction of correlation into the wave function (GVB-SOPP) (right).

occurrences of such delocalization are an indication of the importance of resonance. We will investigate this shortly for the example of PtCO. In passing we remark that for $Ni(CO)_4$ the extent of delocalization of the LMOs and SOPP orbitals is less for than for the $Cr(CO)_6$ molecule.

Turning to the bonding in PtCO, we observe that thus far it has been a reasonable approximation to assume only one resonance structure (or one term in Eq. (1)) is necessary to describe the wave function. Due to the delocalization of the LMOs and SOPP orbitals noted above for $Cr(CO)_6$ it is unlikely that one term (*i.e.*, the GVB-PP method or the MO method) is adequate to describe this molecule. However, rather than explore the consequences of resonance in describing metal-CO bonds for a large molecule such as $Cr(CO)_6$, we will consider the situation for the simpler case of PtCO bonding (*15*).

As mentioned in the introduction, the wave function of CO may be approximated as a superposition of three resonance structures (see Fig. 1a). In the example of $Cr(CO)_6$, we discussed the bonding on the basis of structure I. The only orbitals which seem to incorporate structures II and III are the delocalized *d*-orbitals. This latter situation suggests that M=CO double bond resonance structures may be important in describing the bonding in such molecules. In our recent study of the PtCO molecule we have found that besides the Pt-CO dative bond VBO structure, the Pt=CO structures shown schematically in Fig. 5a contribute quite significantly. The SOPP orbitals for one of the double bond resonance structures are shown in Fig. 5. The bonds between the *d,s,p*-hybrids on Pt and the hybrids on the carbon atom are seen to be bent.

In general, the relative importance in the total wave function of the dative bond VBO structure and the double bond VBO structures will vary from molecule to molecule. A mean-field description of resonance such as molecular orbital theory or GVB-PP for one VBO structure, averages over the resonance structures to yield some delocalized LMOs or SOPP orbitals. This is the origin of the back-bonding description of mean-field theory. Mean-field theory always overestimates the charge fluctuations in a system and provides for no correlation among them. The GVB model reduces local charge fluctuations (*e.g.*, within bonds and lone pairs) and resonance provides a convenient pictorial way to understand the correlations among the charge fluctuations.

Transition metals have the potential to form many different hybrids by combining *s*, *p* and *d* atomic orbitals. In the discussion of molecules with only *s* and *p* orbitals, we have found that approximate sp^3-hybrids arise in the description of single, double and triple bonded molecules, as well as in conjugated systems. In the course of investigating the utility of the polyhedral hybrid concept (*16*), we have considered the bonding of ReH_9^{2-}. The ReH_9^{2-} molecule is a prototype for covalent bonding to a transition metal, in much the same way as H_2 and CH_4 are for bonding to atoms with *s*-orbitals and *s*, *p*-orbitals. Thus, qualitatively nine covalent bonds are expected between the Re atom and the H atoms. The geometrical structure of the molecule is shown in Fig. 6a. The H atoms form a tri-capped trigonal prism with six atoms at the vertices of a trigonal prism and three others capping in three rectangular faces of the

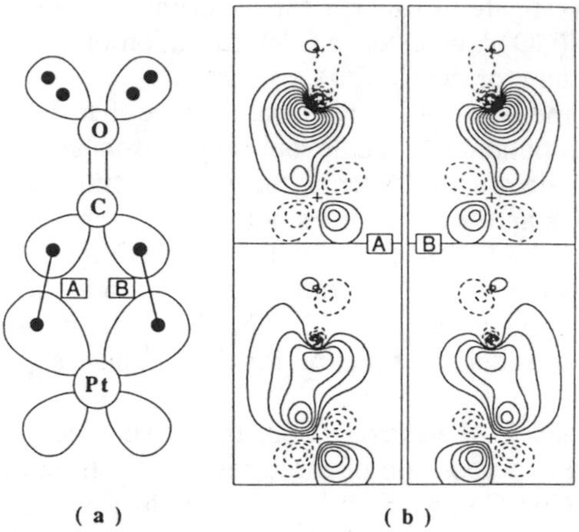

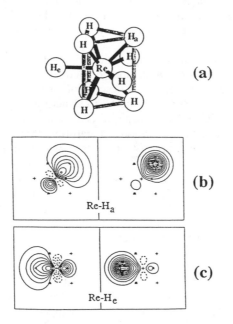

Figure 5. Schematic representation (a) and GVB orbitals (b) or the Pt=CO double bond in the PtCO molecule.

(a)

(b)

Re-H$_a$

(c)

Re-H$_e$

Figure 6. (a) Structure of the ReH$_9^{2-}$ ion. Covalent bonds are indicated by dark lines and the triangonal prism by light lines. (b,c) GVB orbitals of the axial and equatorial ReH bonds.

prism. The capping H atoms we will refer to as equatorial atoms and the six others as axial H atoms. In Figs. 6b and 6c we show orbital contour plots of one of the six equivalent Re-H axial bond-pairs and one of the three equivalent Re-H equatorial bond-pairs respectively. These are SOPP orbitals obtained using the GVB-PP method.

The covalent bonds of ReH_9^{2-} look very much like those of methane: a hydrogen orbital overlapping with a hybrid orbitals of Re instead of C to form a bond. This suggests that metal-metal covalent bonds might be formed much the same way as in carbon-carbon multiple bonds - namely by overlap of hybrid atomic orbitals. To make the analogy clearer, consider Fig. 7. If two methane molecules are oriented as in Fig. 7a and the four hydrogen atoms between the carbon atoms are removed, then the resulting hybrids on the carbons can overlap to form the bent double bonds of C_2H_4. Likewise, if the methane molecules are oriented as in Fig. 7b, the hydrogen atoms between the carbon are removed, and the carbon hybrids allowed to overlap, we can understand the origin of the bent triple bonds of acetylene.

This same procedure can be applied to ReH_9^{2-} to give some insight regarding the valence bond description of Re-Re multiple bonds. Referring back to Fig. 6a, if we imagine this molecule being reflected through a vertical plane and a mirror image appearing to the right, then we may perform the same "Gedanken" experiment discussed for the molecules in Fig. 7. That is, we remove the eight H atoms between the two Re atoms, this allowing one to see how Re-Re quadruple bonds could be formed. Then, if an H^- is removed from the left equatorial position of the molecule in Fig. 6a and from the right equatorial position of its mirror image molecule, and the two resulting moieties are brought together, we could have an $Re_2H_8^{2-}$ molecule. Next if the H atoms in the eight axial bonds (Fig. 6a and its mirror image) are replaced by Cl atoms, we obtain $Re_2Cl_8^{2-}$. The formally empty hybrid orbitals (arising from the removal of an H^- from the two equatorial position along the Re-Re internuclear axis) will be shared by the lone pair electrons of the Cl atoms. Thus one obtains a simple qualitative picture of the bonding in $Re_2Cl_8^{2-}$ and of its relationship to ReH_9^{2-}. Calculations presently being carried out will determine whether this qualitative picture is retained when put to a quantitative test.

Another qualitative applications of polyhedral hybrids illustrates bent single metal-metal bonds in the polynuclear molecules $Os_3(CO)_{12}$ and $Ir_4(CO)_{12}$. In these molecules Os and Ir atoms may be thought of as having three doubly occupied d-orbitals, which do not participate in bonding, and six octahedral hybrids for bonding. Figure 8a shows the octahedral hybrids of Os and Ir in a schematic form in which the dark lines indicate positions where dative bonds to CO can be formed and the light lines terminated with a dot indicate hybrids which are occupied by a single electron and therefore are available to form covalent bonds. Figures 8b and 8c illustrate how the geometries of $Os_3(CO)_{12}$ and $Ir_4(CO)_{12}$ respectively can be understood on the basis of polyhedral hybrids overlapping to form bent metal-metal bonds. There are, in fact, a vast number of molecules and their structures which can be rationalized on this basis.

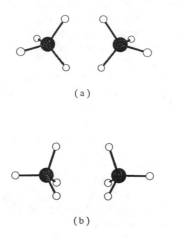

(a)

(b)

Figure 7. Methane molecules oriented so that their bonds point in appropriate directions for (a) ethylene or (b) acetylene molecule formation.

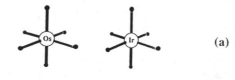

(a)

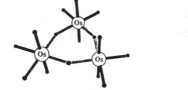

(b)

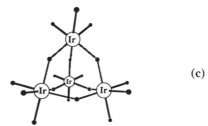

(c)

Figure 8. Polyhedral hybrid representation of Osmium and Iridium carbonyl clusters.

Summary

The major conclusions of the studies described in this contribution are the following: (1) the full GVB model for *sp*-molecules gives bent multiple bonds for the molecules studied thus far, which is in contrast to the results of the GVB-PP model where most of the molecules are found to have σ and π bonds; (2) the localized molecular orbital (LMO) model, using either the Foster-Boys or Edminston-Ruedenberg methods also gives bent multiple bonds; (3) the canonical molecular orbitals can be interpreted as the time-independent probability amplitudes for a hole hopping among the LMOs, which also provides an interpretation consistent with the valence bond model; (4) bonding in transition metal molecules can be interpreted with a model which makes use of polyhedral hybrids and either localized molecular orbitals or valence bond theory -- this should be a useful complement to familiar molecular orbital approaches; (5) although it is yet to be determined whether metal-metal multiple bonds are bent in transition metal molecules using the full GVB model, bent bonds may provide a convenient conceptual framework for interpreting bonding in such molecules.

Finally, if it is not entirely obvious, we should like to make clear that the concepts of bent bonds, resonance, hybrids, etc. have been known for over fifty years. Our contribution has been in testing and evaluating these concepts through computational studies and attempting to synthesize them into a more coherent framework. We believe that having alternative frameworks for discussing electronic structure not only can contribute to the intellectual vitality of the field but also will allow for novel applications and insights.

Acknowledgements

This work was supported in part by the National Science Foundations MRL Program under Grant DMR-8519059 at the Laboratory for Research on the Structure of Matter, University of Pennsylvania and in part by the Office of Naval Research. The authors wish to thank CA Markowski for the preparation of this manuscript.

Literature Cited

(1) Messmer, R. P.; Schultz, P. A.; Tatar, R. C.; Freund, H.-J. *Chem. Phys. Lett.* **1986**, *126*, 176.
(2) Messmer, R. P.; Schultz, P. A. *Phys. Rev. Lett.* **1986**, *57*, 2653.
(3) Schultz, P. A.; Messmer, R. P. *Phys. Rev. Lett.* **1987**, *58*, 2416.
(4) Schultz, P. A. *Ph.D. Thesis*, Physics Department, University of Pennsylvania, 1988.
(5) Schultz, P. A.; Messmer, R. P., to be published.
(6) Ladner, R. C.; Goddard, W. A., III *J. Chem. Phys.* **1969**, *51*, 1073.
(7) Hay, P. J.; Hunt, W. J.; Goddard, W. A., III *J. Am. Chem. Soc.* **1972**, *94*, 8293. Hunt, W.J.; Hay, P.J.; Goddard, W. A., III *J. Chem. Phys.* **1972**, *57*, 738. Bobrowicz, F. W.; Goddard, W. A., III In *Modern Theoretical*

Chemistry; Schaefer, H. F., III, Ed.; Plenum Press: New York, 1977; Vol. 3, Chapter 4.

(8)   Carter, E. A.; Goddard, W. A., III *J. Am. Chem. Soc.* **1988**, *110*, 4077.
(9)   Bauschlicher, C. W., Jr.; Taylor, P. R. *Phys. Rev. Lett.* **1988**, *60*, 859.
(10) Ruedenberg, K.; Schmidt, M. W.; Gilbert, M. M.; Elbert, S. T. *Chem. Phys.* **1982**, *71*, 65.
(11) Messmer, R. P. *J. Mol. Struct. (THEOCHEM)*, **1988**, in press.
(12) Lamson, S. H.; Messmer, R. P., to be published.
(13) Foster, J. M.; Boys, S. F. *Rev. Mod. Phys.* **1960**, *32*, 300.
(14) Edminston, C.; Ruedenberg, K. *Rev. Mod. Phys.* **1963**, *35*, 457.
(15) Wang, H.; Schultz, P. A.; Messmer, R. P. *Bull. Am. Phys. Soc.* **1988**, *33*, 650; to be published.
(16) Messmer, R. P. *Chem. Phys. Lett.* **1986**, *132*, 161.
(17) Patterson, C. H.; Messmer, R. P., to be published.

RECEIVED November 15, 1988

Chapter 15

Transition Metal Diatomic and Monocarbonyl Molecules

An Experimental Viewpoint

W. Weltner, Jr., and R. J. Van Zee

Department of Chemistry, University of Florida, Gainesville, FL 32611

Recent experimental data on the transition-metal diatomics is used to test the "isoelectronic" principle as applied to molecules with the same number of d + s valence electrons. Electron configurations and ground states in the first-row transition-metal monocarbonyl molecules (MCO) are examined on the basis of their CO stretching frequencies, ESR evidence, and theoretical calculations.

Our objective is to discuss two of the simplest classes of transition-metal molecules of mutual interest to theorists and experimentalists and basic to the understanding of clusters. Our desire is to fill in the many gaps among their ground states by extrapolating from the present data.

Reviews of the status of information on the diatomics have been made by us (1), by Shim (2), by Morse (3), by Koutecky and Fantucci (4), and by Salahub (5). Morse's review is recommended here since it considers rather thoroughly both the heteronuclear and homonuclear diatomics. Apparently there is no review of the MCO molecules, and we hope to provide a very brief one here. A more thorough discussion will still be needed of the sophistications in the many theoretical calculations that have been made.

Gas-phase preparation and spectroscopic investigations of these molecules are difficult not only because of the need for reasonable concentrations and low temperature spectra, but also because of the analysis problems. High multiplicities, large masses, large spin-orbit constants, and large nuclear moments contribute to produce low-lying electronic states, second-order spin orbit effects (zero field splittings), small vibrational frequencies, and large hyperfine interactions. Among the diatomic metals only Cr_2 (6-9), CrMo (10), Mo_2 (6, 11), V_2 (12), Fe_2 (13) Ni_2 (14), Cu_2 (15, 16), CuAg (17), CuAu (18) and Ag_2, Au_2 (19) have been investigated in the gas phase. All have been found to have the simpler $^1\Sigma_g^+$ ground state, except V_2, Fe_2, and Ni_2. The ground states of Fe_2 and Ni_2 have not been definitely established, but V_2 demonstrates some of the complications mentioned above in having $^3\Sigma_g^-$ ground state and an

0097–6156/89/0394–0213$06.00/0

extraordinarily large $^3\Sigma_{0^+}$ - $^3\Sigma_{1g}$ spacing of 75 cm^{-1}. Recently Morse, et al., have found NiCu to have a $^2\Delta_{5/2}$ ground state [20] and have also studied Pt$_2$ spectroscopically in the gas phase [21]. (NbO is an example of a "simple" molecule exhibiting many of the complications referred to here; the study by Merer and co-workers represents a real tour de force of spectroscopic analysis [22]). Except for determinations of bond strength there are no gas-phase spectroscopic studies of the MCO molecules.

Almost all of the experimental data on these two classes of molecules has been obtained in rare-gas matrices. It is, in general, a reliable source since these matrices do not produce significant perturbations [23]. Vibrational frequencies and electronic transitions can be observed, but no knowledge of bond distances can be obtained. Most of the electronic and magnetic properties have been obtained from electron-spin-resonance (ESR) spectroscopy of the trapped molecules [24, 25]. Mossbauer spectra have also contributed in the case of Fe-containing species [26, 27].

Because of our reliance here on ESR results, it is perhaps worthwhile to briefly summarize their benefits and limitations: The spectra of randomly-oriented molecules (the general case) in matrices can provide the total spin (S), the g tensor components, the hyperfine tensor components with magnetic nuclei, and sometimes quadrupole coupling constants. Hyperfine splittings can serve to identify the molecule being observed (thus eliminating an objection sometimes made to matrix studies). The spin-rotation contant(s) can be obtained from the g components via Curl's relationship [28]. If linear, only Σ states can be observed since orbital angular momentum so broadens the spectrum as to make it undetectable [29, 24]. Another reason for undetectability, if S $\geqslant$ 1, is a large zero-field splitting parameter (D) relative to X-band quanta; this usually occurs for molecules with large spin-orbit constants and/or through coupling to low-lying electronic states. (V$_2$, with a $^3\Sigma_g$ ground state but D = 75 cm^{-1}, is therefore undetectable in matrix ESR.) These two conditions for not observing an ESR spectrum, even though there is good reason to believe that a magnetic molecule has been matrix isolated [29], will be used as circumstantial evidence supporting some of the ground states to be suggested here.

Transition-Metal Diatomic Molecules
─────────────────────────────────────

Figure 1 shows the array of all possible diatomics derived from the first-row transition metals (except Zn). Ground states are not reliably derived from Mossbauer data [30, 31] so that those given for the seven molecules containing Fe must be considered as suspect. To an earlier version of this Figure [32] we have added a designation of the number of valence 3d + 4s electrons in the isoelectronic molecules occurring in blocks perpendicular to the principal diagonal of homonuclear molecules. Thus, Cr$_2$, VMn, TiFe, and ScCo all have 12 valence electrons. Cr$_2$ is known experimentally (as all the bold borders indicate) to have a $^1\Sigma_g$ ground state. Can we then presume that the other three isoelectronic diatomics also have that ground state? (TiFe and ScCo in that series have been "inferred from experiment" to be $X^1\Sigma$; that will be discussed below.)

Similar groups of isoelectronic molecules are listed in Table I, including those involving Zn.

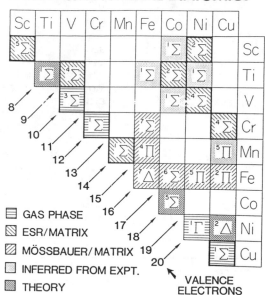

Figure 1. The ground states of possible diatomics formed from
the first-row transition metals, excluding Zn. Those in bold
borders are definitely established, the others are derived as
indicated. "Inferred from experiment" and "valence electrons"
are explained in the text.

Table I. First-Row Transition-Metal Diatomics

Valence Electrons	Experimentally Studied Molecules	Experimental Ground State	Isoelectronic Molecules
6	Sc_2	$^5\Sigma_u^-$	
7			ScTi
8	Ti_2	$^1\Sigma^+$	ScV
9	TiV	$^4\Sigma_g$	ScCr
10	V_2	$^3\Sigma_g^-$	TiCr, ScMn
11			VCr, TiMn, ScFe
12	Cr_2	$^1\Sigma_g^+$	VMn, TiFe‡, ScCo‡
13	TiCo, ScNi	$^2\Sigma_g$	CrMn*, VFe
14	Mn_2	$^1\Sigma^+$	CrFe*, VCo†, TiNi†, ScCu
15	VNi	$^4\Sigma$	MnFe*, CrCo, TiCu, ScZn
16			Fe_2*, MnCo*, CrNi*, VCu, TiZn
17	CrCu	$^4\Sigma$	FeCo*, MnNi*, VZn
18	CrZn	$^7\Sigma$	Co_2*, FeNi*, MnCu*
19			CoNi*, FeCu*, MnZn
20			Ni_2*, CoCu*, FeZn
21	NiCu	$^2\Delta$	CoZn
22	Cu_2	$^1\Sigma_g^+$	NiZn*
23	CuZn	$^2\Sigma_g$	
24	Zn_2	$^1\Sigma_g^+$	

‡ Ground state inferred from experiment - see text.
* Not detected in ESR but believed to have been prepared.

Table II. Mixed-Row Transition-Metal Diatomics

Valence Electrons	Experimentally Studied Molecules	Experimental Ground State	Isovalent Molecules
13	YNi, ScPd, YPd	$^2\Sigma$	ScPt, YPt
15	VPd, NbNi	$^4\Sigma$	VPt, NbPd, NbPt
17	CrAg, CrAu	$^6\Sigma$	
18	MnAg	$^7\Sigma$	MnAu*, CrCd*
19			FeAg*
20			CoAg*
21			NiAg*
22	CuAg, CuAu	$^1\Sigma$	
23	CuCd, CuHg	$^2\Sigma$	
	AgZn, AgCd, AgHg		
	AuZn, AuCd, AuHg		

In Table II our definition of "isoelectronic" has been broadened to just requiring the same number of d + s electrons (regardless of principal quantum numbers), so that mixed-row diatomic molecules can be included, and a more appropriate designation is "isovalent". Table II lists only experimentally-studied molecules and their "isovalent" counterparts, but there are, of course, a large number (465!) of possible transition-metal diatomics. [In these Tables an asterisk (*) indicates a molecule not observed in the ESR but believed to have been prepared (29), while a dagger (†) indicates that the ground state of the molecule was "inferred from experiment", as discussed below.]

Four molecules in column 4 of Table I, TiFe, ScCo with 12 valence electrons and VCo, TiNi with 14 valence electrons (see Figure 1) have been inferred to have $^1\Sigma$ ground state (33). The reasoning is that the addition or subtraction of one electron in the $d\pi^4 d\delta^4 d\sigma^2 s\sigma_1^2 s\sigma_2^1$ configuration of TiCo, ScNi would probably lead to closed shell molecules. ScCu, although also in that category, was considered more doubtful because of the reluctance of Cu to form multiple bonds (however, see below). As Figure 1 indicates, there are no experimental, or theoretical, data on any of these putative $^1\Sigma$ molecules, but the "inferences" are in accord with the known ground states of Cr_2 and Mn_2.

Pursuing the "Isovalent" Principle. The question, admittedly naive, is whether the "isoelectronic" principle of simple MO theory applies among these electronically-complex transition-metal molecules. It seems that exact adherrence to such a rule might be surprising in these cases, but if there is a commonality among the lowest-lying states, it is worth pursuing.

We have already considered two first-row metal series containing 12 and 14 valence electrons where there are hints of adherence to the principle. However, there is a more definite example of "isoelectronic" behavior where the ground states of four 13-electron molecules have been established to be $^2\Sigma$. These molecules are also in the class of what have been referred to, after Gingerich, et al. (34-36), as Brewer-Engel molecules (37); each involves two elements from opposite ends of the periodic table, in this case Groups IIIB and VIII. Such molecules are expected to form strong multiple bonds and therefore to be of low spin. $^{45}ScNi$, $^{45}Sc^{105}Pd$, ^{89}YNi, $^{89}Y^{105}Pd$ not only all have S = 1/2 but the unpaired spin has similar characteristics in the four molecules, as derived from hyperfine interaction with the indicated nuclei (38). The spin is largely (~70%) on the lighter atom and has about 30% s character throughout. Shim and Gingerich (36) have made an all-electron Hartree-Fock calculation for YPd and find a $^2\Delta$ ground state, not in agreement with the ESR result, but $^2\Sigma^+$ and $^2\Pi$ states lie only about 0.2 eV higher in energy. It is likely that improvement in the calculation could change the ordering of these levels. The resulting bonding is unusual (and reminiscent of that in metal carbonyls) in that the d-rich Pd donates electrons to the d-poor Y and Y back donates s electrons. Their similarities indicate that this type of bonding prevails in all four molecules. Extension of the experimental studies to La and Pt would be interesting, and a more extensive ab initio calculation, perhaps on ScNi, would be worthwhile.

More recently in our laboratory, Cheeseman has extended the earlier study of VNi (39) with 15 electrons to VPd, VPt, and NbNi to find that they also have $^4\Sigma$ ground states (40).

"Isoelectronic" Sc_2 ($X^5\Sigma_u^-$) and Y_2 are bothersome, since only the former has an X-band ESR spectrum (41, 42) and theory finds the same ground state for both (43, 44). Walch and Bauschlicher (44) suggest that a more extensive calculation may lower the strongly-bonded $^1\Sigma_g^+$ state below the $^5\Sigma_u^-$ in the case of Y_2, but ESR cannot distinguish between that choice and the possibility of large zfs in a $^5\Sigma_u^-$ ground state.

Surprisingly, the binding of Cu has been puzzling in the two diatomics CrCu and MnCu. CrCu does not appear to have a $^6\Sigma$ ground state, as do CrAg and CrAu, and after considerable agonizing, the spectrum has been interpreted as $^4\Sigma$ (45). Thus it is probably triply bonded with properties intermediate between Cr_2 and Cu_2. An exceptionally large electric field gradient at the Cu nucleus in CrCu supports its anomalous ground state.

The situation in the 18 valence electron series MnCu, MnAg, MnAu is similar but different. Whereas MnAg is easily formed and characterized as a $^7\Sigma$ molecule, MnCu (and MnAu) remains undetected in the ESR (29). Other evidence for a preference for $^7\Sigma$ ground states among this isoelectronic class is that CrZn also has $X^7\Sigma$ (29); however Co_2, even after many trials, was not observed via ESR.

In summarizing this section, one can say that the present meager experimental data are in encouraging support of the application of the "isoelectronic" principle to these diatomics. Both first-row and/or mixed-row diatomics containing 12, 13, 14, 15, and 18 valence electrons show indications of having $^1\Sigma$, $^2\Sigma$, $^1\Sigma$, $^4\Sigma$, and $^7\Sigma$ ground states, respectively. At present there are no definite discrepancies except in the CrCu, CrAg, CrAu series. It is clear that theory has much to contribute here.

Transition-Metal Monocarbonyls, M-CO Molecules

The experimental ground states of most of these molecules are unknown. ESR has established VCO as $^6\Sigma$ and CuCO as $^2\Sigma$, and recently also ScCO as $^4\Sigma$, but it has failed to detect FeCO, which theory indicates has a $^5\Sigma^-$ (or $^3\Sigma^-$) ground state. CrCO($^7\Sigma$) has also apparently been detected via ESR. The present state of affairs is summarized simply in Figure 2, where the origin of a given ground state is indicated as experimental (E), theoretical (T), or suggested (?), usually from ESR evidence. The general scheme of bonding is shown in Figure 3 for VCO (46). The classic Dewar-Chatt-Duncanson model (47,48) involves donation by the CO 3σ orbital into the metal $3d\sigma$ + $4s\sigma$ + $4p\sigma$ orbital to form the 3σ VCO orbital and back donation by the metal $d\pi$ orbitals into the antibonding 2π orbital to form the 2π MO, weakening the CO bond. Experimentally, the effects of the metal-CO interaction are evidenced by lowering of the C-O stretching frequency, ν_{CO}, below 2143 cm^{-1}, the value in the free CO molecule. The trend in the CO stretching frequency varies as shown in Figure 4 for first-row transition metals (49). An unknown frequency in this Figure is that of MnCO (and perhaps ScCO) and it is proposed to be high, i.e., the bonding of Mn to CO to be weak. (This remains to be established by some other experiment.) The variation in ν_{CO}, taken equivalent here to CO bond strength, may also be considered as inversely proportional to the M-C bond strength.

Qualitatively, one can account for the bonding in these MCO molecules by considering the $4s^2 3d^{n-2} \to 4s^1 3d^{n-1}$ promotion energy of

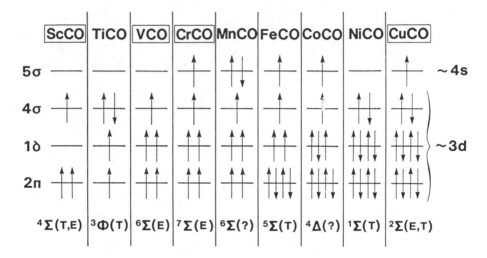

Figure 2.   Molecular orbital schemes and ground states of MCO molecules (M = first-row transition metal).   Boxes here emphasize the only molecules with experimentally established ground states.   Ground states are indicated as: theoretically calculated (T); experimentally determined (E); or suggested (?), based on absence of an ESR spectrum.

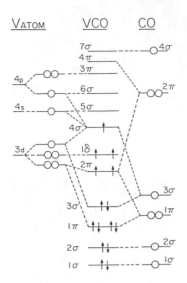

Figure 3.   Molecular orbital scheme for the $^6\Sigma$ VCO molecule (46).

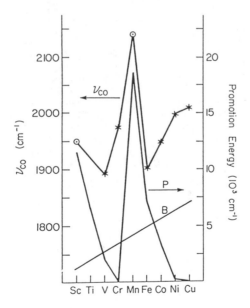

Figure 4.   Plot of the CO stretching frequencies, ν_{CO}, in the first-row transition-metal monocarbonyl molecules MCO (circled points are tentative) (49).   P gives the variation of the atomic energy of promotion corresponding to $4s^2 3d^{n-2} \rightarrow 4s^1 3d^{n-1}$, where n is the number of valence electrons.   Curve B is a crude indication of the expected trend in CO bonding strength due to decreasing $d\pi$ backbonding from Sc to Cu.

the metal atom, the dπ backbonding involved in the DCD theory, and hybridization. The former yields the extreme variations (trace P) shown in Figure 4 and provides a barrier to σ-bond formation. Promotion and/or hybridization of the 4s electrons relieves the antibonding between these metal electrons and the CO 3σ orbital, resulting in a shorter M–C bond and stronger Mdπ back donation. Back donation is expected to generally decrease as one proceeds from left to right because the 3d orbital shrinks in size (44, 50). This effect tends to strengthen the CO bond as one moves across the transition series from left to right, as depicted roughly (trace B) in Figure 4. Then the sum of B + P should have the contours of ν_{CO}, which is approximately true.

What about the ground states? We will consider each of the carbonyls as the 3d shell is filled up, beginning with ScCO.

ScCO. In the IR ScCO was presumably formed since a proper ^{13}CO shift was observed, but the bands were uncomfortably broad (49). Then, although not firm, $\nu_{CO} = 1950$ cm^{-1}, as given in Figure 4, indicating a bound MCO molecule. Jeung and Koutecký (51) have made a MRD SDCI pseudopotential calculation and found a strongly bound $^4\Sigma^-$ lowest state corresponding to the configuration $2\pi^2 4\sigma^1$ shown in Figure 2. A repulsive $^2\Pi$ state is next highest and then a strongly bound $^4\Pi$ state ($2\Pi^1 1\delta^1 4\sigma^1$) about 0.2 eV higher. In these two bound states the Sc 4s electron is polarized away from the CO in the 4σ orbital. Also in both states there is a net electron transfer from Sc to CO.

ESR experiments, after several trials, were successful in detecting ScCO in an argon matrix at 4 K (52). The ground state was found to be $^4\Sigma$, as predicted by theory, with the zero-field splitting parameter $|D| > 1$ cm^{-1}. Hyperfine interaction with the ^{45}Sc nucleus was observed but substitution of ^{13}CO produced no splittings, indicating only a small spin density on the C atom of that ligand. The s character at the Sc atom is estimated to be about 60% from the hyperfine parameters, and the very large $g_\perp$ shift is reasonably accounted for by the low lying $^4\Pi$ state found by Jeung and Koutecký (51).

TiCO. GVB theory (perhaps the oldest theoretical treatment of an MCO molecule) finds the classical σ donation of the CO nonbonding orbital to Ti dσ and delocalization of Ti dπ to acquire CO π* character (53). Nonbonding Ti electrons are polarized away from the CO in 4s-λdp orbitals. In our symbolism the ground state is $^3\Phi$ ($4\sigma^2 2\pi^1 1\delta^1$) where 2π is mainly 4pπ rather than 3dπ. (It is stated that this inducing of 4p character into the valence orbitals should be enhanced in Sc and decrease rapidly in proceeding past V.) Experimentally, ν_{CO} has not been established; in Figure 4 it has been assumed to lie along the line joining ν_{CO} for ScCO and VCO. An ESR spectrum for TiCO was not detected which is in accord with a $^3\Phi$ ground state.

VCO. ^{51}V and ^{13}C hyperfine structure (hfs) in the ESR spectrum established the molecule as VCO and yielded a complete electronic and magnetic picture of a S = 5/2 ground state molecule including g tensor, hf parameters and zero-field-splitting parameter $|D|$ (46). The distribution of the five unpaired electrons is in approximate accord with the $(2\pi)^2(1\delta)^2(4\sigma)^1$ configuration in Figure 3. The σ electron has about 30% s character, the remainder being $3d\sigma$ and $4p\sigma$. The spin density on the CO is quite small. There is the interesting observation of two conformations of VCO; both were observed in argon matrices, one in neon, and the other in krypton. It is reasoned that one of these is bent and that it is most likely the gas-phase form. Earlier infrared studies by Hanlan, et al. (54), had suggested that VCO might be bent, and extended Hückel theory (assuming low spin) indicated that there was a monotonic decrease in energy of the molecule as the angle decreased. Thus, an ab initio calculation would be of real interest. A strong M–CO bond is indicated by a decrease of 240 cm^{-1} (in solid argon) in ν_{CO} (54) (see Figure 4).

CrCO. The CO stretching frequency was observed at 1977 cm^{-1} (49), but there appear to be no other data on CrCO. Unpublished ESR spectra in our laboratory showed uncharacteristic very broad lines when chromium metal was trapped in an argon matrix containing CO in various concentrations. A tentative analysis indicated a $^7\Sigma$ molecule, the reason for the extraordinary breadth of the lines was not understood. In Figure 2, the ground state is tentatively given as $^7\Sigma$ $(2\pi^21\delta^24\sigma^15\sigma^1)$, in agreement with the addition of one electron to the known VCO configuration.

MnCO. Two sets of authors agreed on the infrared spectrum of Mn + CO in matrices (55,49), but the firm identification of a MnCO signal was in doubt. It was suggested by us that the molecule is essentially nonbonded and therefore that $\nu_{CO} \cong 2140$ cm^{-1}. This was rationalized, as indicated in Figure 4, by the high promotion energy of the Mn atom, implying that the 4s electrons provide a repulsive interaction with CO. There is evidence of the existence of MnCO in the gas phase, at least with a lifetime long enough to allow ionization to form MnCO$^+$, if the suggested mechanism for its preparation is correct. That ion and Mn$^+$, Mn$_2^+$ were observed during photofragmentation of Mn$_2$(CO)$_{10}$ and believed to be produced by multiphoton ionization of the neutral MnCO (and Mn, Mn$_2$) photoproducts (56). The signal shows first-order dependence upon the sample pressure. (Note that Mn$_2^+$ is proposed to be formed from Mn$_2$, which is a van der Waals molecule as proposed for MnCO.)

FeCO. Experimentally, IR in matrices finds $\nu_{CO} \cong 1898$ cm^{-1} (57,58), indicating a relatively strong Fe–C bond, but Engelking and Lineberger (59) estimate the bond energy as 1.0 ± 0.3 eV. Also the Mössbauer isomer shift of -0.60 mm/s is close to that of the free atom, -0.75 mm/s (59). Theory has not quite decided whether the ground state is $^3\Sigma^-$ or $^5\Sigma^-$ and is finding it difficult to account for the isomer shift (50,60-65). Thus, Guenzberger, et al. (62), employing the discrete variational method with the Xα local approximation, place the δ orbital below the σ and, on an aufbau basis, obtain a $d\pi^4d\delta^24\sigma^2$, $^3\Sigma^-$ ground state. The calculated isomer shift is -0.12 mm/s, in disagreement with experiment. A recent calculation by Marathe, et al. (65) (using SCF + MP4SDTQ) derived a spin-quintet $(d\pi^4d\delta^24\sigma^15\sigma^1$, $^5\Sigma^-)$ ground state rather than a spin-triplet. $^5\Pi$ states (such as $\pi^3\delta^24\sigma^25\sigma$) are calculated to lie at least 1.4 eV higher in energy. However, with the $^5\Sigma^-$ ground state the calculated isomer shift is -0.11 mm/s, still in poor agreement with experiment, although the calculated C-O stretching force constant is in good agreement with that derived from the experimental frequency. It is suggested that perhaps the matrix effects are large, which is not consistent with the vibrational results. However, another recent calculation by Daoudi, et al. (66) [using SCF with (CIPSI)] finds the $^3\Sigma^-$ state to be lower than the $^5\Sigma^-$ state by 0.4 eV, the Fe-C bond energy to be 1.34 eV and ν_{CO} = 1986 cm^{-1}. It is disappointing that an ESR spectrum for FeCO was not observed since it could resolve the multiplicity problem in the ground state, but if Σ states are lowest, a large zero-field-splitting is implied.

CoCO. ν_{CO} in this molecule has been determined to be 1953 cm^{-1} in solid argon, but its ESR spectrum was not detected at the time even though there was no question that Co atoms and CO were present in the matrix (67). It is probable that the molecule contains at least one unpaired electron whether one reasons from FeCO or NiCO, so that its undetectability in powder ESR spectra is due either to orbital angular momentum or large zfs in the ground state. The former seems more likely so that the $^4\Delta$ state suggested in Figure 2 is reasonable.

NiCO. The dissociation energy has been determined but with rather large uncertainty, 29 ± 15 kcal/mole (68, see also 69), and the CO stretching frequency found to be 1996 cm^{-1} (70,68). The electron affinity has been measured (68). That is essentially the extent of the experimental data except for observation of NiCO over the nickel surface under special conditions (71,72). Early theory predicted a $^3\Delta$ ground state, (73,74,75-81) but Rives and Fenske (82) were the first to show, using a many configuration wavefunction, that the $^1\Sigma$ state is slightly lower (0.15 eV) than the $^3\Delta$, but the bond distance is much shorter in the singlet (1.70 Å) and the binding energy (2.7 eV) and Ni-C vibrational frequency (505 cm^{-1}) much larger. The most recent theoretical studies (83,84,50,85,86-96) have indicated that π bonding is much more important than σ bonding, which is in fact repulsive. This is "softened" by sdσ hybridization. The nickel atom is close to d^9 rather than d^{10}, expected for zero valent Ni because the promotion to the $^1S(d^{10})$ state is energetic whereas the sd^9 state is almost degenerate with the ground state.

CuCO. Kasai and Jones (97) proved that CuCO has a $^2\Sigma$ ground state and observed Cu and ^{13}C hfs. The CO stretching frequency is 2010 cm^{-1} in an argon matrix (98-100). Theory does not predict a bound complex (50,73,74,83-85,101-103) or at least only a "possible weak van der Waals interaction for the $^2\Sigma^+$ state" (104). However, the ESR hf data indicate that the spin density distribution is $\rho(4S)_{Cu}$ = +0.67, $\rho(4p\sigma)_{Cu}$ = +0.08, $\rho(2s\sigma)_C$ = +0.05 with the appreciable fraction remaining probably in the C($2p\sigma$) orbital. Thus, although weakly bound, the spin on Cu is polarized, and more important, perhaps 20% of the spin is on the CO ligand.

Acknowledgments

The authors wish to thank their co-workers who have contributed to the topics discussed here: S. B. H. Bach, C. A. Baumann, M.

Cheeseman, C. A. Taylor, R. L. DeKock, L. B. Knight, Jr., and M. T.
Vala. This research was supported by National Science Foundation
Grant CHE 8514585.

Literature Cited

1. Weltner, W., Jr.; Van Zee, R. J. Ann. Rev. Phys. Chem. 1984,
    35, 291.
2. Shim, I. Mat.-Fyx. Meddr. Danske Vidensk. Salsk. (16 Res.
    Rep. of Niels Bohr Fellows) 1985, 41, 147.
3. Morse, M. D. Chem. Rev. 1986, 86, 1049-1109.
4. Koutecky, J.; Fantucci, P. Chem. Rev. 1986, 86, 539.
5. Salahub, D. R. Adv. Chem. Phys. 1987, 69, 447-520.
6. Efremov, Y. M.; Samoilova, A. N.; Kozhikhovsky, V. B.;
    Gurvich, L. V. J. Mol. Spectrosc. 1978, 73, 430-40.
7. Michalopoulos, D. L.; Geusic, M. E.; Hansen, S. G.; Powers, D.
    E., Smalley, R. E. J. Phys. Chem. 1982, 86, 3914-16.
8. Riley, S. J.; Parks, E. K.; Pobo, L. G.; Wexler, S. J. Chem.
    Phys. 1983, 79, 2577-82.
9. Bondybey, V. E.; English, J. H. Chem. Phys. Lett. 1983, 94,
    443-47.
10. Efremov, Y. M.; Samoilova, A. N.; Gurvich, L. V. Chem. Phys.
    Lett. 1976, 44, 108.
11. Hopkins, J. B.; Langridge-Smith, P. R. R.; Morse, M. D.;
    Smalley, R. E. J. Chem. Phys. 1983, 78, 1627-37.
12. Langridge-Smith, P. R. R.; Morse, M. D.; Hansen, G. P.;
    Smalley, R. E.; Merer, A. J. J. Chem. Phys. 1984, 80, 593-
    600.
13. Leopold, D. G.; Almlöf, J.; Lineberger, W. C.; Taylor, P. R.
    J. Chem. Phys. 1988, 88, 3780.
14. Morse, M. D.; Hansen, G. P.; Langridge-Smith, P. R. R.; Zheng,
    L. S.; Geusic, M. E.; Michalopoulos, D. L.; Smalley, R. E. J.
    Chem. Phys. 1984, 80, 5400.
15. Åslund, N.; Barrow, R. F.; Richards, W. G.; Travis, D. N.
    Ark. Fys. 1965, 30, 171-85.
16. Lochet, J. J. Phys. B. 1978, 11, L55-L57.
17. Ruamps, J. C. Spectrochim. Acta, 1957, Suppl. 11, 329.
    Joshi, K. C.; Majumdar, K. Proc. Phys. Soc. London 1961, 78,
    197.
18. Ruamps, J. C. C. R. Hebd. Seances Acad. Sci. 1954, 239,
    1200.
19. Numerous studies, see reference 3.
20. Fu, Z.; Morse, M. D. J. Chem. Phys. (submitted).
21. Taylor, S.; Lemire, G. W.; Hamrick, Y. M.; Fu, Z.; Morse, M.
    D. J. Chem. Phys. (submitted).
22. Femenias, J. L.; Cheval, G.; Merer, A. J.; Sassenberg, U. J.
    Mol. Spectrosc. 1987, 124, 348-68.
23. See, for example, Jacox, M. E. J. Mol. Spectrosc. 1985, 113,
    286-301.
24. Weltner, W., Jr. Magnetic Atoms and Molecules; Van Nostrand
    Reinhold: New York, 1983.
25. Kasai, P. H.; McLeod, D., Jr. Faraday Symp. Chem. Soc. 1980,
    14, 65-78.
26. McNab, T. K.; Micklitz, H.; Barrett, P. H. Phys. Rev. B
    1971, 4, 3787-97.

27. Montano, P. A. Faraday Symp. Chem. Soc. 1980, 14, 79-86.
28. Curl, R. F. Mol. Phys. 1965, 9, 585.
29. Baumann, C. A.; Van Zee, R. J.; Weltner, W., Jr. J. Phys.
     Chem. 1984, 88, 1815-1820.
30. Guenzburger, D.; Saitovich, E. M. B. Phys. Rev. B 1981, 24,
     2368-79.
31. Nagarthna, H. M.; Montano, P. A.; Naik., V. M. J. Am. Chem.
     Soc. 1983, 105, 2938-43.
32. Weltner, W., Jr. In Comparison of Ab Initio Quantum Chemistry
     with Experiment; Bartlett, R. J., Ed.; Reidel, Dordrecht,
     1985; pages 1-16.
33. Van Zee, R. J.; Weltner, W., Jr. High Temp. Sci. 1984, 17,
     181-191.
34. Cocke, D. L.; Gingerich, K. A. J. Chem. Phys. 1974, 60,
     1958.
35. Gingerich, K. A. Faraday Symp. Chem. Soc. 1980, 14, 109.
36. Shim, I.; Gingerich, K. A. Chem. Phys. Lett. 1983, 101, 528.
37. Brewer, L. Science 1968, 161, 115.
38. Van Zee, R. J.; Weltner, W., Jr. Chem. Phys. Lett. (in
     press).
39. Van Zee, R. J.; Weltner, W. Jr. Chem. Phys. Lett. 1984, 107,
     173.
40. Cheeseman, M.; Van Zee, R. J.; Weltner, W., Jr. (to be
     published).
41. Knight, L. B., Jr.; Van Zee, R. J.; Weltner, W., Jr. Chem.
     Phys. Lett. 1983, 94, 296-99.
42. Knight, L. B. Jr.; Woodward, R. W.; Van Zee, R. J.; Weltner,
     W., Jr. J. Chem. Phys. 1983, 79, 5820.
43. Walch, S. P.; Bauschlicher, C. W., Jr. J. Chem. Phys. 1983,
     79, 3590-91.
44. Walch, S. P.; Bauschlicher, C. W., Jr. In Comparison of Ab
     Initio Quantum Chemistry with Experment; Bartlett, R. J. Ed.;
     Reidel, Dordrecht, 1985; pages 17-51.
45. Baumann, C. A.; Van Zee, R. J.; Weltner, W., Jr. J. Chem.
     Phys. 1983, 79, 5272-79.
46. Van Zee, R. J.; Bach, S. B. H.; Weltner, W., Jr. J. Phys.
     Chem. 1986, 90, 583-88.
47. Dewar, M. J. S. Bull. Soc. Chem. Fr. 1951, 18C, 79.
48. Chatt, J; Duncanson, L. A. J. Chem. Soc. 1953, 2939.
49. Bach, S. B. H.; Taylor, C. A.; Van Zee, R. J.; Vala, M. T.;
     Weltner, W., Jr. J. Am. Chem. Soc. 1986, 108, 7104.
50. Bauschlicher, C. W., Jr.; Bagus, P. S.; Nelin, C. J.; Roos, B.
     O. J. Chem. Phys. 1986, 85, 354-364.
51. Jeung, G. H.; Koutecký, J. Chem. Phys. Lett. 1986, 129, 569-
     76.
52. Van Zee, R. J.; Cheeseman, M.; Weltner, W., Jr. (to be
     published).
53. Mortola, A. P.; Goddard, W. A., III J. Am. Chem. Soc. 1974,
     96, 1-10.
54. Hanlan, L.; Huber, H.; Ozin, G. A. Inorg. Chem. 1976, 15,
     2592-97.
55. Huber, H.; Kundig, E. P.; Ozin, G. A.; Poe, A. J. J. Am.
     Chem. Soc. 1975, 97, 308-14.
56. Lichtin, D. A.; Bernstein, R. B.; Vaida, V. J. Am. Chem.
     Soc. 1982, 104, 1830-34.

57. Moskovits, M.; Ozin, G. A. Cryochemistry (eds. Moskovits and
    Ozin, Wiley, New York, 1976) pages 261-439.
58. Peden, C. H. F.; Parker, S. F.; Barrett, P. H.; Pearson, R.
    G. J. Phys. Chem. 1983, 87, 2329-36.
59. Engelking, P. C.; Lineberger, W. C. J. Am. Chem. Soc. 1979,
    101, 5569-73.
60. Bursten, B. D.; Freier, D. G.; Fenske, R. F. Inorg. Chem.
    1980, 19, 1810-11.
61. Barbier, C.; Del Re, G. Folia Chim. Theor. Lat. 1984, 12,
    27-44.
62. Guenzburger, D.; Saitovitch, E. M. B.; De Paoli, M. A.;
    Manela, H. J. Chem. Phys., 1984, 80, 735-744.
63. Daoudi, A.; Suard, M.; Barbier, C. C. R. Acad. Sc. Paris,
    Ser. II 1985, 301, 911-914.
64. Sawaryn, A.; Aldridge, L. P.; Blaes, R.; Marathe, V. R.;
    Trautwein, A. X. Hyperfine Interact. 1986, 29, 1303-6.
65. Marathe, V. R.; Sawaryn, A.; Trautwein, A. X.; Dolg., M.;
    Igel-Mann, G.; Stoll, H. Hyperfine Interact. 1987, 36, 39-
    58.
66. Daoudi, A.; Suard, M.; Barbier, C. J. chim. phys. 1987, 84,
    795-798.
67. Hanlan, L. A.; Huber, H.; Kuendig, E. P.; McGarvey, B. R.;
    Ozin, G. A. J. Am. Chem. Soc. 1975, 97, 7054-68.
68. Stevens, A. E.; Feigerle, C. S.; Lineberger, W. C. J. Am.
    Chem. Soc. 1982, 104, 5026-31.
69. Blomberg, M.; Brandemark, U.; Johansson, J.; Siegbahn, P.;
    Wennerberg, J. J. Chem. Phys. 1988, 88, 4324-33.
70. DeKock, R. L. Inorg. Chem. 1971, 10, 1205-11.
71. Garrison, B. J.; Winograd, N.; Harrison, D. E., Jr. J. Vac.
    Sci. Technol. 1979, 16, 789-92.
72. Liang, D. B.; Abend, G.; Block, J. H.; Kruse, N. Surf.
    Sci., 1983, 126, 392-6.
73. Itoh, H.; Kunz, A. B. Z. Naturforsch. A 1979, 34, 114-116.
74. Itoh, H.; Kunz, A. B. Fundament. Res. Homogeneous Catal.
    1979, 3, 73-82.
75. Cederbaum, L. S.; Domcke, W.; Von Niessen, W.; Brenig, W. Z.
    Phys. B 1975, 21, 381-88.
76. Walch, S. P.; Goddard, W. A., III J. Am. Chem. Soc. 1976,
    98, 7908-17.
77. Clark, D. T.; Cromarty, B. J.; Sgamellotti, A Chem. Phys.
    Lett. 1978, 55, 482-87.
78. Bullett, D. W.; O'Reilly, E. P. Surf. Sci. 1979, 89, 274-81.
79. Bagus, P. S.; Hermann, K. Surf. Sci. 1979, 89, 588-95.
80. Rosen, A.; Grundevik, P.; Morovic, T. Surf. Sci. 1980, 95,
    477-95.
81. Rives, A. B.; Weinhold, F. Int. J. Quantum Chem., Quantum
    Chem. Symp. 1980, 14, 201-209.
82. Rives, A. B.; Fenske, R. F. J. Chem. Phys. 1981, 75, 1293-
    1302.
83. McIntosh, D. F.; Ozin, G. A.; Messmer, R. P. Inorg. Chem.
    1981, 20, 3640-50.
84. Ha, T.-K.; Nguyen, M. T. J. Mol. Struct. (Theochem) 1984,
    109, 331-338.
85. Bauschlicher, C. W., Jr. J. Chem. Phys. 1986, 84, 260-267.

86. Saddei, D.; Freund, H. J.; Hohlmeicher, G. Chem. Phys. 1981, 55, 339-54.
87. Howard, I. A.; Pratt, G. W.; Johnson, K. H.; Dresselhaus, G. J. Chem. Phys. 1981, 74, 3415-19.
88. Bagus, P. S.; Roos, B. O. J. Chem. Phys. 1981, 75, 5961-62.
89. Dunlap, B. I.; Yu, H. L.; Antoniewicz, P. R. Phys. Rev. A 1982, 25, 7-13.
90. You, X. Jiegou Huaxue 1983, 2, 183-188.
91. Huzinaga, S.; Klobukowski, M.; Sakai, Y. J. Phys. Chem. 1984, 88, 4880-86.
92. Kao, C. M.; Messmer, R. P. Phys. Rev. B 1985, 31, 4835-47.
93. Bauschlicher, C. W. Chem. Phys. Lett. 1985, 115, 387-391.
94. Rohlfing, C. M.; Hay, P. J. J. Chem. Phys. 1985, 83, 4641-49.
95. Madhavan, P. V.; Whitten, J. L. Chem. Phys. Lett. 1986, 127, 354-359.
96. Carsky, P.; Dedieu, A. Chem. Phys. 1986, 103, 265-75.
97. Kasai, P. H.; Jones, P. M. J. Am. Chem. Soc. 1985, 107, 813-18.
98. Huber, H.; Kuendig, E. P.; Moskovits, M.; Ozin, G. A. J. Am. Chem. Soc. 1975, 97, 2097-2106.
99. Ozin, G. A.; VanderVoet, A. Prog. Inorg. Chem. 1975, 19, 105-172.
100. Moskovits, M.; Ozin, G. A. Vibrational Spectra and Structure (Durig, J., Ed. Elsevier, Amsterdam 1975).
101. Bagus, P. S.; Hermann, K.; Seel, M. J. Vac. Sci. Technol. 1981, 18, 435-452.
102. Bagus, P. S.; Nelin, C. J.; Bauschlicher, C. W., Jr. J. Vac. Sci. Technol. 1984, A2, 905-909.
103. Kuźminskii, M. B.; Bagatuŕyants, A. A.; Kazanskii, V. B. Izv. Akad. Nauk. SSSR, Ser. Khim. 1986 <2>, 284-8.
104. Merchán, M.; Nebot-Gil, I.; González-Luque, R.; Ortí, E. J. Chem. Phys. 1987, 87, 1690-1700.

RECEIVED December 9, 1988

Chapter 16

Spin Density Functional Approach to the Chemistry of Transition Metal Clusters

Gaussian-Type Orbital Implementation

J. Andzelm[1], E. Wimmer[1], and Dennis R. Salahub[2]

[1]Cray Research, Inc., 1333 Northland Drive, Mendota Heights, MN 55120
[2]Département de Chimie, Université de Montréal, Montréal, Québec H3C 3J7, Canada

In this contribution we review the accuracy and computational efficiency of the local spin density functional (LSDF) approach using linear combinations of Gaussian-type orbitals (LCGTO) for the calculation of the electronic structures, ground-state geometries, and vibrational properties of transition metal compounds and clusters. Specifically this is demonstrated for (1) bis(π-allyl) nickel where this approach gives an excellent qualitative and quantitative interpretation of the observed photoemission spectrum; (2) chemisorption of C atoms on a Ni(100) surface, where the present computational approach determines the adsorption site of C as the four fold-hollow position above the surface with a calculated C-Ni bond length of 1.79 Å (exp.: 1.75 ± 0.05 Å) and a vibrational frequency of 407 cm^{-1}, (exp.: 410 cm^{-1}); (3) vibrational frequencies of CO on Pd: the calculations reveal that inclusion of surface/subsurface Pd-Pd motions couple significantly to the CO-Pd vibration leading to a reduction of the vibrational frequency by about 20% compared with a rigid substrate model. Inclusion of an external electrical field shows a stiffening of the C-O vibration with increased positive potential of the electrode; and (4) the electronic structure of Zn clusters where it is found that properties such as the first (s-electron) ionization potential converge rather slowly towards the value of the extended system requiring at least 20 transition metal atoms for an accurate description of the surface electronic structure. It is demonstrated that the computation of three-index two-electron integrals can be achieved with a highly efficient vector/parallel algorithm based on recursive integral formulas recently published by Obara and Saika. Furthermore, we present the theoretical framework for LCGTO-LSDF gradient calculations.

0097–6156/89/0394–0228$06.00/0

There is growing evidence that local spin density functional (LSDF) theory (1) provides a unified theoretical framework for the study of electronic, geometric, and vibrational structures of solids, surfaces, interfaces, clusters and molecular systems (2,3) encompassing metallic as well as covalent bonding. Although widely used in solid state physics, including semiconductors, transition metals, lanthanides and actinides, density functional methods are still rather rarely applied to problems in chemistry. One of the reasons appears to be the lack of experience with this method in addressing typical chemical questions such as molecular conformations, vibrations, and reactivity. Such investigations require the capability to calculate accurate analytical derivatives of the total energy with respect to displacements of the atomic nuclei. Evidently, the potential of LSDF gradient calculations has not yet been fully developed. However, if such a goal could be achieved for molecules and large clusters including transition metals, one would not only have an additional theoretical/computational tool to investigate large molecules with first and second row atoms (which is the current domain of Hartree-Fock calculations) but one could also study organometallic compounds, investigate reactions on metallic surfaces, and simulate large and complex systems such as zeolites and enzyme catalysts at an unprecedented level of detail.

In this paper, we present results obtained with a particular molecular orbital implementation of local spin density functional theory using a linear combination of Gaussian-type orbitals (LCGTO's). The results, derived from single-point total energy calculations, illustrate the applicability of the LCGTO-LSDF approach to questions of electronic structure, vibrational properties, and geometries of transition metal clusters and compounds and, in addition, shed light on the computational issues encountered in this kind of large-scale molecular simulation. Furthermore, we demonstrate that the LCGTO implementation allows a compact analytical formulation of energy gradients thus setting the stage for future exploitations of this chemically important feature.

The paper is organized in the following way. First the key features of the LCGTO-LSDF method are reviewed (3); for a more detailed description of LSDF calculations, the reader is referred to recent reviews (2,3). Four examples, discussed next, highlight the performance of the LCGTO-LSDF approach to predict (1) the electronic structure (photoelectron spectrum) of a transition metal complex, bis(π-allyl) nickel; (2) chemisorption geometries of carbon atoms on the Ni(100) surface; (3) the vibrational properties of CO on Pd including the influence of an external electric field; and (4) electronic properties of Zn clusters as a function of cluster size. The last example also illustrates the dependence of computational requirements on the size of the system.

The Linear Combination of Gaussian-Type Orbitals (LCGTO) Implementation

Basis Sets. Since the suggestion of Boys (4), Gaussian-type basis functions have become the standard in quantum chemical ab initio methods. In the Hartree-Fock theory the occurence of four-center two-electron integrals makes this choice a computational necessity. In local density functional theory, on the other hand, a conceptually simpler Hamiltonian gives greater freedom in selecting the variational basis set. For example, in many solid state calculations it has become common practice to use plane waves and augmented plane waves (5) with numerically generated radial functions in a linearized form (6). An elegant alternative consists in the use of numerically generated atomic orbitals as basis for molecular orbitals (7). Both approaches using numerical basis functions, the solid state linearized augmented plane wave (LAPW) method (6,8) and the molecular/cluster approach (7,9) have proven extremely useful in carrying out high-precision local spin density functional calculations for solids, surfaces, clusters, and molecules (2). On the other hand, the accurate evaluation of analytic energy derivatives within these implementations turns out to be a considerable challenge (10).

In the present studies, we use Gaussian-type basis functions for the molecular orbitals to solve the local density functional equations (11). This choice offers several advantages: (i) there is a wealth of experience from numerous ab initio Hartree-Fock calculations in using GTO's for molecular calculations (12), (ii) computationally, this approach can be implemented in a highly efficient way, as will be discussed below, (iii) the analytic nature of the basis functions opens the possibility for accurate analytic calculations of energy gradients for geometry optimizations and density gradients for non-local corrections (13), and (iv) effective core potentials or model potentials can be readily incorporated (14).

Besides the Gaussian basis set for the wavefunctions, there are two other sets of Gaussian expansions used in the present approach, one for the electron density and one for the exchange-correlation potential. The expansion of the electron density is used in the evaluation of Coulomb integrals. Hence the expansion coefficients of the electron density are chosen (11b) such as to minimize the error in the Coulomb energy arising from the difference between the "exact" electron density (i.e. the density originating directly from the wavefunctions) and the fitted electron density. All necessary steps to obtain the expansion coefficients of the electron density can be carried out analytically. On the other hand, the expansion coefficients for the exchange-correlation potential have to be obtained numerically by generating the values of the exchange-correlation potential on a grid, which are then used to fit a Gaussian expansion. After this numerical step, the matrix elements of the exchange-correlation potential operator are calculated analytically.

Integral Evaluation. In contrast to Hartree-Fock methods, the LCGTO-LSDF approach requires evaluation of only three-index integrals, thus representing an N^3 algorithm (with N being the number of basis functions). All examples discussed below, except the Zn clusters, were calculated using the Hermite Gaussian basis originally implemented by Sambe and Felton (11a) and further developed by Dunlap, Connolly and Sabin (11b). In contrast, Cartesian, not Hermite, Gaussians are the most widely used choice in ab initio quantum chemistry. The scheme of recursive computation of four-index cartesian Gaussian integrals, originally developed by Obara and Saika (15) for the Hartree Fock method, has now been reformulated for the three-index integrals needed in the present method. As shown below, a computationally highly efficient scheme results from this approach.

Two kinds of three-index integrals are needed, Coulomb integrals of the form (adopting the notation of Obara and Saika (15))

$$I_C = [a(1)b(1)\|c(2)] \tag{1}$$

where "$\|$" refers to the $1/r_{12}$ operator, and overlap-like integrals to calculate exchange-correlation potentials and energies of the form

$$I_{xc} = [a(1)b(1)c(1)] \tag{2}$$

Here a and b stand for orbital basis functions and c denotes Gaussian functions used in the fitting of the electron density or the exchange-correlation potential. Most of the time is spent in the computation of I_C. We can rewrite the original Obara and Saika formula (15) in a form suitable for computation of three-index integrals:

$$[ab||(c+1_i)]^{(m)} = (W_i - Q_i) \, [ab||c]^{(m+1)}$$
$$+ \tfrac{1}{2} \, \eta \, N_i(c) \, \{[ab||(c-1_i)]^{(m)} - (\rho/\eta) \, [ab||(c-1_i)]^{(m+1)}\}$$
$$+ \tfrac{1}{2}(\zeta+\eta)) \, \{N_i(a) \, [(a-1_i)b||c]^{(m)} + N_i(b) \, [a(b-1_i)||c]^{(m+1)}\} \qquad (3)$$

Here, 1_i is a short-hand notation for a p_x, p_y or p_z function and function $c+1_i$ has an angular momentum one order higher than c. The recursive nature of Equation 3 allows to build, for example, integrals with d-functions from integrals with only s- and p- functions. The superscript (m) refers to the order of the incomplete Γ function. W_i and Q_i are related to geometries and values of exponents of Gaussian functions; η, ζ and ρ depend on these exponents and $N_i(c)$ is a generalized Kronecker delta. For details the reader is referred to the original paper by Obara and Saika (15), Equation 39.

Expansion of the integral with a single function c, instead of a (or b) already cuts the number of arithmetic operations by as much as 30% for d-type integrals. Substantial savings in the time of integral evaluation occur if we calculate entire groups of integrals with shared exponents in various symmetries. Particularly, we can design basis sets for the electron density and exchange-correlation potential with shared exponents without loosing accuracy in the fitting process (16,17). Hegarty and van der Velde (18) analyzed the number of arithmetic operations necessary to calculate 4-index integrals. The best algorithm (18) requires about 22 operations per integral with d functions. In contrast, the present method requires about 8 operations per 3-index integral. The new formulas for the integral calculations can be efficiently programmed on a vector computer and the algorithm is amenable to parallelization as will be shown below.

As for Hartree-Fock calculations, storage of integrals becomes the computational bottleneck for systems with a large number of basis functions. In this case, a "direct SCF" (19) scheme can be adopted (20). For the LCGTO-LSDF method we deal with a smaller number of integrals (of the order N^3 rather than N^4 as in Hartree-Fock calculations) and therefore we may use the standard approach for up to about N=1000, as discussed below for the case of large Zn clusters. There is an additional advantage in the direct scheme as we have to calculate the full Hamiltonian matrix only once and then, in each iteration, add the changes to the matrix which correspond to modifications in the density and the exchange-correlation potential, but only for those matrix elements where this change is greater then a threshhold. During the course of iterations towards self-consistency, a smaller and smaller number of matrix elements needs to be updated.

Gradients. The calculation of energy gradients within the LSDF method using localized basis sets has been investigated by a number of researchers (10,21). However, there has been no published formulation for the case of the LCGTO-LSDF method. We will now give an outline of the formulas that have recently been implemented (22). Full details will be published in due course. In this method, both Coulomb and exchange-correlation energies are calculated analytically once the fitting coefficients for the electron density, ρ_r with $\rho = \Sigma_r \rho_r g_r$, and exchange-correlation energy-density, ε_s, and potential, μ_s, are obtained (11). The total energy (E_{LSDF}) can then be expressed as

$$E_{LSDF} = \Sigma_{pq} P_{pq} \{ h_{pq} + \Sigma_r \rho_r \, [pq||r] + \Sigma_s \varepsilon_s \, [pqs] \} - \tfrac{1}{2} \Sigma_{rt} \rho_r \rho_t \, [r||t] + U_n \qquad (4)$$

Here P denotes the density matrix, h contains kinetic energy and electron-nuclear attraction operators; [pq||r] and [r||t] are Coulomb repulsion integrals with 3 and 2 indices, respectively, [pqs] denote one-electron 3-index integrals and U_n is the nuclear-nuclear repulsion term. The form of Equation 4 ensures (11b) that the Coulomb energies are accurate up to second order in the difference between the fitted density and the "exact" density obtained directly from the wavefunction.

In order to obtain the energy gradient of E_{LSDF} by differentiation of Equation 4 with respect to a nuclear coordinate, x, one can, in the first step, closely follow the procedure used in Hartree-Fock theory (23). In addition, derivatives of the density fitting coefficients occur in the present approach. These terms can be eliminated using the equation for the density fitting together with the normalization condition of the total density. At this point, the intermediate gradient formula is given by

$$\partial E_{LSDF}/\partial x = \Sigma_{pq} P_{pq} \{ \partial h_{pq}/\partial x + \Sigma_r \rho_r \partial[pq\|r]/\partial x + \Sigma_s \varepsilon_s \partial[pqs]/\partial x \}$$
$$- \tfrac{1}{2} \Sigma_{rt} \rho_r \rho_t \partial[r\|t]/\partial x + \partial U_n/\partial x - \Sigma_{pq} W_{pq} \partial[pq]/\partial x$$
$$+ \Sigma_{pq} \partial P_{pq}/\partial x \{ \Sigma_s (\varepsilon_s - \mu_s) [pqs] \} + \Sigma_{pq} P_{pq} \Sigma_s \partial \varepsilon_s/\partial x [pqs] \qquad (5)$$

Here, W_{pq} is an energy-weighted density matrix element as in the Hartree-Fock gradient formula (23). Equation 5 contains two "difficult" terms (the last two), the derivative of density matrix elements and the fitting coefficients ε_s. It turns out that these terms can be eliminated by using the relationship

$$\partial \rho/\partial x (\mu_{xc} - \varepsilon_{xc}) = \rho \ \partial \varepsilon_{xc}/\partial x \qquad (6)$$

which can be obtained by differentiation of the fundamental LSDF formula (1)

$$\mu_{xc} = \partial (\varepsilon_{xc} \rho) / \partial \rho = \varepsilon_{xc} + \rho \ \partial \varepsilon_{xc} / \partial \rho \qquad (7)$$

after multiplying both sides by ρ. In practice, μ_{xc} and ε_{xc} are obtained through a fitting procedure. The fitted quantities do not correspond strictly to the original density and eq. (6) no longer holds exactly. However, with the fitting basis sets currently used, this approximation appears to be reasonable (22). Assuming therefore the validity of Equation 6, we obtain the following LSDF energy gradient formula which is valid also for the spin-polarized case and in the presence of non-local corrections to the Hamiltonian (20))

$$\partial E_{LSDF}/\partial x = F_{HFB} + F_D \qquad (8)$$

with

$$F_{HFB} = \Sigma_{pq} P_{pq} \{ \partial h_{pq}/\partial x + \Sigma_r \rho_r [\partial(pq)/\partial x \| r] + \Sigma_s \mu_s [\partial(pq)/\partial x \ s] \}$$
$$+ \partial U_n/\partial x - \Sigma_{pq} W_{pq} \partial[pq]/\partial x \qquad (9)$$

and

$$F_D = \Sigma_r \rho_r [\partial(r)/\partial x \| \Sigma_{pq} P_{pq} \ p \ q - \Sigma_t P_t \ t] \qquad (10)$$

F_{HFB} is the Hellman-Feynman force plus the correction for the orbital basis set dependence on the nuclear coordinate, x. The term F_D is specific to the present LCGTO-LSDF implementation. Equation 10 is equivalent to

$$F_D = \Sigma_r \rho_r [\partial(r)/\partial x \| (\rho - \rho^f)] \qquad (11)$$

with ρ being the "exact" density and ρ^f the fitted density. Clearly, in the case of a perfect fit F_D vanishes. It is important to realize that within the LCGTO implementation, evaluation of LSDF gradients boils down to computations of 2 and 3 index integrals which can be accomplished by the same efficient technique of integral calculation as described above.

Illustrative Examples

Electronic Structure and Photoelectron Spectrum of bis(π-allyl) nickel. The photoelectron (PE) spectrum of bis(π-allyl) nickel has attracted considerable attention since it was first measured in 1972 by LLoyd and Lynaugh (24). Later experiments (25) established an accurate basis for comparisons with theory. The first theoretical investigation by Veillard et al. (26) using the Hartree Fock approach (cf. Table I) revealed that the interaction between d orbitals of Ni and π^* orbitals of the allyl group is responsible for most of the bonding. Koopmans' theorem was found to be invalid in this case and total energy ΔSCF calculations are required. However, even large-scale CI calculations (27) could not correctly identify the first band as ionization from the ligand π-type orbital ($7a_u$ orbital assuming C_{2h} symmetry of the molecule). A semi-empirical Green function approach (28) (cf. Table I) provided an efficient calculation of all of the ionization potentials (IP's) and gave satisfactory assignment for most of the PE bands. Hancock et al. (29) performed scattered-wave (SW-Xα) calculations of IP's applying the transition state method. Compared with experiment, the calculated spectra are shifted considerably towards lower energies and especially ionizations out of π-type orbitals seem to be in error.

The present calculations were performed with an all electron version of the LCGTO-LSDF method. Triple zeta basis sets for nickel and carbon atoms including polarization functions were employed. Details of basis sets and the method of their optimization are given in Ref. 16. A recent neutron diffraction study (30) revealed a pronounced bending of the anti-hydrogen atoms (by 30°), as a result of their strong repulsion from the nickel atom. First we discuss results obtained assuming planar geometry of the allyl groups as this allows a direct comparisons with the other theoretical calculations of the PE spectra performed so far.

The interaction diagram between the 3d and 4s orbitals of Ni and the $(C_3H_5)_2$ fragment together with the resulting orbital levels of bis(π-allyl) nickel are shown in Fig. 1. The basis set superposition error (BSSE) (17) is corrected for by using the same basis set in all calculations except for the isolated Ni atom. This accounts for the splitting in the $Ni(C_{2h})$ case.

Bonding is caused by the interaction between d-electrons of Ni and the π^* electrons of the $(C_3H_5)_2$ fragment. By symmetry, bonding is allowed within the b_g and a_g manifolds. The donation of electrons from occupied $2b_g$ orbitals of the metal atom to the unoccupied $5b_g$ orbital of the $(C_3H_5)_2$ fragment results in the bonding molecular orbital, $5b_g$. Another bonding orbital, $10a_g$ is formed as a combination of $7a_g$ and $6a_g$ orbitals of bis(π-allyl) and nickel, respectively. This bonding effect is, however, largely cancelled by the antibonding partner, the $13a_g$ orbital. Close to the HOMO ($6b_g$) there are levels of mainly d-character which have slightly antibonding or nonbonding character. One finds, surprisingly at first, a stabilization of the $5a_u$ and $7b_u$ levels of $(C_3H_5)_2$. Inspection of Mulliken charges reveals that the corresponding molecular orbitals ($7a_u$ and $11b_u$) have some admixture of metal 4p states. The coupling with 4p states causes a transfer of the ligand's charge to originally unoccupied 4p orbitals of the metal. This "back donation" mechanism was first discovered in SW-Xα calculations (29) and is confirmed by the present LCGTO-LSDF study.

Table I. Experimental and theoretical results for the photoelectron spectrum of
bis(π-allyl) nickel (in eV)
The principal orbital character of the ionized electron is added in parenthesis

band	exp. (a)	ΔSCF (b)	ΔSCF-CI (c)	GF (d)	SW-Xα (e)	LCGTO-LSDF(f) planar	bent
1	7.8 (π)	5.5	6.4	8.7	2.5	7.8	8.1 (π, $7a_u$)
2	8.2 (d)	5.6	6.6	8.9	4.5	8.0	8.1 (d, $13a_g$)
		6.8	6.7	9.2		8.2	8.4 (d, $12a_g$)
3	8.6 (d)			9.2	5.0	8.4	8.4 (d, $6b_g$)
				9.5	5.1		
4	9.4 (d, π)			10.0	5.5	9.2	9.4 (d, $11a_g$)
					5.6	9.5	9.9 (d, π, $5b_g$)
5	10.4 (π)	11.0	10.8	10.9	6.6	10.7	10.3 (π, $11b_u$)
6	11.6 (π)			12.2	7.9-8.2	11.9	11.2 (π,d, $10a_g$)
7	12.7			13.0-13.2	9.0-9.2	12.1-12.5	11.8- (σ)
							13.3
8	14.2			15.4		13.6	14.3 (σ)
				15.6		13.8	14.6

(a) experimental data : Batich, Ref. 25
(b) Veillard; Rohmer et al., Ref. 26
(c) Moncrieft et al.; Hillier, Ref. 27
(d) semi-empirical Green Function calculation: Bohm and Gleiter, Ref. 28
(e) scattered-wave Xα calculation: Hancock et al., Ref. 29
(f) present calculations: allyl groups are planar or hydrogen atoms are bent

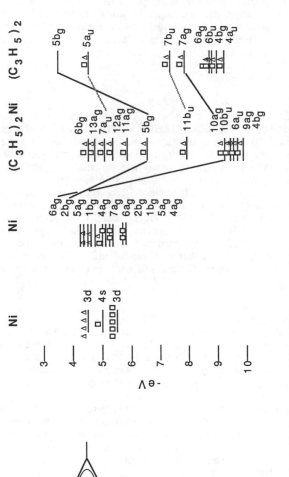

Figure 1. Orbital diagram of the LSDF one-particle energy levels of isolated nickel, nickel(C_{2h}), bis(π-allyl) nickel(C_{2h}) and the $(C_3H_5)_2(C_{2h})$ fragment. □ and △ indicate spin up and spin down electrons. The three-dimensional structure of bis(π-allyl)nickel is shown on the left-hand side.

Using a planar geometry for the H atoms, the calculated LSDF ΔSCF values for the PE agree rather well (within 0.2 eV) with experiment (cf. Table I). Using a bent geometry (30), the total energy of the transition metal complex is lowered by 0.9 eV. The corresponding calculated photoelectron spectrum agrees less convincingly with experiment. The main bonding mechanism, i.e. donation from Ni 3d to allyl π^*, and back donation from allyl π to Ni-4p, remains, however unchanged. The d orbitals of Ni are barely affected by this geometry change, and the important shifts are found in the ionizations from ligand π and σ orbitals. It should be noted that the experimental PE was taken in the gas phase while the "bent" structure was deduced from measurements on the crystalline solid (30). Clearly, a full geometry optimization of the isolated complex would settle the question of crystal packing effects and clarify the details of the photoelectron spectrum.

Chemisorption of Carbon Atoms on the Ni(100) Surface. The next example demonstrates the capability of the LCGTO-LSDF method to predict the geometrical structures and vibrational frequencies of carbon atoms chemisorbed on the Ni(100) surface. The C/Ni system is of fundamental importance in catalytic petrochemical processes and thus has been the subject of many experimental and theoretical studies. Despite these efforts, many aspects such as the equilibrium position of the C atoms (above or below the surface Ni atoms) remained unsettled. Recently, a joint experimental and theoretical study of the chemisorption of carbon on Ni(100) led to a clearer understanding of this system (31). It was found that in the ground state, C is adsorbed in four-fold hollow sites above the surface with a Ni-C distance of 1.79 Å, in agreement with the experimental value of 1.75 ± 0.05 Å, obtained from surface extended energy loss fine structure (SEELFS) measurements. Furthermore, the calculated vibrational frequency of the perpendicular mode of the adsorbed C atom, 407 cm^{-1} is in excellent agreement with the experimental value of 410 cm^{-1}.

Vibrational Frequency of CO Adsorbed on Pd Clusters. In the next example, we present results for the vibrational properties of a CO molecule on a Pd surface, modelled by a cluster of 14 transition metal atoms (32,33). Furthermore, the influence of an external electric field is investigated using a smaller cluster, Pd_2CO (33,34). In both cases, CO is assumed to be bonded in the bridge position. The chemically inactive Pd core electrons are described by a relativistic model potential as described in detail in Ref. 14.

Two coupling modes are considered: for the $Pd_{14}CO$ cluster the first mode (denoted as h) represents vibration of the rigid CO molecule with respect to the transition metal surface. The second mode is either the Pd-Pd vibration within the plane of Pd surface atoms (r) or out-of-plane stretch of the surface/sub-surface Pd-Pd bond (z). The total energy surfaces (h,r) and (h,z) are calculated for discrete points and then fitted to a fourth order polynomial. Variational and Quantum Monte Carlo (QMC) methods were subsequently applied to calculate the ground and first excited vibrational states of each two-dimensional potential surfaces. The results of the vibrational frequences ω using both the variational and QMC approach are displayed in Table II.

If one assumes a rigid substrate which, at first, seems reasonable because of the high mass of Pd compared with the CO molecule, a frequency of almost 500 cm^{-1} is obtained for the vibration of the entire (rigid) CO molecule perpendicular to the surface. This value is in significant disagreement with the experimental value of 340 cm^{-1} (35). As can be seen from Table II, a substantial lowering of this "beating" mode occurs due to anharmonic effects in the coupling of the z and h modes. In other words, the vibration of the surface/sub-surface Pd-Pd bond stretching couples to the CO beating mode and lowers

Table II. Vibrational frequencies (cm^{-1}) of the two coupled modes (h,r) and (h,z) obtained from a harmonic (Har), variational (Var) and Quantum Monte Carlo (QMC) approach

mode		Har	Var	QMC
(h,r):				
	ω_r	240	192	192
	ω_h	498	521	521
(h,z):				
	ω_z	196	65	66
	ω_h	498	402	397

its frequency by about 20%. Taking into account the approximations involved in the present model (movement only of the nearest neighbours of CO, lack of direct coupling between r,h,z modes, and finite cluster effects) we feel that the calculated frequency of about 400 cm^{-1} is in reasonable agreement with experiment. Further investigations of the coupling between the CO stretch and the CO-cluster vibration (h) reveal that this coupling appears to be negligible.

In a model study of Pd$_2$CO, the coupling of the h mode (which is the same as in the 14-atom cluster) and the d mode (CO bond stretching) was investigated in the presence of an external electric field (33,34). In this case, the calculated points on the energy hypersurface extend further away from equilibrium so as to allow the inclusion of second excited vibrational levels. The differences between Variational and QMC approaches are now small, but on the other hand the electric field induced frequency shifts are small as well and therefore a high accuracy of the potential is nedeed . It turns out that the QMC method is less sensitive to the fitting of the potential and thus is preferable. In Table III we present the CO stretching frequency (d) vs. electric field calculated for values of an external electric field appropriate for an electrochemical situation.

Table III. CO stretching frequency (cm^{-1}) vs. electric field (V/cm x10^7)

Electric Field	ω_{CO} Var	QMC	$\Delta\omega_{CO}/\Delta E$ Var	QMC
-5.0	1833	1833		
-2.5	1859	1855	26	22
0.0	1873	1878	14	23
2.5	1908	1900	35	22
5.0	1927	1920	19	20
10.0	1961	1949	17	14.5

Both the variational and the QMC approach show the same qualitative trend: the vibrational frequency of the CO bond-stretching mode increases with the electric field, i.e. when the electrode becomes more positively charged. However, these changes are nearly symmetrical for positive and negative fields only in the case of the QMC treatment. It is interesting to note that the experimental electrochemical curve (36) has a symmetrical sigmoid shape.

A comparison with the calculations of the interaction of the electric field with the dipole moment of the isolated CO molecule shows that the Stark effect is clearly dominant and the chemical contributions account for about 25% of the total shift. This fraction is significantly larger than the chemical contributions found by Miller and Bagus (37) for CO adsorption on Cu.

Electronic Structures of Zn Clusters. In the last example of this short review, we examine the electronic structures of Zn clusters as a function of increasing cluster size starting with the dimer and expanding the size to a total of 61 atoms. In these calculations, symmetry was not used to simplify the problem and all electrons are included in the SCF calculations. This systematic study also provides an opportunity to assess the computational requirements as a function of cluster size.

Dimers. The Zn_2 dimer is the subject of controversy because theoretical calculations performed so far (38,39) yielded no bound ground state, whereas experimental data suggest a weak van der Waals bond (40) with a dissociation energy of about 0.17eV (41) or 0.06eV (42) and a bond distance of 2.35±0.08 Å (41). In Table IV we summarize the results of the present calculations with a (43321/4211*/311+) orbital basis set (16), a (11/5/5) expansion for density and exchange-correlation fit and 4500 grid points per atom using the "best" local spin density potential of Vosko et al. (VWN) (43).

Table IV. Optimized bond distances, R(Å), vibrational frequencies, $\omega(cm^{-1})$, and binding energies, $\Delta E(eV)$, for Zn_2 and Zn_2^+

molecule	R(Å)	$\omega(cm^{-1})$	$\Delta E(eV)$
Zn_2	2.6	207	0.3
Zn_2^+	2.5	229	2.1

The present results were corrected for basis set superposition error (BSSE) (17) which is equal to 0.09 eV at the equilibrium distance of Zn_2. We studied the sensitivity of the results to the choice of basis set. The most compact set (43321/431*/5) + (11/1/2;11/2/1) (see Ref. (16,17) for explanation of the symbols) which we used resulted in a binding energy of 0.2 eV and a bond distance of 2.55 Å. Due to the rather flat potential curve we noticed larger variations of equilibrium distance then in binding energy. We expect our results to be within 0.1 Å and 0.1 eV of the LSDF limit. The d-orbitals of Zn do not participate in the bond and the attraction is a result of a polarization of the 4s orbital. Thus, the inclusion of 4p orbitals is crucial and the dimer would not be bound without them. In fact, inspection of the wavefuction reveals a strong mixing of 4p and 4s states. The Zn_2^+ ion was optimized at the same level of accuracy as the neutral dimer (cf. Table IV). We predict the bond distance to shrink and the bonding to be much stonger than in the case of Zn_2. The reason for this behavior lies in the fact that upon ionization an electron is removed from an antibonding 4σ level. The hole in the valence sp shell is equally shared between two Zn atoms. There remains a discrepancy between calculated (2.5Å) and experimental bond distances (3.0Å) and binding energies (exp.:0.1eV, theory: 2.1eV). Given the bond strengthening upon ionization, Zn_2^+ is expected to have a shorter bond length, larger binding energy and higher vibrational frequency compared with the neutral dimer, as was found by the calculation. Thus, the experimental values for Zn_2^+ deduced indirectly using crude approximations (44) should be reexamined.

Large Clusters. Metal clusters play an important role in the understanding of various aspects of chemisorption and catalysis, but they are interesting species in their own right as well since they bridge the gap between individual atoms and the condensed phase. The following preliminary study provides insight into the convergence of cluster properties to those of surfaces and bulk solids.

First we discuss the dependence of the ionization potential on the cluster size. Clearly, this quantity should convergence to the work function of Zn, known experimentally to be 4.33 eV for a polycrystalline sample and 4.9 eV for the (0001) surface (45), with the Fermi energy falling in the the sp-band. The ionization energy of the isolated Zn atom is 9.8 eV (46). Somewhere between, small clusters of Zn start to resemble the bulk. Earlier Hartree-Fock calculations by Tomonari et al. (39) seemed to indicate that Zn clusters as small as 5 and 6 atoms approach bulk-like properties. This finding is in contrast to recent experiments (40) using ultraviolet photoemission spectroscopy (UPS) which clearly indicate a behavior of small Zn clusters distinct from the bulk material. This experimental result is consistent with our theoretical finding of a rather slow convergence of the ionization potential towards the work function of the extended system.

In the present calculations, the geometries of the Zn clusters up to Zn_6 were taken from Ref. (39). The larger clusters are assumed to have simple hexagonal symmetry (hex) or hexagonal close-packed symmetry (hcp) with the same bond distances as in the bulk (45). We performed self consistent calculations on both neutral and ionized clusters (ΔSCF) in order to account properly for relaxation energies. In the case of ions the calculations were carried out using the spin polarized approach.

From Table V one can see a rather good agreement of ionization potentials (IP's) for the atom and dimer compared with the experimental values. The recent value of 9.0 eV (44) for the Zn_2 dimer was estimated from reactions of Zn_2^+ with alkenes and alcohols. A direct measurement of the ionization potentials should resolve the controversy between the present theoretical and the indirect experimental results (44). The calculated work function decreases gradually as the cluster size increases and it slowly approaches the value of the extended system. The IP's calculated with LSDF theory are much larger then HF results and even for a 6 atom cluster they differ from the bulk value by as much as 2.5 eV. Clusters with at least 20 atoms are necessary to come close to the IP of the extended system indicating the necessity for transition metal clusters in this size range to accurately reproduce the electronic structure on a transition metal surface. This is particularly true for properties involving the sp-band, such as the IP's of zinc clusters. The difference between orbital energies and ΔSCF results (see Table V) should not, of course, be interpreted as a relaxation effect, since the Koopmans' theorem is not valid in LSDF theory (1).

Computational Aspects for Large Transition Metal Clusters. The study of Zn clusters of increasing size provides an opportunity to assess practical computational aspects in the simulations of transition metal systems using the LCGTO-LSDF method. Table VI gives the time in central processor unit (CPU) seconds for the computation of all three-index integrals for various Zn clusters, obtained on one processor of a CRAY X-MP/416 supercomputer (8.5ns clock cycle time). In the first test we used a DZ basis set (43321/431*/41+) for the Zn atom and an (11/3/2) expansion for the density fit. The total number of orbitals and fitting functions per atom was 26 and 32, respectively. No symmetry was employed in the calculations. In the first test we assumed the clusters to have the geometry of a simple hexagonal lattice (hex) . This created a much more open structure then the actual structure of Zn, which is hexagonal close-packed (hcp). In the second test we grew the clusters within hcp symmetry. Now the number of non-zero integrals is much larger and we have to use a direct version of the program already for the 43-atom cluster. An SZ type basis set (43321/431*/5) + (11/2/3;11/0/3) was employed in the second test.

It can be seen from Table VI that for the (hex) geometry the number of significant integrals (larger than our threshold, 10^{-9}) increases quadratically rather than as the third power of the number of orbital basis functions, N_b. The CPU time closely follows this pattern. For the case of a densely packed geometry (hcp) the time of calculation is significant and it scales more as the third power. Analysis of the CPU time shows that about 30% of the time is spent in the routine which compresses primitive integrals into contracted ones and writes them onto external storage (disk or solid state storage device). The formation of the contracted integrals is a bottleneck of the computations, since in general it requires gather and scatter operations simultaneously, which is not vectorized. Using an unrolling technique, this problem can be solved, resulting in a sustained average speed of the integral package of about 80 MFLOPS per processor. This value may increase with the size of the basis set. The same integral routine running in the direct scheme delivers about 110 MFLOPS per processor. Obviously, one can form the Hamiltonian matrix within an uncontracted basis set first and at the end (once) couple contributions from integrals over primitive Gaussians.

An important aspect of an efficient implementation of any program on current and future high-performance computers is the level of parallelism. Our tests show that for the Zn_{14} cluster 96% of the code is parallel and for larger clusters this ratio increases. Taking into account Amdahl's law (**48**) we can expect a factor of at least 3.5 improvement in performance (total time) on a four procesor machine. In the last column of Table VI the expected wall clock time is presented.

In view of these improvements it now appears that integral evaluation (in standard or direct mode) no longer represents the bottleneck for systems in the size range of about one-hundred transition metal atoms.

Summary and Conclusions

In this contribution we have reviewed the applicability, accuracy and computational efficiency of the local spin density functional approach to the chemistry of transition metal complexes and clusters using a linear combination of Gaussian-type orbital basis set for the calculation of electronic structures, ground state geometries and vibrational properties.

This has been demonstrated for bis(π-allyl) nickel, where the current approach provides a detailed understanding of the ligand-transition metal interactions (in the form of Ni-d to allyl-π^* donation and allyl-π to Ni-p backdonation) as well as a quantitative

Table V. Ionization Potentials of Zn Clusters (in eV)

Cluster Size	Orbital Energy	Δ SCF	experiment	HF(d)
1	-6.2	9.8	9.8(a)	7.6
2	-5.2	7.9	8.4(b) - 9.0(c)	6.0
3	-5.5	8.1		6.5
4	-4.2	6.5		4.6
5	-4.3	7.6		4.6
6	-3.3	6.9		4.9
9 hex	-3.6	5.7		
14 hex	-3.6	5.3		
21 hex	-3.5	5.2		
13 hcp	-4.0	5.7		
25 hcp	-3.8	5.1		
43 hcp	-3.8	5.0		
polycrystalline			4.3(e)	
(0001) surface			4.9(e)	

(a) Ref. 46; (b) Ref. 47; (c) Ref. 44; (d) Ref. 39; (e) Ref. 45.

Table VI. Time (in CPU seconds of one processor of a CRAY X-MP supercomputer) of Coulomb integral calculations (I_C) vs. number of integrals (N_I) and size of the orbital basis set (N_b)

Cluster Size	N_b	N_I 10^6 total	non zero	Computing time (seconds) CPU(a)	CPU(b)
hex:(c)					
7	182	4	1	2	
14	364	30	12	11	3
19	494	74	18	21	
37	962	548	82	91	24
hcp:(d)					
13	260	14	9	12	
25	500	100	41	54	
43	860	509	210	238(168)	64(45)
61	1220	1453	416	601(391)	158(103)

(a) the time of an integral run in direct scheme is given in parenthesis
(b) the theoretical wall clock time on a 4 processor machine
(c) (43321/431*/41+) + (11/2/3;11/3/2)
(d) (43321/431*/5) + (11/2/3;11/3/0)

account of the photoelectron spectrum. In contrast to earlier approaches using Hartree-Fock, CI and scattered-wave Xα theories, the current LSDF calculations yield all observed photoemission lines to within about 0.2 eV.

Chemisorption bond lengths such as those of C atoms on a Ni(100) surface are reproduced with the experimental error bar, i.e. to within a few percent. In addition, the calculation allow a determination of adsorption sites: in the case of C on Ni(100) it is found that C is adsorbed *above* the Ni surface in four-fold hollow sites settling a controversy of C being above or below the Ni surface atoms.

The study of vibrational frequencies of CO molecules on a Pd(100) surface reveals a rather intriguing lowering of the CO/Pd frequency due to the coupling of surface/sub-surface Pd-Pd vibrations to the CO/Pd beating mode. In addition, an electric field causes an increase of the C-O stretching frequency when the potential of the electrode becomes positive and the calculated shift rate (assuming an inner layer thickness of $5*10^{-8}$ cm) of 32 cm^{-1} per volt is in reasonable agreement with experimental value of about 37 cm^{-1}/V (**36**).

A systematic study of the first ionization potentials of Zn clusters as a function of cluster size with up to 43 atoms reveals that at least 20 transition metals are needed to approach surface (bulk) electronic structure properties. In addition, for the first time LSDF theory predicts a stable Zn dimer with a realistic bond length and binding energy.

The investigation of computational requirements and their dependence on cluster size show the LCGTO implementation as highly efficient, particularly if the integrals are calculated using a method based on the recursive formulas of Obara and Saika (**15**). For example, on a CRAY X-MP supercomputer integrals can be calculated at a rate of 1 million integrals per second per processor with a high level of vectorization and parallelism. In addition to the integral evaluation, three other parts needed in the LCGTO-LSDF approach, namely density fit, exchange-correlation fit and diagonalization, have been implemented here in an efficient manner resulting in a balanced program, as will be discussed elsewhere (**49**). Cluster sizes of 50 transition metal atoms (without symmetry and with all electrons) are thus readily tractable with present technologies. Inclusion of pseudopotentials and/or symmetry and progress in computational hardware (faster clock, more processors) will enable the treatment of systems with 100 transition metal atoms or more.

In this paper, we presented explicit formulas for LCGTO-LSDF gradient calculations which require just the calculation of integral derivatives. Hence, the same highly efficient integral package can be applied.

In conclusion, the LCGTO-LSDF method promises to be an accurate and efficient approach for the study of the chemistry of complex systems including transition metal compounds and surfaces.

Acknowledgments

JA and DRS are grateful to Marc Bénard and to Mark Wrighton for stimulating discussions about bis(π-allyl) nickel and adsorbed CO molecules in an electric field, respectively. Work at the Université de Montréal was supported by the National Sciences and Engineering Research Council of Canada (operating and supercomputer grants), by the Fonds FCAR of the government of Québec and by the Institut Francais du Petrole. Part of the computational resources were supplied by the Centre de Calcul de l'Université de

Montréal. JA and EW would like to thank the staff of the Corporate Computer Services of Cray Research, Inc. in Mendota Heights for their competent and friendly service during the code development and extensive testing phase and for their support during the calculations on the Zn clusters. Furthermore, JA and EW would like to express their gratitude to their colleagues at Cray Research, in particular to John Larson and Jong Liu.

Literature Cited

1. (a) Hohenberg, P.; Kohn, W. Phys. Rev. 1964, **136**, B864. (b) Kohn, W.; Sham, L. J. Phys.Rev. 1965, **140**, A1133.

2. Wimmer, E.; Freeman, A. J.; Fu C. -L; Cao P. -L; Chou, S. -H.; Delley B. In **Supercomputer Research in Chemistry and Chemical Engineering**; Jensen, K. F.; Truhlar, D. G., Eds.; ACS Symposium Series No. 353, 1987; p. 49.

3. (a) Salahub, D. R. In **Ab Initio Methods in Quantum Chemistry Part II**; Lawley, K. P., Ed.; **Adv. Chem. Phys.** Vol. 69; Wiley - Interscience, Chicester, 1987; p. 447, and references therein; (b) Rosch, N.; Knappa, P.; Sandl, P.; Gorling, A.; Dunlap, B. I. In this volume, and references therein.

4. Boys, S. F. **Proc. Roy. Soc.** 1950, A**200**, 542.

5. Slater, J. C. Phys. Rev. 1937, **51**, 846.

6. (a) Andersen, O. K. **Phys. Rev.** 1975, B**12**, 3060; (b) Koelling, D. D.; Arbman, G. O. **J. Phys.** 1975, F**5**, 2041.

7. Rosen, A.; Ellis, D. E.; Adachi, H.; Averill, F. W.; **J. Chem. Phys.** 1976, **65**, 3629.

8. Wimmer, E.; Krakauer, H.; Freeman, A. J. In **Advances in Electronics and Electron Physics**; Hawkes, P. W., Ed.; Vol 65, Academic Press, New York, 1985; p. 357, and references therein.

9. Delley, B.; Ellis, D. E. **J. Chem. Phys.** 1982, **76**, 1949.

10. (a) Satoko, C. **Chem. Phys. Lett.** 1981, **83**, 111; (b) Satoko, C. **Phys. Rev.** 1984, B**30**, 1754.

11. (a) Sambe, H.; Felton, R. H.; **J. Chem. Phys.** 1974, **61**, 3862. (b) Dunlap, B. I.; Connolly, J. W.; Sabin, J. R. **J. Chem. Phys.** 1979, **71**, 3396.

12. Hehre, W. J.; Radom, L.; v. Schleyer, P.; Pople, J. A. **Ab initio molecular orbital theory**; John Wiley & Sons, New York, 1986, and references therein.

13. Becke, A. **J. Chem. Phys.** 1988, **88**, 1053, and references therein.

14. Andzelm, J.; Radzio, E.; Salahub, D. R. **J. Chem. Phys.** 1985, **83**, 4573, and references therein.

15. Obara, S.; Saika, A. **J. Chem. Phys.** 1986, **84**, 3963.

16. (a) Andzelm, J.; Radzio, E.; Salahub, D. R. **J. Comput. Chem.** 1985, **6**, 520; (b) Andzelm, J.; Russo, N.; Salahub, D. R. **Chem. Phys. Lett.** 1987, **142**, 169.

17. Radzio, E.; Andzelm, J.; Salahub, D. R. **J. Comput. Chem.** 1985, **6**, 533.

18. Hegarty, D.; van der Velde, G. **Int. J. Quant. Chem.** 1983, **23**, 1135.

19. Almlof, J.; Faegri, Jr., K.; Korsell, K. **J. Comp. Chem.** 1982, **3**, 385.

20. Andzelm, J.; unpublished.

21. (a) Bendt, P.; Zunger, A. **Phys. Rev. Lett.** 1983, **50**, 1684; (b) Averill, F. W.; Painter, G. S.; **Phys. Rev.** 1985, B**32**, 2141; (c) Versluis, L.; Ziegler, T. **J. Chem. Phys.** 1988, **88**, 322; (d) Feibelman, P. J.; **Phys. Rev.** 1987, B**35**, 2626.

22. Fournier, R.; Andzelm, J.; Salahub, D. R., to be published; in Xα theory, eq. (5) can be further simplified by using the linear relationship between the exchange-correlation energy density, ε_{xc}, and the exchange-correlation potential, μ_{xc}. In general LSDF theory, this simple relationship does not apply.

23. Pulay, P. **Mol. Phys.** 1969, **17**, 197.

24. LLoyd, D. R.; Lynaugh, N. In **Electron spectroscopy**; Shirley, D. E., Ed.; North-Holland, Amsterdam, 1972; p. 443.

25. (a) Batich, C. D.; **J. Am. Chem. Soc.** 1976, **98**, 7585; (b) Bohm, M. C.; Gleiter, R.; Batich, C. D. **Helv. Chim. Acta** 1980, **63**, 990.

26. (a) Veillard, A. **Chem. Commun.** 1969, 1022, 1427.; (b) Rohmer, M. M.; Demuynck, J.; Veillard, A. **Theor. Chim. Acta** 1974, **36**, 83.

27. (a) Moncrieft, D.; Hillier, I. H.; Saunders V. R.; von Niessen, W. **Inorg. Chem.** 1985,
    24, 4247; (b) Hillier, I. H. In **Quantum Chemistry; The challenge of Transition Metals and Coordination Chemistry**; Veillard, A. Ed.; Reidel, D. 1986, p. 143.

28. Bohm, M. C.; Gleiter, R. **Chem. Phys. Lett.** 1986, **123**, 87.

29. Hancock, G. C.; Kostic, N. M.; Fenske, R. F. **Organometal.** 1983, **2**, 1089.

30. Goddard, R.; Kruger, C.; Mark, F.; Stansfield, R.; Zhang, X. **Organometal.** 1985, **4**, 285.

31. Chiarello, G.; Andzelm, J.; Fournier, R.; Russo, N.; Salahub, D. R. **Surf. Sci. Lett.** 1988, **202**, L621.

32. (a) Andzelm, J.; Salahub, D. R. **Int.J.Quant.Chem.** 1986, **XXIX**, 1091; (b) Andzelm, J.; Salahub, D. R.**Proc. of the NATO Adv.Work. on Physics and Chemistry of Small clusters**; 1986, in press.

33. Caffarel, M.; Claverie, P.; Mijoule, C.; Andzelm, J.; Salahub, D. R. **J.Chem.Phys** in press.

34. Andzelm, J.; Salahub, D. R. to be published.

35. Behm, R. J.; Christmann, K.; Ertl, G.; van Hove, M. A.; Thiel P. A.; Weinberg, W. M. **Surf.Sci.** 1979, **88**, L59.

36. Kunimatsu, K. **J. Phys. Chem.** 1984, **88**, 2195.

37. Muller, W.; Bagus, P. S.; **J. Elec. Spect. Rel. Phen.** 1986, **38**, 103.

38. (a) Hay, P. J.; Dunning, Jr. T. H.; Raffenetti, R. C. **J. Chem. Phys.** 1976, **65**, 2679; (b) Bender, C. F.; Rescigno, T. N.; Schaefer, H. F.; Orel, A. E. **J. Chem. Phys.** 1979, **1122**, 71.

39. (a) Tomonari, M.; Tatewaki, H.; Nakamura, T. **J. Chem. Phys.** 1984, **80**, 344; (b) Tatewaki, H.; Tomonari, M.; Nakamura, T. **J. Chem. Phys.** 1985, **82**, 5608.

40. Schroeder, W.; Wiggenhauser, H.; Schrittenlacher, W.; Kolb, D. M. **J. Chem. Phys.** 1987, **86**, 1147.

41. Carlson, K. D.; Kuschnir, K. R. **J. Phys. Chem.** 1964, **68**, 1566.

42. Carlson, K. D.; Kuschnir, K. R. **J. Chem. Phys.** 1984, **81**, 11.

43. Vosko, S. H.; Wilk L.; Nusair, M. **Can. J. Phys.** 1980, **58**, 1200.

44. Buckner, S. W.; Gord, J. R.; Freiser, B. S.; **J. Chem. Phys.** 1988, **88**, 3678.

45. (a) West, R. C.; Astle, M. J. **Handbook of Chemistry and Physics;** CRC West Palm Beach,Florida, 1978-79; (b) Michaelson, H. B. **J. Appl. Phys.** 1977, **48**, 4729.

46. Moore, C. E. **Atomic Energy Levels;** Nat.Bur.Stand., GPO Washington,1952.

47. Cooper, W. F.; Clarke, G. A.; Hare, C. R. **J. Phys. Chem.** 1972, **76**

48. Amdahl, G. **Proc. 3 AFIPS SJCC;** 1967, p. 483.

49. Andzelm, J.; Wimmer, E. to be published.

RECEIVED September 23, 1988

Chapter 17

f Electron States in Condensed Matter

Jaime Keller and Carmen de Teresa

División de Ciencias Básicas, Facultad de Química, Universidad Nacional
Autónoma de México, Apartado 70–528, 04510 México D.F., Mexico

In condensed matter the d and f electron states give
rise to a large variety of properties due to their
possibility of being either atomic–like states or
band–like states with small changes in chemical composition, temperature, pressure, etc. Both types
of states are discussed both from the theoretical
(and computational) point of view and from the observed spectroscopic, electric and magnetic properties
point of view with special emphasis in f electrons.
Electron–electron correlation is shown to play a
dominant role in these series of phenomena.

Atomic–like f electron states in condensed matter were first
studied in rare–earth and actinide metallic or non metallic compounds. There the multiplicity of the f states and related properties like magnetic moment, Curie–Weiss susceptibilities and
spectra (where the crystal field splitting is measured) indicate
that for most of the rare–earth series (RE) it is a good approximation indeed to consider those f electrons as atomic–like
states. Then for the calculation of properties we can treat the
f electrons in those compounds within the same approximations
as for the core electrons and assume that the interaction between
f electrons in different sites is carried through the conduction
or the valence electrons.

The magnetic interaction between the ions in the magnetic
metals for example, can then be considered as carried by the conduction electrons in the well known Rudermann–Kittel–Kasuya–
Yoshida (1) interaction. The physical origin of this interaction
is a point like polarization of the conduction electrons (CE),
at the atomic sites, by the magnetic moments of the f electrons,
resulting in an oscillation of the spin density of the CE. The
point like approximation is useful because the maxima of the f
wave functions are found well inside the atomic core, in radii
smaller than 0.7 atomic units. This polarization is carried from
ion to ion by the generated polarization oscillation of the con-
duction electron spins, which has a wave length $\lambda_f = 2\pi/\varepsilon_f^{1/2}$ ($\varepsilon_f =$

0097–6156/89/0394–0246$06.00/0

Fermi energy relative to the bottom of the CB). The Fermi level
wave lengths are in general incommensurable with the crystal's
interatomic or interplanar distances; as a result, the magnetic
ground states of the rare-earth intermetallics and pure metals
have complicated spatial distributions, ferromagnetic, antiferromagnetic, helical, etc.

There are, nevertheless, non metallic or metallic elements
and compounds containing rare-earths or actinides, where the f
electrons are clearly not properly described as localized atomic-
like states. A wide range of special properties are found associated with phenomena like electronically induced phase transitions, valence fluctuations, mixed valency (MV), disappearance of
Kondo resistivity and heavy electron behaviour. As we will see
below, there is a common origin to all these properties which
arise from the f electrons forming f bands which are highly hybridized with the s, p and d valence or conduction electrons, presenting new types of correlation between the f electrons themselves, between the f electrons and the conduction electrons and
between f electrons in different atomic states. As a result we
find a whole series of new electronic ground states. A further
complication arises at high temperatures, because it is usually
the case that the peculiar electronic ground state, which gave
rise to the special properties, is destroyed and the materials
behave again as if the f electrons were localized, atomic-like
states. The main reason for this electronic transition is that
the correlations are weakened by the thermal disorder and that
the entropy ΔS_a of the localized, disordered, atomic like magnetic
moments adds a term $\Delta F_a = -\Delta S_a T$ to the free energy F.

At low temperatures, there are at least two ways in which
the f electrons, and their magnetic moments, behave: either they
order spontaneously in ferromagnetic, antiferromagnetic or complicated magnetic structures, or the f electrons can form a heavy
fermion state, strongly correlated with the conduction electrons.

Some of these materials change again their ground state,
at temperatures very close to absolute zero, to a non conventional
superconducting state.

The first example of a heavy electron system $CeAl_3$, is a
good material to study the heavy fermion systems because it presents an extreme case of these properties (it does not become
superconducting at low temperatures). Then it should be useful
to study the basic properties of these materials.

There is a series of materials with very interesting properties showing large qualitative and quantitative changes in
their electronic structure arising from small composition, temperature or pressure changes. Among them the rare earth and
actinide mixed valence compounds are currently the subject of
extense experimental and theoretical studies (2-5). Boppart and
Wachter (3) have presented a very clear example of this behaviour
in their studies of the moment formation in $TmSe_{1-x}Te_x$ under pressure where both mixed valency, or intermediate valence (IV), and
semiconductor to metal transition are found. The particular interest of this case is not only that these materials have been
extensively characterized, but also because they show, from comparison of valences determined by two different experimental
methods, that a unique picture which considers only one type of

electronic structure change, cannot, at the same time, explain
the volume and the magnetic properties change as a function of
composition or pressure. The substitution of Te by Se is of course
a "compositional" pressure, to be added to the external pressure
inducing the valence instabilities. The valence of a material
$\nu = v - n_f^{atomic}$, the difference between the number v of electrons
outside the core and the number n_f^{atomic} of localized f electrons
is determined from the analysis of two different experimental
properties, say the change in the lattice constant and the change
of the molar Curie constant (see Fig. 4 of Ref.(3) and it has
been found that they do not coincide for a given set of com-
positions and external conditions. In fact, a linear interpolation
of the lattice constant and the Curie constant between those of
Tm^{3+} and Tm^{2+} give different valences. This can be understood
if one realizes that the Curie constant reflects more the orbital
character, while the lattice constant reflects also the real space
extension of charges. As has been pointed out by Schoenes (6),
the degree of delocalization of the f states can be correlated
with the oscillator strength of the f → d transition. Table 1
in ref. (7) gives a selection of optically determined f → d oscillator strengths for various Ce and U compounds.

I.   Real Space Wave Functions of f Electrons in Condensed Matter.

The total wave function of a material can be written as a sum
of fixed configuration (f^n) states which will be mixed by the
actual electronic correlation. This procedure is usually known
as configuration interaction (CI), being a standard method in
many body techniques (8-10) where it is usual that each configuration is represented by one determinant constructed from appropriate single particle wave functions of the different angular
momenta. Each single particle wave function ψ is, on the other
hand, constructed as a linear combination of (usually) an atomic
like and a series of polarization wave functions. This LCAO procedure is suitable to construct a given single particle wave
function for the actual geometry of the material.
    The calculation of the single particle 4f-electron wave
functions in atomic problems shows that even if these states are
occupied only after the 5s, 5p and 6s orbitals, the charge distribution of the 4f is such that most of it is inside the sphere
of maximum charge density of the 5s and 5p. This is also the
case for the 5f electrons, in the actinides, relative to the 6s
and 6p states. The physical reason for this behaviour is the
dominant role played by the effective f potential $V_{\ell=3}(r)$, a radial
potential well confining the f-wave function to a small region
of space, see Figure 1. The resultant f-wave function ψ_f should be
appropriately called atomic-like. When the rare earth or actinide
atom is in condensed matter, the boundary conditions and the
potential well changes drastically, mainly in the outermost part
of the Wigner-Seitz cell, through the superposition of the atomic
potential wells, with a lower potential in the interstitial region.
The change in the potential makes it necessary to include an extra
contribution $\psi_{f'}$, in the bonding region to construct an appropriate f wave function $\tilde{\psi}_f$ in this material as:

$$\psi_{\tilde{f}} = (a_n \psi_f + \sqrt{1 - a_n^2}\ \psi_{f'})A \quad \text{where in general:}$$

$$\int |\psi_{f'}|^2 d\tau = 1, \quad \int |\psi_f|^2 d\tau = 1 \tag{1}$$

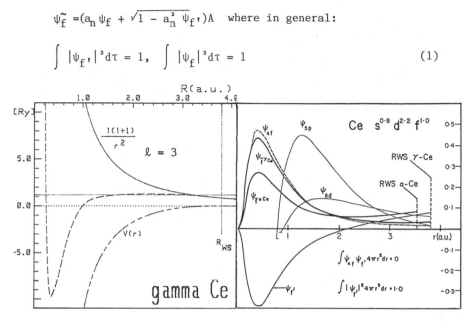

Figure 1. The radial potential well $V_{\ell=3}(r)$ for f-electrons in solid γ-cerium and the resulting bottom of the band f-wave function. Note the deviation from atomic character at the Wigner-Seitz radius giving rise to a wide f-band in these materials. Correlation will contract the f wave function to an atomic like distribution reducing the band character even to the extent of producing atomic-like behaviour.

The auxiliary $\psi_{f'}$ wave function is not of atomic character, on the contrary it must have contributions outside the sphere where the 5s, 5p and 5d (or 6s, 6p and 6d for actinides) have their maxima. The $\psi_{f'}$ is usually taken to be orthogonal to ψ_f. The angular part is of course the same as the 4f (or 5f); but in molecular and solid state calculations the $\psi_{f'}$ could also be termed "polarization" wave functions. Because the two components ψ_f and $\psi_{f'}$ are orthogonal, the normalization integral

$$A^2 \int |(a_n \psi_f + \sqrt{1-a_n^2}\psi_{f'})|^2 d\tau = A^2 \int \left[a_n^2\ |\psi_f|^2 d\tau + (1-a_n^2)|\psi_{f'}|^2 d\tau \right]$$

$$= \left[a_n^2 + (1-a_n^2) \right] = 1 \tag{2}$$

will have two main contributions: the atomic-like part a_n^2 and the delocalized part $(1 - a_n^2)$. For the complex constant A we require $A^2 = 1$.

The coefficient a_n is given the subscript n because it strongly depends on the total amount of f-character for a given configuration f^n being considered. The lower the average energy of the f-electron band in the f^n configuration for that atom in condensed matter, the larger the average a_n is expected to be ($a_n^2 \sim 1$). But in general $a_n = a_n(E)$ with $a_n^2(E_1) < a_n^2(E_2)$ if $E_1 < E_2$ and E_1, E_2 within an f band, that is bonding states at the bottom of the band are less localized than the antibonding states at the end of the band. If the average eigenvalue ε_{fn}, the center of the f-band, is close or above the fermi energy the set a_n will decrease its value. Because the f-electrons eigen-values are found to be highly occupation dependent, shifting by 1.5 to 5 eV with a unit change in occupation, $\varepsilon_{fn} < \varepsilon_{fn+1}$ and and $a_n^2 > a_{n+1}^2$, thus the average amount of atomic like character, per electron, in the f^{n+1} state should in general be taken to be smaller than in the f^n configuration.

Equation 1 is an example of the expansion of one type of orbital into a linear combination of other sets, or LCAO procedure. A discussion of the real space wave functions for RE and actinides, showing the large differences with the atomic like wave functions can be found, for example, in Freeman's introductory paper to the Physics of Actinides and Related 4f Materials (11), (especially Figure 8 of that paper for $\gamma - U$). See also (12). In our analysis here we have concentrated on the rare earths.

Here it should be stressed that for condensed matter a con-venient analysis of the wave functions is made in the form of cellular orbitals, these are defined inside the Wigner-Seitz cells around each atom, a band can be formed for each angular momentum beginning at the energy at which these wave functions have zero slope at the surfaces of the Wigner-Seitz cells up to the energies where the wave functions have zero value at the same points. In a sense we are speaking of a linear combination of cellular orbitals (LCCO) procedure.

The real space wave functions $\psi_{\tilde{f}}$ are the solutions of a Schrödinger like equation

$$\hat{H}_i \psi_i = \varepsilon_i \psi_i \tag{3}$$

where the operator $\hat{H}_i$ is obtained from a multiconfiguration (MC) total energy E expression

$$E[\{\psi^*\}, \{\psi\}] \tag{4}$$

when the E is minimized subject to the constraint

$$\int |\Psi|^2 d\tau_1 \ldots d\tau_N = 1 \tag{5}$$

where the MC wave function Ψ is a linear combination of determi-nants of the ψ_i. Then

$$\delta\left\{ E[\{\psi^*\}, \{\psi\}] - \varepsilon\left(\int |\Psi|^2 d\tau_1 \ldots d\tau_N - 1\right)\right\} = 0 \tag{6}$$

contains a set of equations, one of which, defining $\hat{H}_i$,

$$\frac{\delta E[\{\psi^*\}, \{\psi\}]}{\partial \psi_i^*} - \varepsilon\psi_i = 0 \tag{7}$$

"Equation 7" reduces to the Hartree Fock and CI if only one determinant is used and $E \rightarrow E^{HF}\left[\{\psi*\}, \{\psi\}\right]$. But in highly correlated systems the use of Equation 7 is mandatory because correlation largely changes the ψ_i from the Hartree-Fock value, then

$$\frac{|\psi_i - \psi_i^{HF}|}{|\psi_i^{HF}|} = f_i(r) \text{ with } f_i(r) \gg 0.0 \text{ in a range of } r.$$

II.   Configuration Space Wave Functions.

In configuration space the wave function for a given n occupancy of the f band will be $|f^n\rangle$, usually a determinant with n single f-states. The actual wave function for a particular experimental situation h, for example an initial state h = i or a final state h = f, is

$$|h\rangle = \sum_n C_{nh} |f^n\rangle \qquad (8)$$

this being the so called configuration interaction (CI) or multi-configuration (MC) schemes (see, for example (8-10), where the total number of f electrons n_f will not, in general, be an integer value.

The configuration wave function $|f^n\rangle$ is usually represented by a determinant for the "active" space where only the v valence electrons contribute to one column each: n columns for the f electrons, v-n-1 columns for the d electrons and one column for the (s-p) conduction electrons for each RE atom $n - n_f$.

In this case the amount of atomic-like f-character:

$$n_f^{atomic} = \sum_n n C_{hn}^2 \qquad a_n^2 \leq n_f \qquad (9)$$

There should be some additional contributions to n_f^{atomic} in Equation 9, from the cross terms in $\langle h|h\rangle$ which can be important, mostly if several C_{hn} are of the same order of magnitude and at least two of these terms have n > 0.

For a given h the parameter n_f^{atomic} is a suitable quantity to measure the amount of very localized, atomic-like f-character, and its relation to the properties of the system.

n_f^{atomic} is not unique because it depends on the choice of ψ_f and $\psi_{f'}$, but a consistent scheme will lead to avoid this, well known, shortcoming of population analysis.

The configuration interaction (in the MC scheme) wave function $|h\rangle$ is needed in the condensed matter problem because the reduced symmetry of the crystal and, mainly, the effect of the scattering wave boundary conditions for energies above the interstitial potential or the existence of bonding states between the anions and the heavy metal ions, will allow several types of coupling or correlation between the f level and the valence or conduction electrons. This coupling, usually denoted by Δ, is in fact the sum of several contributions which are responsible for either MV, Kondo or other effects. In general there will be a dominant configuration n contribution ($C_n \simeq 1.0$) and then a second with lower amplitude C_{n+1} (or C_{n-1}) $\simeq \Delta/(E_{n+1}^h - E_n^h)$, E_n^h being the total energy computed for a fixed f^n configuration.

For example in the different phases of the Ce problem the ground state is usually described as in a f^1 configuration, but if a core hole is present or for the case of BIS spectroscopy the f^2 configuration will then take a larger weight. In the ground state of α–Ce the increased interaction between the f electrons and the conduction band will bring in the f^0 configuration because in this case the energy difference $E_{4f^1 5d^2 6s^1} - E_{4f^0 5d^3 6s^1}$ is small; this being one of the reasons for the complicated behaviour of Ce metal and Ce compounds. For these materials the CI method has been extensively used by Gunnarson and Schönhammer (13) (and references therein).

The coefficients C_n^h will in general depend on the hybridization of the f electron wave functions which is the origin of the formation of an f band. But, as already pointed out by Haldane (14), these materials tend to behave as "low density" or impurity like systems at intermediate temperature. This consideration is important because in a material such as $CeAl_3$, at intermediate temperatures, transport properties correspond to incoherent scattering by each RE atom at the maximum resonance scattering cross section (15), as expected from an impurity like Kondo–resonance, and the coherence between the RE atoms, although fundamental for the understanding of the low temperature regime, can be introduced a posteriori.

We can describe the process of pressure induced valence transitions (as in Fig. 1 of Ref.(3)) as a three step phenomenon. The first, in the pressure range $P \leq P_1$ only qualitative changes in the wave functions (as described by Equation 1) are induced, for a given ground state f^{n+1} configuration, $C_{n+1} \simeq 1.0$, because of the total electronic energy for that configuration $E((n + 1),P) < E(n,P)$ and transitions to f^n states are not favoured. The second, where on increasing P, for some range $P_1 < P < P_2$, the total energies, $E(n,P) \simeq E((n + 1),P)$, are similar and a real mixed valence regime is set on with a corresponding decrease of the bulk modulus of the material which arises because the atomic radii in the f^n configuration are smaller (by some 6%) than in the f^{n+1} states. Afterwards, a third step, for a pressure $P > P_2$ where the dominant configuration will be f^n because now $E(n,P) < E(n +1,P)$ or $C_n(P) \simeq 1.0$, and in general, for this pressure the center of the f^n band would have dropped to $\varepsilon_f(n) = \varepsilon_f(n + 1) - \frac{\partial \varepsilon_f}{\partial n}|_{n+1/2} < E_F$ and the wave functions recover some atomic character. Our calculations for the RE show that for a unit change in f–atomic–like character the radii where the 5d wave functions have their maxima, change by 6% approximately. On the other hand a change in the real space distribution of the f wave function will not change the magnetic moment but by the amount given by the change in hybridization.

In terms of the Equation 7 above the correlation between the n and the n+1 states will result in large attractive or repulsive effective potentials arising from the kinetic energy of the d electrons, as far as terms like $-\int C_{n+1}^* C_n \psi_f^* \nabla^2 \psi_d \, d\tau$ appear in the expression for the total energy (Equation 4), which may even not contribute to the total energy but that will contribute in Equation 7 with a term $-C_{n+1}^* C_n \nabla^2 \psi_d = V_{df}$ as an effective attractive potential if V_{df} is negative which will induce a larger band character or as a repulsive potential if V_{df} is positive which will induce a larger atomic like character. The reason for this can be seen in Figure 1 because ψ_d is "outside" of the atomic like part of the f wave function and there is where V_{df} will act.

One particular example of the application of these ideas would be the analysis of CeO which from both band structure calculations (Koelling, Boring & Wood (1983) (16)) and core level spectroscopy, has about $n_f = 0.5$, the f electron is, however, bonded to the oxygen p-electrons. The real space analysis for these bonding, delocalized character, f-orbitals show that $n^{atomic} \simeq 0.4$ $n_f \simeq 0.2$, a value too low to show intraatomic $f \rightarrow f$ or $f \rightarrow d$ transitions with final state splittings.

III. Considerations for the Analysis of the Heavy Fermions Systems.

For normal metals most of the electronic and magnetic properties are dominated by the density of electronic states at the Fermi level $N(E_F)$. One of the conclusions of the nearly free electron theory is that at low temperatures, the ratio of the magnetic susceptibility χ due to independent magnetic electrons, to the electronic specific heat parameter γ, should be a constant (in principle universal) number

$$R = \pi^2 k_B^2 \chi / g_J^2 J(J+1) \mu_B^2 \tag{10}$$

An important feature of the heavy electron systems which have been discussed by Ott in the previous paper of this Symposium is that, even if the specific heat parameter and the paramagnetic susceptibility are both very large compared to the normal free electron metals, their ratio is not very different from what is found for normal metals (a review by Lee et al. of the theories, and experimental facts, of heavy electron systems, can be found in (17)).

There are at least two conclusions to be gained from this fact: the high specific heat parameter and the high magnetic susceptibility at low temperatures should have their origin in a common feature of the electronic structure as in the case of the nearly free electron metals. Second, that either the concept of Fermi level should be modified from its usual formulation, or that there is a new type of mechanism allowing that the Fermi level coincides with the high density of electronic states without the system becoming unstable. That is, it could be that we should abandon the picture where the Fermi level corresponds (for a metal) to a series of an electron like quasiparticles. There is another property which makes the study of these materials a series of exciting discoveries: some of them form a superconducting state at very low temperatures, where the characteristics of these superconducting states should be considered unusual in the sense that the energy parameter is not evidently related to the Debye temperature T_D of the solid but to an order of magnitude smaller energy parameter

T_h. In our study of these materials we have found some evidence that this temperature could be related to a spin flip scattering mechanism which is also responsible for the Kondo resistivity observed in these materials.

There are at least three types of materials presenting heavy electron properties:

-Heavy fermion systems becoming superconductors at low temperatures, $CeCu_2Si_2$, UBe_{13}, U_2PtC_2.

-Heavy fermion systems with a magnetic low temperature ground state UCd_{11}, U_2Zn_{17}.

-Heavy electron systems with a "normal" low temperature ground state $CeAl_3$, UAl_2.

It is thought that the same electron-electron couplings which give rise to the heavy electron behaviour for the "normal" heavy electron systems, should be responsible for the magnetic or superconducting ground states. Then it is important to analyze the properties of a material which remains "normal" at all temperatures, like $CeAl_3$. It should be stressed that the known examples of a low temperature magnetic ground state: UCd_{11}, $NpBe_{13}$, U_2Zn_{17}, $NpIr_2$, $NpOs_2$, correspond to actinide compounds and only one case of a cerium compound $CeAl_2$, is included in this list; this could be related to a stoichiometry and lattice parameter effect.

The study of the electrical resistivity ρ may be the first property where some conclusions can be drawn. In ordinary transition or rare earth metals, the presence of the d and f bands is clearly seen in the analysis of the resistivity (for example (18)).

In many studies of the properties of valence and conduction electrons in the different materials, an approximate theoretical result to fit experimental data is obtained through the introduction of an effective mass m* using a nearly free electron picture and formulae with the mass m parametrized accordingly. In most metals, where the ratio m*/m varies in the range of $0.1 < m*m/m < 10$, this value of the parameter m* can be explained considering an actual electron density of states $N(E)$ modified from the free electron value $N_o(E) = aE^{1/2}$ for two main reasons: resonant states corresponding to s, p, d, f bands of atomic origin and, second, splitting of those bands by crystal field or chemical bonding interactions (19). Each one of these contributions will tend to create a sub-band, generally narrow, which will change m*(E). A transition metal for example, (see reference (19)), will have from resonant d atomic scattering a large $N(E)$ in the energy range corresponding to the d band, that d band will be split, in cubic crystals, into E_g and T_{2g} sub-bands, corresponding to the crystal field symmetry, and, finally, each sub-band will in turn be split into (a series of) bonding and antibonding sub-bands from the chemical bonding with the rest of the material, mainly the nearest neighbors. Nevertheless the sharpest of those sub-bands will not explain an effective mass larger than say m* = 20 m.

Heavy electron systems (see for example Ott, Rudigier, Fisk & Smith (20)) are known since the discovery of the unusual properties of $CeAl_3$, to show from measurements of the specific heat C_p, a large effective mass ranging from 150 m for UAl_2 to 1180 m for UBe_{13}; simultaneously a very large magnetic susceptibility ranging from a spin fluctuation behaviour (where the susceptibility is at least two orders of magnitude higher than the transition metal usual spin enhanced susceptibility) to a high temperature local moment behaviour, with a typical Curie-Weiss law and a localized moment corresponding to an orbital and spin parts very similar to the f electron atomic values, and finally, a resistivity maximum at low temperatures, suggesting a Kondo effect contribution (as mentioned above, it was found that these heavy electron systems, being spin fluctuators, could become superconducting as in the case of $CeCu_2Si_2$, UPt_3, UBe_{13}, URe_2, showing unconventional super-conductivity; the experimental data suggesting from the very beginning the possibility of s = 1 and p-pairing. A theoretical discussion can be found, for example, in Ott, Rudigier, Rice, Ueda, Fisk and Smith (21), where the analogy suggested by Pines of a behaviour similar to superfluid liquid ^3He is further analyzed).

At this point we want to stress that the heavy fermion systems (22) are related to a very large spin enhanced susceptibility or localized magnetic moments, very narrow bands of elementary excitations at the Fermi level, and a new type of pairing, in the case where they become superconducting, at low temperatures even if they were spin fluctuators above the critical superconducting transition temperature T_s.

As it was presented in our studies of the f bands, in terms of the actual f-wave functions in the atomic and in the condensed matter case (7) we found that the width of the band is related to a distortion, in real space, of the radial part of the f-wave function to adapt to the condensed matter potential and boundary conditions: if the distortion is large, (α Ce for example) the band width is also large, otherwise if the f-wave function remains very close to atomic like the band width is smaller (γCe, for example). Then it is possible that the peculiar behaviour of heavy fermion intermetallics is related to a mechanism which will contract the f-wave functions to an atomic, core like form.

The question of these materials being spin fluctuators on the other hand, similar to the problem of the itinerant $TiBe_2$ is related to both a large conduction electron density of states at the Fermi level and to a large enhancement of the spin susceptibility. The approach we have followed for these materials is to analyze the spin susceptibility density in real space to show those regions of the material with a large electron density of electrons at the Fermi level $\gamma(r)$ and the local enhancement β those electrons will have in that point. The spin fluctuators are those materials for which β is large either in an atomic cell (with localized moment behaviour) or in the interstitial region (with itinerant magnetism behaviour like in $TiBe_2$). We have found that the rare earth and the actinides compounds of Ce and U present a large spin susceptibility in the boundary regions between atomic cells, similar to early transition metals, but larger.

Suppose now that some mechanism, which will be explained in the self consistent approximation below, contracts the f-wave functions of the rare earth or actinides metals to the point where the orbital angular momentum is not quenched, as in broad d-band transition metals, then the orbital magnetism will act as a strong local magnetic moment H_0 which will induce a (local) spin magnetic moment on the enhanced susceptibility conduction band electrons.

This will contribute an additional energy $\Delta E_{o-s} = - 1/2\chi_s H_o^2$ to the system, with the following consequences: first an additional, magnetic, scattering of the conduction electrons will be found, increasing the resistivity in a Kondo like mechanism, second, the susceptibility to an external field will have two contributions, one from the spin enhanced susceptibility and the second from the giant localized moments consisting of the f-electron moment and the induced local spin moment. Third, at low temperatures, the additional ion-conduction electrons magnetic scattering, for electrons of the same spin, will add a p-pairing (because of the total spin s of the conduction electrons pair being $S = 1$) to the weak, normal, s-pairing of the Cooper mechanism. It is known that when p-pairing is present the s-pairing will tend to be negligible. Then, a mechanism which changes the hybridized f-electron wave functions back into atomic like will, in a high spin susceptibility material, induce the observed behaviour now known as heavy electrons in metals, if at the same time the coupling, through magnetic interaction, of the conduction electrons at the Fermi level and the electrons in the f-band, is accounted for.

Because of the coupling between Fermi level conduction electrons and the f-band (allowing the hopping between both types of states) it could be thought that the one electron f band will always be at the Fermi level, but it is known that this is not the actual case because the occupation of the f-band cannot be deduced from the energy eigenvalue (because a removal of an f-electron will drastically change the electronic structure and the relaxation corrections are large and cannot be neglected). Then, the narrowed f-band occupation has to be found from the study of the lowest total free energy configurations allowing for integer n_f.

The mechanism we propose in the present paper is just a self consistent analysis of all the previous considerations and their introduction into the, otherwise typical, calculational procedure.

The magnetic coupling between conducting electrons and f-electrons as well as other correlation effects will modify the kinetic energy of the f-electron. For this the conduction electrons have to be thought as the source of the magnetic field and the field momenta incorporated into the Schroedinger equation (P_i contains correlation effects of the type described in the previous section where terms like $-C_n^* C_{n-1} \nabla^2 \psi_\alpha$ act as an additional kinetic energy):

$$\frac{1}{2m} (P_i - \frac{e}{c} A_i)^2 \simeq \frac{1}{2m^+} P_i^2 \tag{11}$$

(here m^+ is a dynamic effective mass, not the heavy fermion mass).

The effective magnetic field H_o , from the orbital part onto the conduction electrons, can be approximated from the spin orbit splitting but it should also be remembered that the effective field could be larger because the forces between the f-electron spin and the conduction electron spins, will add to the dipole field of the orbital moment. To illustrate the effect of this dynamic effective mass m^+ we present the change in the f-band width of Ce in YCe as a function of the effective mass, Figure 2 . Other calculations supporting this analysis will be published elsewhere.

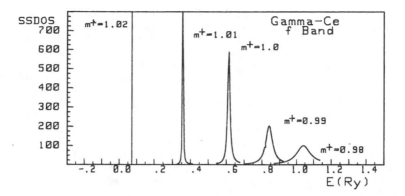

Figure 2.   The change in the energy of the center of the 4f bands
$\bar{\varepsilon}_f$ and in the band widths Γ_f as a function of the electron dynamic
effective mass m^+ in γ-cerium metal. Energy in Rydberg units.
For $m^+/m > 1$ the state becomes almost atomic like, and the f bands
for different $J = L + S$ values do not overlap in energy, $\Delta E_{SL} \gg \Gamma$

We found the contraction of the f-wave functions to an atomic
like character. This has an important consequense, because we
see the onset of a cooperative effect, the orbital magnetic moment
to spin conduction band electrons interaction and other correlations
will generate an effective mass for the f-electrons which, in turn,
will, by contraction of the f-wave function, enhance the atomic
like character and then enhance the full process, hence the process
will not be smooth or show linear behaviour; the process will
tend to present a sudden change of the f-electrons from a band
character to an almost atomic like behaviour. This many-body effect
is responsible for further scattering between the conduction electrons and the rare earth or actinide f electron states.

IV.   Correlation Among Localized States.

Many materials can be studied within the Hartree-Fock scheme because
the atomic, molecular or Bloch orbitals in the independent particle
HF approximation are a very good first approximation to the actual
electron-like quasiparticles of the system. The scheme breaks
down in at least three special cases:

First, when the states are so localized that the Coulomb correlation is larger than the separation of adjacent levels or the
widths of the bands. In this case the main failure of the Hartree-
Fock scheme is caused by the eigenvalues being strongly different
from the ionization potentials (in an energy dependent way). This
is because two states of approximately the same eigenvalue have
ionization potentials which are different among them by a quantity
larger than their difference in energy to the Fermi level.   As
a result the sum of the eigenvalues is largely different from the
total energy and the states cannot be occupied in order of increasing energies.

<u>Second</u>, when sets of states of the occupied part of the spectrum lie very close to the Fermi level and other sets of empty states, also near the Fermi level, can be correlated with the first set through the Coulomb or other interactions. This is the case of aromatic compounds in organic chemistry (correlation dominated properties, as opposed to the aliphatic compounds, molecular orbital dominated properties) and of the heavy fermion systems.

<u>Third</u>, in cases similar to the previous one, when another source of correlation, like the electron phonon coupling, generates the large correlation between the occupied and the empty states. This is the case of the vibronic. states in molecules or of. the superconducting states in solids. Because it is not the Coulomb correlation alone the mechanism responsible for the behaviour, the correlation leads to an actual pairing of electrons and either boson states are formed or this boson states can condense into a boson liquid like in superconductors.

Heavy fermion materials correspond to the second case. In $CeAl_3$ for example there is a $f_{5/2}$ band just below the Fermi level, the crystalline field splits this band into sub-bands of different symmetry and the interaction with the rest of the material further splits each symmetry sub-band into a bonding and antibonding part. In the cerium metal and its compounds only the bonding part of the lowest lying symmetry sub-band is occupied. It can then correlate both to the electron states at the Fermi level through a Kondo mechanism and to the lowest lying f empty states through a Coulomb correlation. When the temperature of the material is decreased, pairs of f states sitting on different atoms form a very weak through bond (a through bond is formed with the aluminum atoms as mediators in $CeAl_3$). Then f electrons in adjacent cerium sites develop a π bond-like structure similar to that of graphite. An equivalent to an aromatic compound has been formed. There is nevertheless a large difference with the case of graphite, because in graphite the energy of delocalization is very large and the bonding π states are separated of the antibonding π states by energies of the order of electron volts and, as a result, the correlation energy is large but smaller than the difference between the eigenvalues of the π states and the Fermi level. In $CeAl_3$, on the other hand, the f levels constitute a very narrow, atomic-like, band which, when forming the extended delocalized states, has a very small splitting between the bonding and antibonding f electrons pair bands. As a result, as the Fermi level sits in the minimum between the sub-bands, the Coulomb correlation is very strong and a Fermi liquid can form at low enough temperatures, when the thermal disorder will not destroy the extended state.

Our conclusion is in agreement with the measurement and the narrow band model for these series of compounds proposed by Wachter, Marabelli and Travaglini (23), and the conclusions of our analysis coincide with those of the splitted narrow band model of those authors.

Acknowledgements.

The technical assistance of Mrs. Irma Aragón is acknowledged. This work was supported by CONACYT, México, project PCEXCNA-022702.

Literature Cited

1. Kittel, C., Quantum Theory of Solids, John Wiley & Sons, Inc., New York, 1963.
2. The Physics of Actinides and Related 4-f Materials, Ed. P. Wachter, North-Holland, 1980.
3. Boppart, H. and Wachter, P., Moment Formation in Solids, W.J.L. Buyers, Ed., Plenum Press, New York, 1984, 229.
4. Wohlleben, D., Moment Formation in Solids, W.J.L. Buyers, Ed., Plenum Press, New York, 1984, 171.
5. Wohlleben, D., Valence Fluctuations in Solids, L.M. Falicov, W. Hanke and M.B. Mapple, Eds., North Holland, 1981, 1.
6. Schoenes, J., Moment Formation in Solids, W.J.L. Buyers, Ed., Plenum Press, New York, 1984, 237.
7. Keller J., de Teresa, C., Schoenes, J., Solid State Comm., 1985, 56 [10], 871.
8. McGlynn, S. et al., Introduction to Applied Quantum Chemistry, Holt, Reinhart and Winston, Inc., U.S.A., 1972.
9. Pilar, F.L., Elementary Quantum Chemistry, McGraw Hill, 1968.
10. Lowe, J., Quantum Chemistry, Academic Press, 1978.
11. Freeman, A.J., Physica 1980, 102B, 3.
12. Keller, J., Castro, M., Amador, C., Physica 1980, 102B, 129.
13. Gunnarsson, O. and Schonhammer, K., Phys. Rev. 1983, B28 [8], 4315.
14. Haldane, F., Valence Instabilities and Related Narrow-Band Phenomena, R.D. Parks, Ed., Plenum Press, New York, 1977, 191.
15. Andres, K., Groebner, J. and Ott, R., Phys. Rev. Lett. 1975, 35, 1779.
16. Koelling, D., Boring, A. and Wood, I., Solid State Comm. 1983, 47, 227.
17. Lee, P., Rice, T.M., Serene, J., Sham, L. and Wilkins, J., Comments on Condensed Matter Physics, 1986, 12, 99.
18. Keller, J., Hyperfine Interactions, 1979, 6, 15.
19. Keller, J., Int. J. Quantum Chemistry, 1975, 9, 583; Amador C., de Teresa, C., Keller, J. and Pisanty, A., Ins. Phys. Conf. Ser., 1981, 55, 225; Keller, J., Amador, C. and de Teresa, C., Rev. Mex. Física, 1984, 3, 447; Keller, J., J. of Molec. Structure, 1983, 93, 93-110; de Teresa, C., Amador, C. and Keller, J., Physica 1985, 130B, 37; Pisanty, A., Orgaz, E., de Teresa, C. and Keller, J., Physica 1980, 102B, 78.
20. Ott, H., Rudigier, H., Fisk, Z. and Smith, J., Physica 1984, 127B, 359.
21. Ott, H., Rudigier, H., Rice, T., Ueda, K., Fisk, Z. and Smith, J., Phys. Rev. Lett. 1984, 52, 195.
22. Ott, H., Rudigier, H., Fisk, Z. and Smith, J., J. Appl. Phys. 1985, 57, 3044.
23. Marabelli, F., Travaglini, G. and Wachter, P., Solid State Comm. 1986, 59, 381; Marabelli, F. and Wachter, P., Conference on Valence Fluctuations, Bangalore, 1987.

RECEIVED April 19, 1988

Chapter 18

Influence of Chemical Composition on Heavy-Electron Behavior

H. R. Ott[1] and Z. Fisk[2]

[1]Laboratorium für Festkörperphysik, ETH-Hönggerberg, 8093 Zürich, Switzerland
[2]Los Alamos National Laboratory, Los Alamos, NM 87545

The drastic influence of small changes in chemical composition on the low-temperature properties of heavy-electron materials is demonstrated. Prevention of the heavy-electron state itself or suppression of phase transitions out of this state are among the most interesting experimental observations.

The occurrence of a heavy-electron state in metals is most distinctly observed in compounds where one of the chemical constituents is an element of the rare-earth (4f) or actinide (5f) series. Within these series, it is the elements at the beginning or the end of the res-8ective row of the periodic system that are most likely involved in this effect (Ce, Yb, U, Np).

The heavy-electron state manifests itself at low temperatures by a large and temperature-dependent ratio c_p^{el}/T, where c_p^{el} is the specific heat due to conduction-electron excitations. This ratio increases with decreasing temperature and, for $T \to 0$ K, reaches values that are one to three orders of magnitude larger than observed in usual metals. These large ratios imply a very high effective density of electronic states at the Fermi energy $N(E_F)$ due to a large enhancement of the effective mass m^* of the itinerant electrons. This is consistent with the large magnetic susceptibilities that are observed in these materials at low temperatures.

Another distinct feature of these materials is the temperature dependence of the electrical resistivity $\rho(T)$ and a typical example $(CeAl_3)$ is shown in fig. 1. Characteristic is a relatively large resistivity at room temperature which stays almost constant or even increases with decreasing temperature, followed by a distinct drop of ρ in a fairly narrow temperature range and low values of ρ at very low temperatures where a T^2 variation is observed [see inset of fig. 1].

This kind of behaviour must be traced back to strong interactions in the electronic subsystem and hence these materials are ideally suited to study many-body effects among electrons which form some kind of Fermi liquid. There are not many general rules to pre-

0097–6156/89/0394–0260$06.00/0

dict the observation of a heavy-electron state. A favourable condition is certainly a distinct separation of the f-electron carrying
atoms within the crystal lattice, both in distance and with respect
to the coordination of the ligands. It should be stressed, however,
that this requirement only appears to be a necessary but by no means
a sufficient condition. This paper is intended to show that changes
in chemical composition may affect the formation of the heavy-electron state itself or, if it is preserved, influence its ground-state
properties drastically (1).

Chemical Composition, Crystal Structure and Heavy-Electron State

In order to elucidate some of the points raised in the introduction, we first mention a few interesting facts related to the chemical composition and the crystal structure of compounds that show
heavy-electron behaviour. Certain crystal structures seem to favour
the formation of a heavy-electron state. We note that compounds
adopting the cubic Cu_3Au structure have no strong tendency in this
respect, with the exception of $U(In_{1-x}Sn_x)_3$ where, with varying
x, the low-temperature properties change quite drastically (2) and
$NpSn_3$ (3). URh_3 and UIr_3, for example, show essentially non-magnetic
behaviour because of a strong hybridization of the 5f wave functions
with wave functions of other symmetries (4), but no features that are
of interest here. If the anion is chosen from the neighbouring column
of the periodic table, UPt_3 crystallizes in the hexagonal stacking of
the same structure, i.e., the Ni_3Sn structure, and is an outstanding
example of heavy-electron materials (5), just as is the aforemention-
ed $CeAl_3$(6) which adopts the same crystal structure. As a side remark
we mention that UPd_3 crystallizes in a combination of cubic and hexagonal crystal structure, the double-hexagonal closed-packed structure, and undergoes structural phase transitions at low temperatures
without any trace of heavy-electron behaviour (7).

Another structure which seems favourable for the formation of
the heavy-electron state is the cubic $AuBe_5$ structure which is shown
in fig. 2. In this respect it is instructive to discuss the properties of some compounds of the form UM_5, where M = Ni, Cu, Pd and Pt.
With the exception of UPd_5, all members of this group of compounds
crystallize in the $AuBe_5$ structure. The crystal structure of UPd_5 is
not clearly established because it is difficult to prepare this compound in single phase form, confirming the view that Pd is an element
that often plays an extraordinary role. With respect to the $AuBe_5$
structure it is important to realize that there are two inequivalent
anion sites with a ratio of 1 : 4 and forming large and small tetra-
hedra respectively. One of the anion sites is crystallographically
equivalent to the cation site. It is interesting to note that the
occupation of a single element on these equivalent sites results in
the cubic Laves-phase structure. UAl_2, which crystallizes in this
latter structure, is one of the earliest examples for which features
of heavy-electron behaviour was discovered (9).

A common feature of the compounds that we are discussing here is
the large distance between the U atoms which ranges from 4.79 Å for
UNi_5 to 5.29 Å for UPt_5. The physical properties of UNi_5 are similar
to those of the compounds with the Cu_3Au structure that were mentioned above and in this sense not of interest here (10). UCu_5, however,
orders antiferromagnetically at T_N = 15 K and enters a heavy-elec-

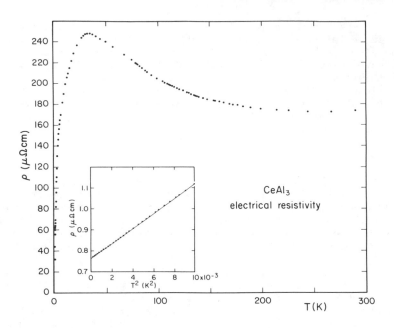

Figure 1.   Temperature dependence of the electrical resistivity
$\rho(T)$ of CeAl$_3$ below room temperature. (Reproduced with permission
from ref. 1. Copyright 1987 Elsevier.)

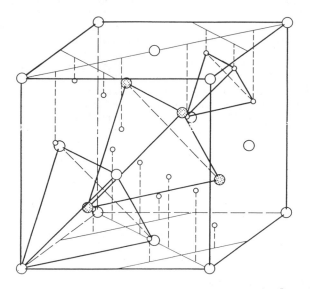

Figure 2. The AuBe$_4$ structure. The large open circles denote Au
sites; the dotted large circles and the small open circles denote
Be sites (see also ref. 8). (Reproduced with permission from ref.
13. Copyright 1987 American Physical Society.)

tron state at temperatures far below T_N (12). With a further decrease in temperature, this state again becomes unstable and a further phase transition of still not well established nature occurs around 1 K. For our purposes it is now interesting to mention how these features described for UCu_5 are varying by manipulating the chemical composition.

At first we note that small amounts of Ni replacing Cu are devastating for both the magnetic ordering and the formation of the heavy-electron state (1,13). This is shown in fig. 3 where it may be seen that the anomaly of the specific heat at T_N is quenched very effectively by replacing a few atomic % of the Cu atoms with Ni. The characteristic upturn of c_p^{el}/T below 4 K that is observed in UCu_5 is also wiped out with only 1% of Ni for Cu (see fig. 4). With a further increase in Ni content, a rather abrupt drop of the low-temperature $c/_p^{el}T$ ratio is noted if the Ni content exceeds 20% (10). This all occurs without any drastic changes in the lattice constants, not to mention the structure. While the drastic effects at low Ni concentrations are probably related with disorder effects on the Ni sublattice, the loss in electronic specific heat at larger Ni concentrations might be due to band-structure effects.

Quite different is the influence of substituting Cu with Ag. If one Cu atom per formula unit is replaced by Ag, T_N is raised to 16 K and the formation of the heavy-electron state below 5 K persists (13). What is different now is that the low-temperature phase transition is quenched as is demonstrated in the $c_p(T)$ plot of fig. 5. The specific heat varies linearly with T down to the lowest temperatures investigated and $\gamma = c_p^{el}/T$ is roughly one hundred times larger than that observed in the equivalent amount of Cu metal. This clearly demonstrates that Ni or Ag substitutions for Cu in UCu_5 induce quite different changes in the low-temperature behaviour but the reasons for it are not yet clear.

In UPt_5, the shortest distance between U atoms was mentioned above and, at more than 5 Å, must be regarded as very large. Nevertheless, the expected magnetic ordering among localized 5f-electron moments is not observed above 1.5 K. Indeed the temperature dependence of the magnetic susceptibility χ gives no evidence for well defined moments on the U ions. This compound again behaves more like a metal whose properties are dominated by electron energy bands of average width with a slight tendency to heavy-electron behaviour at the lowest temperatures (14). This tendency may now be greatly enhanced by replacing Pt with Au. This may be seen in fig. 6 where c_p/T versus T^2 is shown for UPt_5 and $UAuPt_4$. The c_p/T ratio of $UAuPt_4$ increases dramatically at very low temperatures and reaches a value of more than 700 mJ/mole K^2 as $T \to 0$ K (see inset of fig. 6) (13). Concomittantly, the low-temperature magnetic susceptibility χ of $UAuPt_4$ is distinctly larger than that of UPt_5. No phase transition is observed above 0.1 K. A further enhancement of the Au content results in a suppression of the specific-heat enhancement and the γ-value of UAu_2Pt_3 is considerably lower than that of UPt_5 (15). On the other hand, χ still increases with increasing Au concentration but no magnetic ordering is observed for UAu_2Pt_3 above 1.5 K. Again it is not obvious what exactly causes these distinct changes in the low temperature behaviour. From band-structure calculations it may be concluded that no drastic changes are induced in the electronic excitation spectrum (16) and therefore again, many-body effects that are

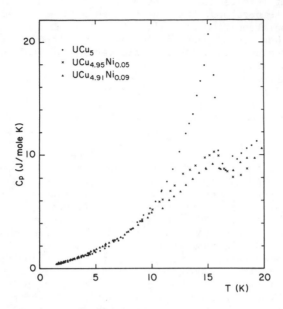

Figure 3. Temperature dependence of the specific heat of UCu_5 and $UCu_{1-x}Ni_x$ with small values of x. (Reproduced with permission from ref. 1. Copyright 1987 Elsevier.)

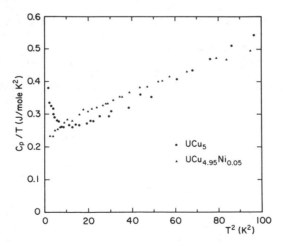

Figure 4. c_p/T versus T^2 for UCu_5 and $UCu_{4.95}Ni_{0.05}$ between 1.5 and 10 K. (Reproduced with permission from ref. 13. Copyright 1987 American Physical Society.)

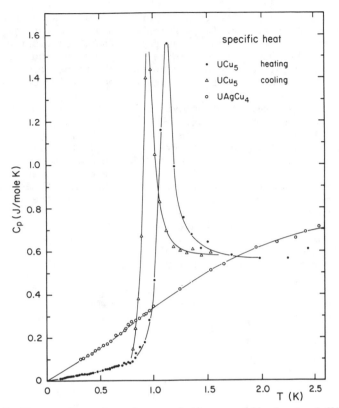

Figure 5. Temperature dependence of the specific heat of UCu_5 and $UAgCu_4$ below 2.5 K. (Reproduced with permission from ref. 12. Copyright 1985 American Physical Society.)

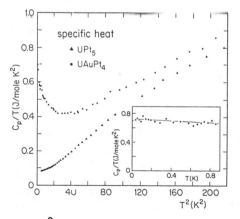

Figure 6. c_p/T versus T^2 for UPt_5 and $UAuPt_4$ below 15 K. The inset is an expanded display of the data for $UAuPt_4$ below 0.9 K. (Reproduced with permission from ref. 13. Copyright 1987 American Physical Society.)

evidently much less easy to identify are probably responsible for the
experimental observations. It is somewhat surprising that in none of
the compounds just discussed is any phase transition observed at low
temperatures. We suspect that this is mainly due to disorder on the
(Pt-Au) sublattice. For the higher Au concentrations, a well defined
moment of more than 3 μ_B per U ion is suggested by the Curie-Weiss
temperature dependence of χ (13).

A similarly interesting behaviour is observed in compounds that
are based on $CeCu_5$. Unlike UCu_5, $CeCu_5$ crystallizes in the hexagonal
$CaCu_5$ structure and is of interest, because $CeCu_6$, an orthorhombic
compound, is also a prototype heavy-electron material (17). $CeCu_5$,
however, undergoes two phase transitions around 4 K but only about
0.2 K apart (18). Neither above nor below these transitions does
$CeCu_5$ show any tendency to heavy-electron behaviour. Again, quite
drastic changes may be induced by replacing Cu with other elements,
without changing the crystal structure.

The hexagonal crystal structure of $CeCu_5$ contains two inequivalent Cu sites with a ratio of 2 : 3. It has been found that Al can
be sub-stituted for Cu on the threefold site (19). Specific heat measurements have shown that replacing one Cu atom per formula unit by
Al leads to an extreme heavy-electron state where the c_p^{el}/T ratio
reaches 2.2 J/mole K^2, so far a record value, and no phase transition
is observed above 0.15 K (18) . The temperature dependence of the
specific heat $c_p(T)$ below 1 K is shown in fig. 7. Increasing the Al
content to obtain $CeCu_3Al_2$ results in a decrease of this giant γ
value by a factor of about two.

It is also possible to sythesize the ordered compound $CeCu_2Zn_3$
in the same crystal structure. Its physical properties are quite similar to those of $CeCu_5$ and it also orders magnetically, with only
one phase transition at about 6 K (18). Again it seems possible to
introduce Al, in this case on the Zn sites. As before, the magnetic
ordering in $CeAlCu_2Zn_2$ is suppressed and c_p^{el}/T reaches values of more
than 2 J/mole K^2 below 1 K. In this case, however, c_p^{el}/T passes over
a maximum of about 2.4 J/mole K^2 at 0.4 K and the extrapolated γ
value for T = 0 K is only about 1.4 J/mole K^2 (20). This type of
behaviour is similar to that observed in other heavy-electron
compounds, the prime example being $CeAl_3$.

It is quite surprising, that Al substitutions induce such drastic effects in both $CeCu_5$ and $CeZn_3Cu_2$, because it is certainly not
to be expected that the electronic structure is the same in both compounds. Therefore it is suspected that the Ce-Al interaction might be
responsible for the formation of the heavy-electron ground state.

With these few examples we intended to demonstrate how changes
in the chemical composition may affect the low-temperature properties
of materials that are likely to show heavy-electron behaviour, without severely affecting other parameters like the lattice constant or
even changing the crystal structure. In the next section we concentrate on changes of the ground state of heavy-electron systems that
are induced by small variations of the chemical composition, without
affecting too much the heavy-electron state itself.

Superconductivity and Magnetism of Heavy-Electron Systems

In this section we briefly discuss how the ground states of the
superconducting heavy-electron materials UBe_{13} and UPt_3 are affected

by only slight changes in the chemical composition of these materials. The occurrence of superconductivity out of a heavy-electron state, although at very low temperatures of 0.9 K (UBe$_{13}$) (21) and 0.5 K (UPt$_3$) (22),respectively, is by itself a quite surprising fact and it is argued that this superconducting state involving heavy-mass quasiparticles must be different from conventional superconductivity observed in simple metals. It cannot be denied that these materials are close to magnetic instabilities and therefore it is conjectured that magnetic interactions are responsible for this superconducting state rather than the usual electron-phonon interaction, and this would almost inevitably lead to electron pairing with other symmetries than assumed in the original BCS theory (23). Such states are known to be very sensitive to any kind of impurities and are easily destroyed by potential scattering. We recall that, in conventional superconductors, it is mainly magnetic impurities with well developed moments that act as pairbreakers.

First reports on the observation of superconductivity of UPt$_3$ already mentioned the extreme sensitivity of this superconducting state to any shortening of the electronic mean free path (22). The superconducting state of UBe$_{13}$ is apparently somewhat more stable but it is also clear that non-magnetic impurities have quite a drastic influence as well. In fig. 8 we show the temperature dependence of c_p below 1 K (24) of pure UBe$_{13}$ and of material doped with small amounts of non magnetic impurities introduced on the U sites. For UBe$_{13}$ we observe a distinct anomaly of c_p at T_c as is expected. It may be seen, however, that with the exception of Th, a case to be discussed separately below, all these impurities shift T_c to distinctly lower temperatures and also almost annihilate the expected c_p anomaly. This result is most likely due to inducing immediately a gapless state of superconductivity which would, on that level of impurity concentration, be somewhat surprising for a conventional superconductor but is actually expected for unconventional anisotropic superconducting states where gap nodes on the Fermi surface are easily smeared out.

Th impurities have quite a different but even more intriguing effect. This is shown in fig. 9 where I_c of $U_{1-x}Th_xBe_{13}$ is plotted versus x. For very small x, T_c is also drastically suppressed but a distinct discontinuity of c_p at T_c is still observed. Around 1.8 % Th, T_c reaches a minimum and increases again with still increasing x. Concomitantly, a second phase transition in the superconducting state, which does not destroy superconductivity, is observed. The critical temperature of this transition is almost independent of x and both transitions merge into one for x > 0.05 (25).

The exact nature of the lower transition is still unclear. Microscopic µSR measurements indicate that some very small magnetic moments of the order of 0.01 µ$_B$ or less might develop below 0.4 K (26), but this has to be confirmed by additional investigations. This latter finding does not necessarily imply a magnetic phase transition. Theoretical investigations (27) actually found particular superconducting states that carry a moment and hence it cannot be ruled out that the tentative phase lines in the diagram shown in fig. 9 separate unconventional superconducting states with different symmetries. Further investigations to check this possibility are being carried out at present. As an additional remark we may mention that about 5% Cu substituting for Be quenches both transitions completely

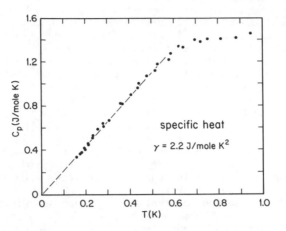

Figure 7. Temperature dependence of the specific heat of CeAlCu$_4$ below 1 K. (Adapted from ref. 18.)

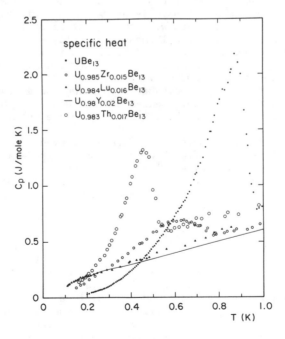

Figure 8. Influence of various non-magnetic impurities on the specific heat of superconducting UBe$_{13}$ below 1 K. (Reproduced with permission from ref. 1. Copyright 1987 Elsevier.)

but leaves the heavy-electron state, as evidenced by the c_p^{el}/T ratio, almost unchanged (28).

Studies of similar nature on UPt_{13} have shown that small amounts of other elements on either the U or the Pt sites may induce a magnetic phase transition in this material (29,30). As mentioned above, even tiny amounts of impurities completely quench the superconducting state in UPt_3. If about 5% of U is replaced by Th or similar amounts of Pd or Au are substituted for Pt, antiferromagnetic ordering is observed between 5 and 6 K. From neutron-diffraction experiments it has been deduced that ordered moments of about 0.65 μ_B/U ion are involved (31). It should be noted, however, that the cooperative magnetic transition is only observed in a narrow range of impurity concentration which shows that also magnetic order is very sensitive with respect to chemical composition. Subsequently a most spectacular result was obtained in pure UPt_3. After some preliminary μSR experiments (32) it was possible to demonstrate, with neutron diffraction again, that even in pure UPt_3, a spontaneous order among tiny moments of only 0.02 μ_B/U ion develops below 5 K (33). Even more intriguing is the fact that the increase of the ordered moment with decreasing temperature is abruptly stopped at the onset of superconductivity below 0.5 K (33). This clearly demonstrates the interference between magnetic order and superconductivity in heavy-electron compounds with interactions of magnetic character.

Finally we should like to show that magnetic order in a pure heavy-electron compound is also easily wiped out by adding small amounts of impurities. For this we discuss the onset of magnetic order in U_2Zn_{17} (34). In fig. 10 we show the anomaly of the specific heat associated with this antiferromagnetic transition at 9.7 K in the form of a c_p^{el}/T versus T^2 plot. The large and constant value of c_p^{el}/T above T_N indicates the heavy-electron state and the sharp discontinuity of c_p^{el}/T at T_N is consistent with a mean-field type transition. The non-zero c_p^{el}/T ratio for $T \to 0$ indicates that not the entire Fermi surface is involved in the phase transition. Also shown in fig. 10 is c_p^{el}/T that is obtained for a sample where 2% of the Zn atoms have been replaced by Cu (35). It is evident that the heavy-electron state itself is hardly affected in this case but the magnetic transition is completely quenched and the heavy-electron state persists to below 1 K. This extreme sensitivity on small amounts of impurities is clearly not typical for a conventional antiferromagnet and microscopic measurements using neutron scattering have confirmed the unusual nature of this transition (36).

Conclusions

In this short review we have given examples of chemically induced changes in the behaviour of heavy-electron materials. First we noticed that the mere occurrence of the heavy-electron state seems almost unpredictable if only the chemical composition and the crystal structure of a compound are given. Then we pointed out that the heavy-electron state itself is quite sensitive to small alterations of the chemical composition in the sense that it can be destroyed or created by minor variations of the constituents of respective materials. Finally we emphasized that possible ground states of heavy-electron systems, like superconductivity or magnetic order, may be

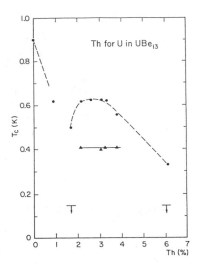

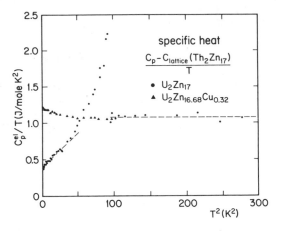

Figure 9. Critical temperatures T_{c1} and T_{c2} of $U_{1-x}Th_xBe_{13}$ as determined from specific-heat measurements. The arrows indicate that no second transition was observed above this temperature. (Reproduced with permission from ref. 39. Copyright 1987 Elsevier.)

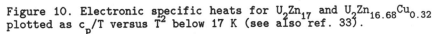

Figure 10. Electronic specific heats for U_2Zn_{17} and $U_2Zn_{16.68}Cu_{0.32}$ plotted as c_p/T versus T^2 below 17 K (see also ref. 33).

changed or even destroyed rather easily with small amounts of impurities. In all known cases of heavy-electron superconductivity this is accomplished in particular with non-magnetic impurities, thus giving further support to ideas that some kind of unconventional and anisotropic superconducting state is adopted in these cases. The considerable sensitivity of magnetic ordering in these materials to small amounts of impurities points to drastic changes of the interactions that trigger these phase transition. These observations are not inconsistent with suggestions that both types of collective phenomena have similar roots in these substances.

Considering all these experimental observations it may seem almost hopeless to tackle these systems on theoretical grounds. In view of all the difficulties that arose in dealing with the Kondo problem, any treatment of interacting magnetic ions and itinerant electrons together will not be simple. It appears that even the most sophisticated model will not be successful if it is not combined with highly advanced calculations of the electronic excitation spectrum. Many-body approaches to the understanding of solids are notoriously difficult but it seems that here one cannot do without them. Readers interested in these aspects should consult refs. 37 and 38.

Acknowledgments

Financial support from the Schweizerische Nationalfonds is greatly appreciated. Work at Los Alamos was done under the auspices of the US-Department of Energy.

Literature cited

1. A compilation of typical properties of heavy-electron systems may be found in: Ott, H.R. in Progress in Low-Temperature Physics, vol. XI, Brewer, D.F. Ed.; Elsevier, Amsterdam 1987; p. 215.
2. Lin, C.L.; Zhou, L.W; Crow, J.E; Guertin, R.P.; Stewart, G.R. J. Magn. Magn, Mat. 1986, 54-57, 391.
3. Trainor, R.J.; Brodsky, M.B.; Dunlap, D.B.; Shenoy, G.K.; Phys. Rev. Lett. 1976, 37, 1511.
4. Arko, A.J.; Koelling, D.D.; Schirber, J.M. in Handbook of the Physics and Chemistry of the Actinides, vol. 2, Freeman A.J. and Lander, G.H. Eds.; North Holland: Amsterdam, 1984; p.175.
5. Frings, P.H.; Franse, J.J.M.; de Boer, F.R.; Menovsky, A. J. Magn. Magn. Mat. 1983, 31-34, 240.
6. Andres, K.; Graebner, J.E.; Ott, H.R. Phys. Rev. Lett, 1979, 35, 1179.
7. Andres, K.; Davidov, D.; Dernier, P.; Hsu, F.; Reed, W.A.; Nieuwenhuys, G.J.; Solid State Commun. 1978, 28, 405.
8. Misch, L.; Metallwirtschaft, 1935, 14, 897.
9. Trainor, R.J.; Brodsky, M.B.; Culbert, H.V. Phys. Rev. Lett., 1975, 34, 1019.
10. van Daal, H.J.; Buschow, K.H.J.; van Aken, P.B.; van Maaren, M.H.; Phys. Rev. Lett. 1975, 34, 1457.
11. Murasik, A.; Ligenza, S.; Zygmut, A. Phys. Status Solidi 1974, a23, K163.
12. Ott, H.R.; Rudigier, H.; Felder, E.; Fisk, Z.; Batlogg, B.; Phys. Rev. Lett. 1985, 55, 1595.
13. Ott, H.R.; Rudigier, H.; Felder, E.; Fisk, Z.; Thompson, J.D. Phys. Rev. B 1987, 35,
14. Frings, P.H. Ph.D. Thesis, University of Amsterdam, 1984.
15. Ott, H. R.; Felder, E.; Fisk, Z. unpublished.
16. Albers, R. C.; Boring, A. M. private communication.
17. Ott, H. R.; Rudigier, H.; Fisk, Z.; Willis, J. O.; Stewart, G. R. Solid State Commun. 1985, 53, 235.

18. Willis, J. O.; Aiken, R. H.; Fisk, Z.; Zirngiebl, E.; Thompson, J. D.; Ott, H. R.; Batlogg, B. Theoretical and Experimental Aspects of Valence Fluctuations and Heavy Fermions; Gupta, L. C. and Malik, S. K., Eds.; Plenum: New York, 1987; p. 57.
19. Takeshita, T.; Malik, S. K.; Wallace, W. E. Solid State Commun. 1978, 23, 225.
20. Ott, H. R.; Felder, E.; Fisk, Z., unpublished.
21. Ott, H. R.; Rudigier, H.; Fisk, Z.; Smith, J. L. Phys. Rev. Lett. 1983, 50, 1595.
22. Stewart, G. R.; Fisk, Z.; Willis, J. O.; Smith, J. L. Phys. Rev. Lett. 1984, 52, 679.
23. Anderson, P. W. Phys. Rev. B 1984, 30, 1549 and 4000.
24. Ott, H. R.; Rudigier, H.; Felder, E.; Fisk, Z.; Smith, J. L. Phys. Rev. B 1986, 33, 126.
25. Ott, H. R.; Rudigier, H.; Fisk, Z.; Smith, J. L. Phys. Rev. B 1985, 31, 1651.
26. Heffner, R. H.; Cooke, D. W.; Hutson, R. L.; Schillaci, M. E.; Smith, J. L.; Willis, J. O.; Mac Laughlin, D. E.; Boekema, C.; Lichti, R. L.; Denison, A. B.; Oostens, J. Phys. Rev. Lett. 1986, 57, 1255.
27. Sigrist, M.; Rice, T. M., preprint.
28. Ott, H. R. Theoretical and Experimental Aspects of Valence Fluctuations and Heavy Fermions, Gupta, L. C. and Malik, S. K.; Plenum: New York 1987,; p. 29.
29. de Visser, A.; Klaasse, J. C. P.; van Sprang, M.; Franse, J. J. M.; Menovsky, A.; Palstra, T. M. J. Magn. Magn. Mat. 1986, 54-57, 375.
30. Batlogg, B.; Bishop, D. J.; Bucher, E.; Golding, B.; Ramirez, A. P.; Fisk, F.; Smith, J. L.; Ott, H. R. J. Magn. Magn. Mat. 1987, 63-64, 441.
31. Goldman, A. I.; Shirane, G.; Aeppli, G.; Batlogg, B.; Bucher, E. Phys. Rev. B, 1986, 34, 6564.
32. Heffner, R. H.; Cooke, D. W.; Mac Laughlin, D. E. Theoretical and Experimental Aspects of Valence Fluctuations and Heavy Fermions, Gupta, L. C. and Malik, S. K., Eds.; Plenum: New York 1987, p. 319.
33. Aeppli, G.; Bucher, E.; Broholm, C.; Kjems, J. K.; Baumann, J.; Hufnagl, J. Phys. Rev. Lett. 1988, 60, 615.
34. Ott, H. R.; Rudigier, H.; Delsing, P.; Fisk, Z. Phys. Rev. Lett. 1984, 52, 1551.
35. Willis, J. O.; Fisk, Z.; Stewart, G. R.; Ott, H. R. J. Magn. Magn. Mat. 1986, 54-57, 395.
36. Broholm, C.; Kjems, J. K.; Aeppli, G.; Fisk, Z.; Smith, J. L.; Shapiro, S. M., Shirane, G.; Ott, H. R. Phys. Rev. Lett. 1987, 58, 917.
37. Lee, P. A.; Rice, T. M.; Serene, J. W.; Sham, L. J.; Wilkins, J. W.; Comments on Condensed Matter Physics 1986, Vol. XII, 99.
38. Fulde, P.; Keller, J.; Zwicknagl, G.; to appear in Solid State Physics, Seitz, F.; Turnbull, D.; Ehrenreich, H., Eds.
39. Freeman, A. J. and Lander, G. H., Eds. Handbook of the Physics and Chemistry of the Actinides, vol. 5; North Holland: Amsterdam, 1987; p. 85.

RECEIVED January 18, 1989

Chapter 19

Response of Local Density Theory to the Challenge of f Electron Metals

M. R. Norman[1] and A. J. Freeman[2]

[1]Materials Science Division, Argonne National Laboratory, Argonne,
IL 60439
[2]Department of Physics and Astronomy, Northwestern University, Evanston,
IL 60208

A brief review is given on the application of local density
theory to the electronic structure of f-electron metals,
including various ground state properties such as observed
crystal structures, equilibrium lattice constants, Fermi
surface topologies, and the electronic nature of known
magnetic phases. A discussion is also given about the
relation of calculated results to the unusual low energy
excitations seen in many of these metals.

Local density theory is an approximate method used to calculate ground
state properties of a many-body system. The local density
approximation (LDA) involves replacing the exact (unknown)
exchange-correlation functional (representing the non-direct part of
the Coulomb interaction) by that of a homogeneous electron gas (the
condition being invoked at each point in space). Such an approximation
would appear to be so gross as to be inapplicable in practical
circumstances, but the LDA turns out to work remarkably well for most
systems (1). One way to see why this is so is to note that the exact
exchange energy density can be represented by a shape function times
the density to the one-third power (2). The LDA simply replaces this
shape function by a constant, which also occurs for the exact
expression when the energy density is integrated to yield the total
energy. On the other hand, the effective potential is obtained by
functionally differentiating the energy, and thus samples the shape
function. This in turn implies that one must be careful when
interpreting one-particle eigenvalues obtained from the effective
potential. In particular, it is now known that the LDA places valence d
levels about 20 mRy too low at the beginning of the transition metal
series (3), and valence f levels about 50 mRy too low for La, and 40
mRy too high for Tm (4).
 There is a deeper question that has been raised, however, in the case
of f-electron metals. Given the large degree of localization of these
electrons and the strong correlations among them, it has been
suggested that the LDA is a poor starting point for understanding the

0097–6156/89/0394–0273$06.00/0

electronic nature of these materials (5). This is a legitimate question, and it is our purpose here to address this problem.

Bulk Properties

Evaluation of the total energy as a function of atomic position can lead to such useful structural information as the equilibrium lattice constant, bulk modulus, and preferred crystal structure. Reviews of some of this work can be found in Refs. 6 and 7. For the 4f elemental metals from Eu to Lu, the observed equilibrium lattice constants are well represented by LDA calculations which treat the f electrons as core states (7) (The inapplicability of Fermi-Dirac statistics for treating highly localized electrons requires the imposition of a fixed occupation number). This need of forcing the f electrons into the core represents a well-known limitation, namely, that the LDA functional does not minimize the f count per site with integer occupation values in the localized limit. In atomic and molecular cases, this deficiency can be remedied by self-interaction corrections (8-9), but the latter terms are conceptually difficult to handle for extended systems. For the 4f elements Pr to Sm, there is about a 4-5% overestimation of the lattice constant, indicating the presence of some f bonding. For Ce metal, the lattice constant of γ-Ce is given fairly well by treating the f electron as a core state, whereas that for α-Ce is underestimated by about 5% when treating the f electron as a band state. Thus, the LDA calculations support the notion that the α to γ transition involves a localization of the f electron, whereas the above mentioned underestimate indicates that the LDA overestimates f bonding for itinerant f metals.

As for the bulk modulus, the LDA well represents its trend across the 4f series, although the absolute errors are quite large (7). A careful analysis of the bulk modulus for the mixed valent metal TmSe indicates that the LDA error in that case can be corrected if the f levels are shifted down 40 mRy (4). This problem is connected to the fact that the LDA does progressively worse for higher angular momentum states (due to probable misrepresentation of the shape of the exchange-correlation hole). Self-interaction corrections may be able to explain this error (9-10).

A great deal of work has been done for the 4f elemental metals concerning the relative stability of various crystal structures. For example, LDA predicts the correct fcc structure for Ce in the 4f band case (11), whereas in the later lanthanides, the crystal structure is determined by the d states. The reader is referred to Refs. 11 and 7 for a further discussion of these results.

The situation improves for the 5f elemental metals due to the greater delocalization of the 5f as compared to the 4f electrons. In fact, the equilibrium lattice constants of the 5f elemental metals are well reproduced even with the f electrons treated as band states, as long as spin polarization is taken into account (12). The spin polarization "turns on" at Am, and correlates with an inferred localization of the f electrons. Even for such strongly correlated metals as the heavy fermion superconductors UPt_3 and UBe_{13}, the equilibrium lattice constant and bulk modulus are well-given by LDA f band calculations (13-14).

Fermi Surfaces

One of the best ways to determine whether an itinerant or localized
picture is applicable is to compare experimental and theoretical
determinations of the Fermi surface topology and effective masses. It
has been known for many years that the Fermi surfaces of most
elemental 4f metals can be represented by an f core calculation (15).
In the case of Pr, the calculated effective masses are a factor of four
too small, a result which can be attributed to a polaronic drag of the
conduction electrons caused by induced virtual excitations between f
crystal field levels (16). This is a dynamic correlation effect, and is
therefore not given when performing a ground state calculation, but can
be represented by adding a self-energy correction to the (ground state)
band results. It is now known that the heavy fermion metal CeB_6 is
also describable by this picture, with the mass renormalization being
of the order of fifty instead of four. This observation is of some
importance, as it implies that large effective masses and f character
in the Fermi surface are not equivalent statements. Also, calculations
indicate that the Fermi surface topologies of both LaB_6 and CeB_6 can be
improved by shifting the unoccupied f levels upwards by about 50 mRy.
This error is analogous to the 20 mRy shift needed for d band metals
such as V and Nb (3).
   These cases can be contrasted by most uranium intermetallics, which
have Fermi surfaces in good agreement with LDA calculations which
treat the f electrons as band states (17). In the one case where a
mixed valent Fermi surface is known ($CeSn_3$), it is also in excellent
agreement with an LDA f band calculation (18-19), with a mass
renormalization of five due to a self-energy correction resulting from
virtual spin fluctuation excitations (20). Notice the different dynamic
correlations used to explain the mass renormalizations in the f core
and f band cases.
   Of great relevance is the recent experimental determination of the
Fermi surface of the heavy fermion superconductor UPt_3 (21), which is
found to be in excellent agreement with LDA f band calculations
(22-23), despite the fact that the Fermi surface is composed of five f
bands. This, along with the above-mentioned structural information,
indicates that the static charge correlations are handled quite
adequately by the LDA. Again, there is a large mass renormalization (a
factor of twenty in this case), attributable to a spin fluctuation
self-energy correction (20).
   It is interesting to contrast the case of hcp UPt_3 to the case of UPd_3,
which has a closely related crystal structure (dhcp). It turns out that
the UPd_3 Fermi surface is well-represented by an f core calculation
(24). Thus one could say that UPt_3 is on the verge of a localization
(Mott) transition. In conclusion, it is clear that the LDA calculations
have been very helpful in sorting out the nature of the ground states for
those f electron metals which show anomalous behavior.

Magnetic Properties

The existence of strong antiferromagnetic correlations in heavy
fermion metals has been confirmed by extensive neutron scattering
studies (25), and these correlations are now thought to be responsible
for both the large mass renormalizations and exotic superconductivity

seen in these metals (Ref. 20 and references therein). It is interesting to note that an antiferromagnetic tendency was predicted from band structure calculations before the neutron scattering experiments were done (13). In reality, however, many heavy fermion metals, although showing strong magnetic fluctuations, exhibit either a small or zero ordered moment. This fact has been used to question the validity of the local spin density (LSD) approximation (5).

The importance of the above-mentioned criticism has led us to undertake polarized calculations on the following metals: the heavy fermion magnets $NpSn_3$ (26), TmSe (27), and UCu_5 (Norman, M. R.; Min, B. I.; Oguchi, T.; Freeman, A. J. Phys. Rev. B, in press.), and the heavy fermion superconductors UPt_3 and URu_2Si_2 (Norman, M. R.; Oguchi, T.; Freeman, A. J. Phys. Rev. B, in press.). At the outset, it must be stated that there does not exist a proper LSD treatment in the presence of spin-orbit coupling, due to the fact that spin is no longer a good quantum number. The approximation we employ is to include the spin-orbit terms as a perturbation within each self-consistent iteration. The orbital contribution to the magnetic moment can then be determined by the method suggested by Brooks and Kelly (28). We note, however, that the orbital magnetic moment does not make any contribution to the energy functional in this formalism, which would not be true in an exact theory. Despite this, the formalism works quite well in determining the total magnetic moment for various uranium compounds (29), and shows the proper Hund's rule effect that the spin and orbital moments are antiparallel in the first half of a series, and parallel for the second half. In the cases of $NpSn_3$ and UCu_5, magnetic moments in good agreement with experiment were found, whereas the discrepancy with experiment in TmSe can be attributed to multiplet effects (which are not included in the LDA). Moreover, in all these cases, one finds strong reductions of the electronic specific heat coefficients at the magnetic phase transition which are also explained quite well by the calculations. Therefore, a standard LSD treatment of magnetism seems to work quite well.

This can be contrasted with the case of the above-mentioned super-conductors. We calculate a magnetic moment of 0.8 μ_B for UPt_3 and 1.2 μ_B for URu_2Si_2 as compared to the 0.02 μ_B (30) and 0.03 μ_B (31) values actually observed. This is the origin of the justifiable criticism mentioned in Ref. 5. The essential difference in the band calculations between the magnets and the superconductors is that in the former, the Stoner criteria for magnetism is greatly exceeded, whereas in the latter cases it is only weakly exceeded, indicating that in these cases, some other effect occurs which acts to greatly suppress the LSD predicted moment. We believe that this is the Kondo effect and note that the band structure moment is recovered by either doping the metal or subjecting it to a strong magnetic field. The inclusion of this "Kondo suppression" effect in an LSD calculation is a major challenge. This suppression is of extreme importance, as without it, these heavy electron metals would be strong magnets instead of superconductors.

Conclusions

Despite several problems, the local density approximation has turned out to be quite useful in elucidating information about the ground state properties of f electron metals. For further information, the reader is referred to reviews given in Refs. 3, 6, 7, 12, 15, 17, and 19. For a further discussion of some problems with local density theory as regards to f electron metals, see Ref. 32. Finally, an excellent discussion of the relation of band calculations to observed low energy excitations seen in heavy electron metals has been given by Pickett (33).

Acknowledgments

Work at Argonne National Laboratory was supported by the U. S. Dept. of Energy, Office of Basic Energy Sciences, under Contract No. W-31-109-ENG -38, and at Northwestern University by the National Science Foundation (DMR Grant No. 85-18607).

Literature Cited

1. Theory of the Inhomogeneous Electron Gas; Lundqvist, S.; March, N., Eds.; Plenum: New York, 1983.
2. Harris, J. Phys. Rev. A 1984, 29, 1648.
3. Koelling, D. D. In The Electronic Structure of Complex Systems; Phairseau, P.; Temmerman, W. M., Eds.; Plenum: New York, 1984; p. 183.
4. Jansen, H. J. F.; Freeman, A. J.; Monnier, R. Phys. Rev. B 1986, 33, 6785.
5. Lee, P. A.; Rice, T. M.; Serene, J. W.; Sham, L. J.; Wilkins, J. W. Comments on Cond. Mat. Phys. 1986, 12, 99.
6. Norman, M. R.; Koelling, D. D. J. Less-Common Metals 1987 127, 357.
7. Freeman, A. J.; Min, B.I.; Norman, M. R. In Handbook on the Physics and Chemistry of Rare Earths; Gschneidner, K. A., Jr.; Eyring, L.; Hufner, S., Eds.; North Holland: New York, 1987; Vol. 10, p. 165.
8. Perdew, J. P. In Density Functional Methods in Physics; Dreizler, R. M.; daProvidencia, J., Eds.; Plenum: New York, 1985; p. 265.
9. Norman, M. R.; Koelling, D. D. Phys. Rev. B 1984, 30, 5530.
10. Harrison, J. G. J. Phys. Chem. 1983, 79, 2265.
11. Skriver, H. L. Phys. Rev. B 1985, 31, 1909.
12. Johansson, B.; Eriksson, O.; Brooks, M. S. S.; Skriver, H. L. Physica Scripta T 1986, 13, 65.
13. Sticht, J.; Kubler, J. Solid State Comm. 1985, 58, 389.
14. Pickett, W. E.; Krakauer, H.; Wang, C. S. Physica B 1985, 135, 31.
15. Liu, S. H. In Handbook on the Physics and Chemistry of the Rare Earths; Gschneidner, K. A., Jr.; Eyring, L., Eds.; North Holland: New York, 1978; p. 233.
16. Fulde, P.; Jensen, J. Phys. Rev. B 1983, 27, 4085.
17. Arko, A. J.; Koelling, D. D.; Schirber, J. E. In Handbook on the Physics and Chemistry of the Actinides; Freeman, A. J.; Lander, G. H., Eds.; North Holland: New York, 1985; Vol. 2, p. 175.
18. Koelling, D. D. Solid State Comm. 1982, 43, 247.
19. Norman, M. R. In Theoretical and Experimental Aspects of Valence Fluctuations and Heavy Fermions; Gupta, L. C.; Malik, S.; K., Eds.; Plenum: New York, 1987; p. 125.
20. Norman, M. R. Phys. Rev. Lett. 1987, 59, 232.
21. Taillefer, L.; Lonzarich, G. G. Phys. Rev. Lett. 1988, 60, 1570.

22. Oguchi, T.; Freeman, A. J.; Crabtree, G. W. J. Magn. Magn. Matls. 1987, 63-64, 645.
23. Wang, C. S.; Norman, M. R.; Albers, R. C.; Boring, A. M.; Pickett, W. E.; Krakauer, H.; Christensen, N. E. Phys. Rev. B 1987, 35, 7260.
24. Norman, M. R.; Oguchi, T.; Freeman, A. J. J. Magn. Magn. Matls. 1987, 69, 27.
25. Goldman, A. I. Jap. Jour. Appl. Phys., Suppl. 1987, 26-3, 1887.
26. Norman, M. R.; Koelling, D. D. Phys. Rev. B 1986, 33, 3803.
27. Norman, M. R.; Jansen, H. J. F. Phys. Rev. B 1988, 37, 10050.
28. Brooks, M. S. S.; Kelly, P. S. Phys. Rev. Lett. 1983, 51, 1708.
29. Brooks, M. S. S. Physica B 1985, 130, 6.
30. Aeppli, G.; Bucher, E.; Broholm, C.; Kjems, J. K.; Baumann, J.; Hufnagl, J. Phys. Rev. Lett. 1988, 60, 615.
31. Broholm, C.; Kjems, J. K.; Buyers, W. J. L.; Matthews, P.; Palstra, T. T. M.; Menovsky, A. A.; Mydosh, J. A. Phys. Rev. Lett. 1987, 58, 1467.
32. Norman, M. R. In Condensed Matter Theories; Vashishta, P.; Kalia, R. K.; Bishop, R., Eds.; Plenum: New York, 1987; Vol. 2, p. 113.
33. Pickett, W. E. In Novel Superconductivity; Wolf, S. A.; Kresin, V. Z., Eds.; Plenum: New York, 1987; p. 233.

RECEIVED October 24, 1988

Chapter 20

Correlations in d and f Electron Systems

G. Stollhoff and P. Fulde

Max-Planck-Institut für Festkörperforschung, D–7000 Stuttgart 80, Federal Republic of Germany

A discussion is given of electron correlations in d- and f-electron systems. In the former case we concentrate on transition metals for which the correlated ground-state wave function can be calculated when a model Hamiltonian is used, i.e. a five-band Hubbard Hamiltonian. Various correlation effects are discussed. In f-electron systems a singlet ground-state forms due to the strong correlations. It is pointed out how quasiparticle excitations can be computed for Ce systems.

Electron correlations in d- and f-electron systems are presently in the center of considerable activity in theoretical solid-state physics. There are two distinct classes of materials which require a better understanding of electron correlations for their intriguing physical properties. To one class belong the recently discovered high-T_C superconducting materials (1) like $La_{2-x}M_xCuO_4$ (M=Ba, Sr), $YBa_2Cu_3O_{7-\delta}$ or $Bi_2Sr_2CaCu_2O_{8-\delta}$. The characteristic structure of these materials involves layers of $(CuO_2)_n$ with approximately one hole per formula unit, depending on the degree of doping or oxygen deficiency. From experiments it is known that the electrons in the layers are strongly correlated. For example, La_2CuO_4 is not a metal despite of the fact that it should have a half-filled band. Instead it is a Mott-Hubbard insulator. To the other class of materials belong the heavy- fermion systems (2-6). Examples are Ce compounds like $CeAl_3$, $CeCu_2Si_2$ or U compounds like UPt_3 or UBe_{13}. In this case the 4f or 5f electrons are responsible for the unusual physical properties of the systems. They are only weakly coupled to the remaining electrons and due to the strong on-site Coulomb

0097–6156/89/0394–0279$06.00/0

repulsions one is again dealing with strong electron correlations.

The most studied d electron systems are the transition metals. Here the on-site Coulomb interactions and the d-band width are of similar size and therefore correlations are moderately strong. We shall concentrate on these systems and describe our present understanding of the groundstate and of effects caused by correlations (7). The latter include reductions of charge fluctuations, a build up of atomic magnetic moments due to Hund's rule correlations, as well as modifications in the Stoner-Wohlfarth criterion for the onset of magnetic order. At present the correlated ground-state wave function can be computed only for a model Hamiltonian, for which we choose a five-band Hubbard system. This approach contrasts present day band-structure calculations based on the local density (LDA) or spin density (LSD) approximation to the density-functional theory (8). In LDA or LSD calculations one does not attempt to compute a ground-state wave function but instead calculates directly certain ground-state properties such as the density distribution or energy. Exchange and correlation are dealt with by using results from the homogeneous electron gas. The advantages of that approach are its conceptual and computational simplicity and a number of outstanding successes. But the price one has to pay are uncertain approximations. Furthermore there is no way of determining the many-body wave function.

In the second part of the lecture we want to demonstrate in the simplest possible way the problem of strongly correlated f-electrons. It is intimately connected with the Kondo lattice problem, or alternatively with the formation of a singlet state (5,6).

Correlations in Transition Metals

We start from a model Hamiltonian which describes the five canonical d bands with dispersions $\varepsilon_\nu(\underline{k})$ ($\nu=1,..5$) and includes on-site interactions

$$H = \sum_{\nu\sigma\underline{k}} \varepsilon_\nu(\underline{k})n_{\nu\sigma}(\underline{k}) + \sum_l H_I(l) \tag{1}$$

$$H_I(l) = \frac{1}{2}\sum_{ij\sigma\sigma'} [\ U_{ij}a_{i\sigma}^+(l)a_{j\sigma'}^+(l)a_{j\sigma'}(l)a_{i\sigma}(l) +$$

$$+J_{ij}[a_{i\sigma}^+(l)a_{j\sigma'}^+(l)a_{i\sigma'}(l)a_{j\sigma}(l)+$$

$$+a_{i\sigma}^+(l)a_{i\sigma'}^+(l)a_{j\sigma'}(l)a_{j\sigma}(l)]]$$

The only free parameter in the one-particle part of the Hamiltonian is the total band width W.   The $n_{\nu\sigma}(\underline{k})$ are the number operators for the Bloch states.   The $a_{i\sigma}^+(\underline{1})$ ($a_{i\sigma}(\underline{1})$) create (destroy) electrons with spin σ in atomic-like d orbital i at site 1.   The on-site interaction matrix elements U_{ij} have the form

$$U_{ij} = U + 2J - 2J_{ij} \tag{2}$$

where U and J are average Coulomb and exchange interaction constants and J_{ij} contains the anisotropies. We shall assume that $J = 0.2U$ and $\Delta J = 0.15\ J$ where J and ΔJ are the isotropic and anisotropic part of J_{ij}, respectively.   Further details can be found in Ref. (9,10).   The Hartree-Fock (HF) ground state is written as

$$|\phi_0\rangle = \prod_{\substack{\underline{k}\nu\sigma \\ \varepsilon_{\underline{k}\nu} \leq \varepsilon_F}} c_{\nu\sigma}^+(\underline{k})\ |0\rangle \tag{3}$$

where the operators $c_{\nu\sigma}^+(\underline{k})$ refer to Bloch states and $|0\rangle$ denotes the vacuum.   By expanding

$$c_{\nu\sigma}^+(\underline{k}) = \sum_{il} \alpha_i(\nu,\underline{k})\ a_{i\sigma}^+(1)\ e^{i\underline{k}\underline{R}_1} \tag{4}$$

one can decompose $|\phi_0\rangle$ into different configurations. In figure 1 two of them are shown with different interaction   energies.   One   configuration   (a)   is favourable while the other (b) is unfavourable with respect to the Coulomb repulsion energy.   One notices that in (b) the electron numbers at different sites fluctuate much more around the chosen average number of 2.5 than in (a).   Electron correlations decrease the weight of unfavourable configurations which is too large in the HF ground state.   Stated differently, their effect is to reduce charge fluctuations at the different sites as compared with the ones contained in $|\phi_0\rangle$. Electron   correlations   are   introduced   into   the ground-state wave function by the ansatz

$$|\psi_0\rangle = e^S\ |\phi_0\rangle \tag{5}$$

The operator S is of the form

$$S = - \sum_{ijl} \eta_{ij}\ O_{ij}(1) \tag{6}$$

with

$$O_{ij}(1) = \begin{cases} n_{i\uparrow}(1) \ n_{i\downarrow}(1) \\ n_i(1) \ n_j(1) \\ \underline{s}_i(1) \ \underline{s}_j(1) \end{cases} \tag{7}$$

The $n_{i\sigma}(1)$, $\underline{s}_i(1)$ are the electron number and spin operators for the different sites, i.e. $n_{i\sigma}(1)=a_{i\sigma}^+(1)a_{i\sigma}(1)$ and $n_i=\sum_\sigma n_{i\sigma}$. Furthermore, in order to exclude single-particle excitations, contractions within the operators $O_{ij}(1)$ are excluded when expectation values are evaluated. For more details see e.g. Ref. (9). The parameters η_{ij} follow from minimizing the energy

$$E = \frac{<\Psi_0|H|\Psi_0>}{<\Psi_0|\Psi_0>} \quad = <e^S \ H \ e^S>_c \tag{8}$$

The subscript c indicates that only connected diagrams have to be taken. The last equation follows after applying a linked cluster theorem. Furthermore the abbreviation $<....> = <\phi_0|...|\phi_0>$ has been used. The energy is evaluated after expanding

$$E = <H> - 2 \sum_{ijl} \eta_{ij} <O_{ij}(1)H>_c +$$
$$+ \sum_{ijmn} \sum_{ll'} \eta_{ij} <O_{ij}(1) \ H \ O_{mn}(1')>_c \ \eta_{mn} \tag{9}$$

The expectation values are evaluated within the R=0 approximation introduced by Friedel and coworkers (see e.g. Ref. (11)) in which only terms with l=l' are kept. With these approximations calculations become very simple. In the following a number of results are presented. The reduction of charge fluctuations as a function of d band filling is shown in figure. 2 for a bcc lattice and U/W = 0.5. Plotted is the mean square deviation of the electron number

$$\Delta n^2 = < \Psi_0|n^2(1)|\Psi_0> - <\Psi_0|n(1)|\Psi_0>^2 \tag{10}$$

where $n(1) = \sum_i n_i(1)$. For comparison we also show the corresoponding results for the HF ground-state. Hund's rule correlations can be studied by computing

$$S^2 = <\Psi_0|\underline{S}^2(1)|\Psi_0> \tag{11}$$

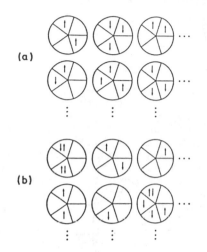

(a)

(b)

Figure 1. (a) "favourable" and (b) "unfavourable" configuration contained in the nonmagnetic HF ground state $|\phi_o\rangle$. The circles symbolize atoms and the five segments the different d-orbitals. The d-electron occupancy per atom is chosen to be 2.5.

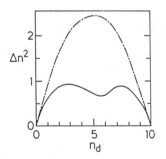

Figure 2. Charge fluctuations as function of d-band filling n_d (bcc structure). Upper curve without and lower curve with electron correlations included. $U/W=0.5$. (Adapted from ref. 9.)

where $\underline{S}(1) = \sum_i \underline{S}_i(1)$. Results are shown in figure 3
where $\bar{S}^2$ is compared with the corresponding expectation
value in HF approximation (S_{HF}^2) and in the localized
limit (S_{loc}^2), respectively. It is seen from this figure that the relative spin alignment

$$\Delta S^2 = \frac{S^2 - S_{HF}^2}{S_{loc}^2 - S_{HF}^2} \tag{12}$$

is almost constant and approximately 0.5 for a ratio
$U/W \hat{=} 0.5$. When computed as function of U, ΔS^2 is found
to be a monotonously increasing function rather than a
function which is different from zero only above a
threshold value U_c. This implies that a fluctuating
local moment is building up continuously as function of
U at the different sites. This is in contrast to the
findings of the coherent potential approximation, when a
static local moment is the only way of accounting for
electrons correlations (12). In that case a static
magnetic moment is found provided that $U>U_c$.
     Electron correlations influence strongly the energy
difference between nonmagnetic and magnetic states,
leading to drastic changes of the Stoner-Wohlfarth criterion for the onset of ferromagnetic order. The reason
is that electronic charge fluctuations are smaller in a
ferromagnetically ordered than in a nonmagnetic state.
Therefore electron correlations decrease the energy by a
larger amount of a state which is nonmagnetic, than of a
ferromagnetic state. For example, in Fe the energy gain
due to ferromagnetic order is 0.56 eV/atom when the HF
approximation is made and a ratio $U/W=0.44$ is assumed.
When density- and in addition spin correlations are
included, this energy reduces to 0.22 eV/atom and 0.15
eV/atom, respectively (10). In the LDA to the density
functional theory, Hund's rule correlations are not
taken into account, because they are not present in an
unpolarized homogeneous electron gas from which the
exchange-correlation potential is taken. When the LSD
approximation is applied instead, they are partially
included. Spin correlations, however, modify the
generalized Stoner parameter strongly (13). The latter
can be related to the exchange correlation energy
$E_{xc}(M_0)$ for fixed magnetization M_0 by writing

$$E_{xc}(M_0) = E_{xc}(0) + \frac{1}{4} \int_0^{M_0} dM^2 I_{xc}(M) \tag{13}$$

where $I=I_{xc}(0)$ is the original Stoner parameter. Figure
4 displays the magnetic field dependence of the Stoner
parameter as obtained within three different approximations for the case of Co. I_1 is the result of a

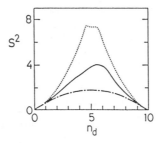

Figure 3. Atomic spin correlations S^2 as function of d-band filling n_d (bcc structure). Upper curve: atomic limit; lower curve: without correlations. The solid line represents the results for $|\psi_o\rangle$. $U/W=0.5$. (Adapted from ref. 9.)

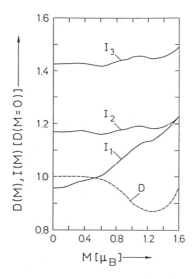

Figure 4. Stoner parameter $I(M)$ and loss of kinetic energy $D(M)$ (dashed line) for Co as functions of magnetization. I_1 - full correlation calculation, I_2 - neglecting spin correlations, I_3 - neglecting all correlations, resulting from J_{ij}. (Adapted from ref. 13.)

complete correlation calculation. I_2 is the Stoner parameter when spin correlations are neglected. Since spin correlations play no role for fully magnetic states, both curves become equal at $M_{max}=1.6$. While I_2 does not display sizeable magnetic field dependencies and compares in this respect with results of LSD computations, I_1 increases by 20% from M=0 to M_{max}. I_3 finally is the curve obtained when all exchange contributions ($\sim J_{ij}$) to the interaction part of the Hamiltonian (Eq. 1) are treated in HF approximation. I_2 and I_3 may be considered as lower and upper limit of the deficiencies of LSD. Although a change of I by 20% may seem small, it has the effect of changing the Curie temperature T_C by approximately a factor of two, because $T_C \sim (IN(0)-1)^{1/2}$ and $IN(0) \simeq 1$, However, even that is not sufficient in order to bring the larger calculated values for T_C in agreement with experiments. This is due to the fact that Stoner theory does not contain fluctuations of the order parameter. For improved calculations of T_C see e.g. Ref. (14).

Another point of considerable importance is the nonlocal character of the exchange. The latter always favours non-uniform distributions of electrons (or holes) among the different d orbitals, e.g. e_g and t_{2g} orbitals, when the system is cubic. Direct Coulomb interactions as well as correlations favour uniform occupations of the different atomic orbitals and therefore counteract the effect of nonlocality of the exchange. Despite this, the anisotropies caused by exchange are important, in particular for bulk Ni (10) as well as for its surface (15).

Finally it is of interest to compute spin correlations between neighboring sites, i.e.

$$S^2(\delta) = <\psi_0|\underline{S}(0)\underline{S}(\delta)|\psi_0>, \qquad\qquad (14)$$

where δ denotes a nearest neighbor of site 0. $|\psi_0>$ is calculated by starting from the nonmagnetic SCF-state as before, but by including in S (see Eq. 6) also operators of the form $O_{ij}(1,1+\delta)=\underline{S}_i(1)\underline{S}_j(1+\delta)$. The effect of these operators is that additional ferromagnetic correlations betweeen electrons on neighboring sites are built into $|\psi_0>$, except for band fillings close to $n_d=5$. Despite this one still finds in the regime $6 \leqslant n_d \leqslant 8$ net antiferromagnetic correlations between neighboring sites to prevail (15), although in this region the ground state is a ferromagnet. This suggests that correlations between more than two sites (clusters) are necessary in order to arrive at ferromagnetic alignments of electrons. One may conclude from the findings that the strong magnetic fluctuations above T_C originate from extended order-parameter fluctuations and not from atomic local moments.

Correlations in f-Electron Systems

Because of the good localization of the f-electron orbital the overlap of it with atomic orbitals of nearest neighbors is rather small.   But the Coulomb repulsion of electrons in a f-orbital is large.   For that reason one is in the strong correlation limit.   In order to demonstrate the effects of the correlations we consider the simplest possible model, which contains all the important ingredients.   It consists of two orbitals l and f with energies $\varepsilon_l > \varepsilon_f$.   The index l stands for ligand orbital and we assume that it is rather extended so that we may neglect Coulomb repulsions within that orbital.   The Hamiltonian then reads

$$H = \varepsilon_l \sum_\sigma l_\sigma^+ l_\sigma + \varepsilon_f \sum_\sigma f_\sigma^+ f_\sigma + V \sum_\sigma (f_\sigma^+ l_\sigma + l_\sigma^+ f_\sigma) + U n_\uparrow^f n_\downarrow^f$$

(15)

with $n_\sigma^f = f_\sigma^+ f_\sigma$ and U very large.

We want to discuss the solutions of the eigenvalue problem for two electrons.   First we set V=0.   In that case, because $\varepsilon_l > \varepsilon_f$ the ground state is a quartet with energy $E_0 = \varepsilon_l + \varepsilon_f$, i.e. one electron is in the f orbital and the other is in the l orbital.   The excited state is a singlet with $E_s = 2\varepsilon_l$, i.e. both electrons are in the l orbital.   When $V \neq 0$ is taken into account, the ground-state quartet splits into a low lying singlet

$$|\psi_0\rangle = \frac{1}{\sqrt{2}} (1 - (V/\Delta\varepsilon)^2)(f_\uparrow^+ l_\downarrow^+ - f_\downarrow^+ l_\uparrow^+)|0\rangle - \frac{\sqrt{2}V}{\Delta\varepsilon} l_\uparrow^+ l_\downarrow^+ |0\rangle$$

(16)

with energy $E_0 = \varepsilon_l + \varepsilon_f - 2V^2/\Delta\varepsilon$ and a triplet

$$|\psi_{\varepsilon 1}\rangle = (1 - (V/\Delta\varepsilon)^2) l_\uparrow^+ l_\downarrow^+ |0\rangle + \frac{V}{\Delta\varepsilon} (f_\uparrow^+ l_\downarrow^+ - f_\downarrow^+ l_\uparrow^+)|0\rangle$$

(17)

$$|\psi_{\varepsilon 2}\rangle = f_\uparrow^+ l_\uparrow^+ |0\rangle \quad ; \quad |\psi_{\varepsilon 3}\rangle = f_\downarrow^+ l_\downarrow^+ |0\rangle$$

with energy $E_t = \varepsilon_l + \varepsilon_f$.   We have set $\varepsilon_l - \varepsilon_f = \Delta\varepsilon$ (see figure 4). The formation of the singlet $|\psi_0\rangle$ with an excited triplet state of excitation energy $E_{ex} = 2V^2/\Delta\varepsilon$, is a characteristic feature of strongly correlated f-electron systems.   When the f orbital is embedded in a sea of conduction electrons, the energy gain due to the singlet formation becomes

$$\Delta E = D \exp[-\frac{|\varepsilon_f|}{2N(0)V^2}]$$

(18)

instead of $\Delta E = 2V^2/\Delta\varepsilon$, as in the case of two orbitals. Here D is the conduction-electron band width and 2N(0) is their density of states.   We have set the Fermi energy equal to zero.   The energy gain is usually identified with a characteristic temperature $k_B T_K = \Delta E$,

the Kondo temperature.   The low lying excitations lead
to heavy-fermion behaviour when the ions with f electron
form a lattice.
        The above calculation suggests that a singlet
formation due to strong correlations with a triplet
excited state should be found in appropriate molecules.
The effect requires an even total number of valence
electrons.   In order to detect it one should search e.g.
for molecules containing Ce, which are diamagnetic, but
which show a f-electron count close to 1, when photoemission experiments are performed.
        The formation of a singlet state due to strong
correlations implies also a new kind of electron-phonon
coupling.   The energy gain ΔE due to singlet formation
depends on the hybridization V, which in turn depends on
pressure P or volume Ω.   In particular in a solid this
dependence $\Delta E(V)$ is very strong (see Eg. (18)),
resulting in a strong electron phonon coupling.   Its
strength can be characterized by an electronic Grüneisen
parameter

$$\eta = \frac{- d\ln T_K}{d\ln \Omega} \tag{19}$$

Measured values of η are as large as 100-200 in heavy
fermion systems (17).
        One important problem, which is presently under
intense investigations is that of the Fermi surface of
strongly correlated f-electron systems.   It was a surprise, at least to the present authors, that the
measured Fermi surface of the heavy-fermion system UPt_3
(18) is very much in accord with the one computed within
LDA (19).   There is no a priori reason why the topology
of the Fermi surface should come out correctly when
electron correlations are strong and a LDA is made.   But
for UPt_3 it does come out surprisingly well, although
the measured effective masses are off by a factor of
order 20 as compared with the calculated ones.   Detailed
investigations have shown (20) that the good agreement
in the case of UPt_3 is due to a large spin-orbit
splitting and a crystal-field (CEF) splitting, which is
much less than $k_B T_K$, i.e. the energy gain due to singlet
formation.   In that case, the theory becomes a
one-parameter (which is the effective mass) theory, and
the topology of the Fermi surface due to the heavy
quasiparticles is completely determined by the geometry
of the unit cell.   In cases in which the CEF splitting
is larger than $k_B T_K$, one expects differences between the
measured Fermi surface and the one which follows from
applying the LDA.   In order to improve the computation
of the Fermi surface one can proceed as follows, at
least for Ce compounds.   One applies the LDA to the
density functional theory for all electrons, except the
f-electrons.   The potential acting on the latter is

described by an energy dependent phase shift $\delta_{l=3}(\epsilon)$, for which a simple, phenomenological ansatz is made. Only these channels within the l=3 manifold obtain a phase shift different from zero, which have the symmetry Γ of the crystal-field ground state. The latter is usually known from inelastic neutron scattering experiments. The slope α of the f-electron phase shift at the Fermi energy ϵ_F is put into the theory as a parameter, which is determined by fitting the measured linear specific heat coefficient. Thus one sets

$$\delta_{\Gamma(l=3)}(\epsilon) = \delta_0 + \alpha(\epsilon-\epsilon_F) \tag{20}$$

where δ_0 is fixed by Luttinger's theorem. For more details we refer to Figure 5. For U compounds with two or three f-electrons it is not clear yet, how an analoguous calculation can be performed. There seems to be a long way to go from such a semiphenomenological approach to a truly ab initio type of calculation.

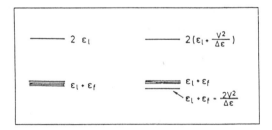

Figure 5.   Eigenvalues of the two electron model for U=∞. Left hand side: V=0; right hand side V≠0.

Acknowledgments

A number of helpful discussions with Dr. J. Keller and Dr. G. Zwicknagl are gratefully acknowledged.

Literature Cited

1. Bednorz, J.G.; Müller, K.A.   Z. Phys. 1986, B64, 189
2. Stewart, G.R.   Rev. Mod. Phys. 1984, 56, 755
3. Steglich, F. In Theory of Heavy Fermions and Valence-

Fluctuations, Kasuya, T.; Saso,T., Ed. Springer
Verlag, Berlin, 1985, p.23
4. Ott, H.R. In Progr. Low Temp. Physics 1988, Vol. XI,
in press
5. Fulde, P.; Keller, J.; Zwicknagl, G. In Solid State
Physics; Ehrenreich, H.; Turnbull, D., Ed.
Academic Press, New York, 1988; Vol. 41. p. 1
6. Lee, P.A..; Rice, T.M.; Serene, J.W.; Sham, L.J.;
Wilkins, J.W. Comments on Condensed Matter Physics
1986, 12, 99
7. For a review see e.g. Fulde, P. ; Kakehashi, Y.;
Stollhoff, G. In Metallic Magnetism; Capellmann,
H., Topics in Current Physics, Ed., Springer
Verlag Berlin, 1987; Vol. 42, p.159
8. For a review see e.g. Moruzzi, V.L.; Janak, J.F.;
Williams, A.R. Calculated Electronic Properties of
Metals, Pergamon Press, New York, 1978
9. Stollhoff, G.; Thalmeier, P. Z. Phys. 1981, B43, 13
10. Oleś, A.M.; Stollhoff, G. Phys. Rev. 1984, B29,
314
11. Kajzar, F.; Friedel, J. J. Physique 1978, 39, 379
12. Heine, V.; Samson, J.H.; Nex, G.M. J. Phys. 1981,
F11, 2645
13. Oleś, A.M.; Stollhoff, G. Europhys. Lett. 1988, 5,
175
14. Kakehashi, Y. Phys. Rev. 1988, B38, 474
15. Oleś, A.M.; Fulde, P. Phys. Rev. 1984, B30, 4259
16. Stollhoff, G. J. Magn. Magn. Mat. 1986, 54, 1043
17. Takke, R.; Niksch, M.; Assmus, W.; Lüthi, B.; Pott,
R,; Schefzyk, R.; Wohlleben, D.K. Z. Phys. 1981,
B44, 33
18. Taillefer, L.; Lonzarich, G.G. Phys. Rev. Lett.
1988, 60, 1570
19. Oguchi, T.; Freeman, A.J.; Crabtree, G.W. J. Magn.
Magn. Mat. 1987, 63, 645
20. Zwicknagl, G. In Proceedings ICCF6, J. Magn. Magn.
Mat. 1988, in press

RECEIVED March 8, 1989

Chapter 21

Ab Initio Relativistic Quantum Chemistry of Third-Row Transition Elements and Actinides

G. L. Malli

Department of Chemistry, Simon Fraser University, Burnaby, British Columbia V5A 1S6, Canada

Ab initio <u>fully relativistic</u> Dirac-Fock (DF) self-consistent field (SCF) calculations for a large number of diatomics of sixth-row elements and actinides (Th,U,Pu) are reported. Our methodology has been delineated recently in the <u>first</u> ab initio Dirac-Fock (DF) as well as relativistic configuration interaction (RCI) calculations (with DF as reference) for AuH. The effects of both <u>relativity</u> and <u>electron correlation</u> on bonding, dissociation energy, bond length, vibrational frequency, dipole moment, etc., of diatomic species involving atoms with $Z \geq 75$ are discussed. It is concluded that the <u>relativistic</u> (via the Dirac equation) and <u>electron correlation</u> effects <u>must</u> be included in all reliable ab initio calculations for these systems. Less rigorous and approximate methods are <u>unreliable</u> even though they can yield results which <u>fortuitously</u> agree with the experimental results.

The theory of relativity and quantum mechanics constitute the two basic foundations of theoretical physics. It is also well known that quantum mechanics based upon the Schrödinger equation has been used for decades to investigate atomic and molecular structure by physicists and chemists. However, the Schrödinger equation is non-relativistic; i.e., it is not Lorentz-invariant as it does not obey the special theory of relativity.

Since the behaviour of fast-moving bodies according to special relativity (with a finite velocity of light) is significantly different from that predicted by non-relativistic (NR) (which assumes velocity of light to be <u>infinite</u>) Newtonian mechanics (which Schrödinger's equation assumes), it is safe to conclude that the use of non-relativistic quantum mechanics (NRQM) to study the behaviour of electrons in atoms and molecules would not be appropriate if the electrons in these systems moved at velocity comparable to that of light. Indeed, even the simplest Bohr model of one-electron atom with nuclear charge Z predicts that this will happen for an atomic system with $Z \geq 70$. These simple considerations would therefore a priori suggest that for a proper understanding of the electronic

0097–6156/89/0394–0291$06.00/0

structure of atoms and molecules involving atoms of heavy elements
(Z>75) the relativistic quantum mechanics (RQM), i.e., quantum
mechanics which is in conformity with the special theory of
relativity is mandatory. Various attempts were made to wed quantum
mechanics and special theory of relativity (1-3); but it was
Dirac (4) who in 1928 discovered the famous Dirac equation
(linear in momentum and kinetic energy) for an electron, which not
only is Lorentz-invariant but also explains naturally the existence
of electron spin which was added as an ad hoc hypothesis in NRQM.
Dirac's theory has been found very successful and indeed its
predictions agree very well with experimental results (apart from
Lamb shift and higher order quantum electrodynamical corrections)
and forms the basis of relativistic treatment for one-electron
atomic (and molecular) systems except that it should be remarked
that Dirac's equation even for the simplest one-electron molecular
system $\underline{viz}$ H_2^+ cannot be solved $\underline{exactly}$, although exact solutions
for this system were obtained by Burrau (5) in 1927 using
Schrödinger non-relativistic equation. It was around this time that
Dirac (6) made the prophetic statement: "...the general theory of
quantum mechanics is now almost complete, the imperfections that
still remain being in connection with the exact fitting in of the
theory with relativity ideas... The underlying physical laws
necessary for the mathematical theory of a large part of physics and
the whole of chemistry are thus completely known...."

However, relativistic quantum mechanics was ignored by chemists
for decades because of the erroneous belief that in all atoms, the
valence electrons (in the outermost shells) which are primarily
responsible for the chemistry moved so slowly that their dynamics
was not significantly modified by relativity, although there was no
evidence to support this premise, especially in the case of valence
electrons of atoms of heavy elements (Z>75).

RELATIVISTIC EFFECTS ON VALENCE ELECTRONS IN HEAVY ATOMS

The first contrary evidence demonstrating conclusively that
relativity significantly modifies even the behaviour of valence
electrons in heavy atoms came from the earliest relativistic
calculation on the Hg atom by Mayers (7) who found that whereas the
binding energy of an electron in 6s shell was increased by
relativity and its mean radius was decreased significantly; the
electrons occupying the 5d relativistic orbitals were less strongly
bound as shown by their smaller binding energy and larger mean
radius compared to the corresponding non-relativistic values. Since
it was known that relativity stabilizes all one-electron atomic
states irrespective of the orbital occupied by the electron, Mayers
(7) pointed out that the destabilizing of the 5d electrons in the Hg
atom was due to an indirect effect, viz, relativity contracts
inner-shell orbitals thereby shielding the 5d electrons more
efficiently from the effect of the nuclear charge thereby weakening
the binding energies of electrons in these (i.e., 5d) orbitals and
increasing their mean radii. This indirect effect, which always
expands the orbital, although unimportant for s electrons, increases
in the order $s<\bar{p}<\underline{p}<\bar{d}<d$, being most important for $\bar{d}$ and d
electrons, where $\bar{\ell}$ and ℓ designate the relativistic sub-shells with
the total angular momentum $j = \ell-1/2$ and $j = \ell+1/2$, respectively, in
the notation introduced by Swirles (8). A few years later

Boyd et al. (9) found for the uranium atom also the destabilization of 5f electrons arising due to the indirect effect.

Therefore, electrons occupying penetrating valence orbitals having low angular momenta (s and $p_{1/2}$ electrons) are substantially stabilized by the direct relativistic effect due to the fact that they spend an appreciable amount of time near the heavy nucleus (with Z≥75) where their velocity is very appreciable compared to that of light, and are therefore affected very significantly by relativity. The direct relativistic effect decreases in importance with increasing total angular momentum of an electron becoming progressively smaller in the order s>$\bar{p}$>$\bar{d}$>d. hence, as a result of these two relativistic effects, the d and f electrons are destabilized while the s and p electrons are stabilized so that the electrons in the 5d (and 6d) and 5f outer atomic orbitals of the third-row transition elements and the actinides can no longer be ignored in the chemistry of these elements. In addition, the spin-orbit interaction splits an ℓ>0 level into two sub-levels with j = ℓ ±1/2. These effects increase as Z^2 and hence are very significant for atoms of heavy elements (Z≥75). Furthermore, as the number of electrons (and especially valence electrons in outer orbitals) in atoms of heavy elements is very large, not only the relativistic but in addition the electron correlation effects become very significant for systems involving heavy elements. Therefore the computational quantum chemists must face the formidable double challenge of the accurate calculation of both the effects of relativity and electron correlation for systems of heavy elements involving 5d, 6d and 5f electrons which happens to be the main theme of this symposium. RELATIVISTIC EFFECTS IN HEAVY ATOM MOLECULES Although a variety of approximate calculations for heavy atom molecules have been carried out (10-13), there is an obvious need for benchmark calculations using reliable ab initio fully relativistic methods for such systems. A computer program, called Relativistic Integrals Program (RIP), capable of performing fully relativistic ab initio calculations for heavy atom diatomics was recently used by Malli and Pyper (14) to discuss bonding in AuH. The RIP can handle any type of basis function which is centered on a nucleus and has the full symmetry of a relativistic atomic orbital. In all the calculations so far performed with the RIP, the numerical Dirac-Fock atomic orbitals (DFAO) of each isolated atom are used as the basis set. The DFAOs are computed using the Oxford Dirac-Fock program (15). This choice of basis set has the advantage of maximizing the insight into the molecular wavefunction in that the bonding can be easily interpreted in terms of the concepts traditionally and successfully used by quantum chemists. It has the additional advantage of avoiding certain technical difficulties involving the admixture into the wavefunction of the negative energy solutions of the Dirac equation, which describe the behaviour of positrons (16-17).

In computations with the RIP, it is found that many relativistic molecular orbitals (RMOs) are almost identical to the DFAOs of the constituent atoms in that the coefficient of one DFAO is very close to unity while the remaining coefficients are small. These orbitals and the electrons they contain are termed the core, and the electrons not included in the core are accommodated in the

valence RMOs constituting the valence wavefunction which, in
general, are composed of several DFAOs of the constituent atoms.
The basis set used to express the valence RMOs can consist of up to
four different types of functions (14):

(I)   DFAOs that are not completely filled in the isolated
        atoms;

(II)   DFAOs completely filled in the isolated atoms that might
        be significantly affected by the formation of the
        molecule;

(III)   Excited DFAOs unoccupied in the isolated atoms but which
        might contribute significantly to the formation of the
        molecule;

(IV)   Functions, called augmenting functions (AF), which are
        needed to describe small residual distortions of the
        valence charge distribution on formation of the
        molecule.

Although the basis set must include all the functions of type
(I), AFs are only needed for quantitatively accurate work. Since
the AFs are constructed to be orthogonal to DFAOs belonging to the
same atom, AFs make only small contributions to the RMOs. Hence the
essential features of the bonding can be understood by using a
basis, called the chemical basis (CB), consisting of functions
selected from types (I), (II) and (III). The role played in bonding
by functions of type (II), of which the $5\bar{d}$ and 5d DFAOs in the gold
atom are discussed in detail elsewhere (14). The importance of
hybridization involving orbitals belonging to type (III), for
example, the $6\bar{p}$ and 6p DFAOs in the gold atom, is deduced by
examining the coefficients with which they contribute to the
valence RMOs and by calculating the fraction of the binding energy
lost by excluding them from the basis set.

CORRELATION EFFECTS: THE RELATIVISTIC CONFIGURATION INTERACTION
The defects of RMO theory of overestimating molecular ionicities and
predicting incorrect dissociation products are remedied by using the
RIP to perform RCI calculations (14). This is achieved by using the
unoccupied RMOs resulting from an RMO calculation to construct more
accurate wavefunctions. Both the RMO and RCI wavefunctions can and
have been used to calculate molecular properties other than the
total energy, e.g., dipole moments, charge distributions, etc.

The details of our methodology as well as the computational
techniques are fully described elsewhere (14) and because of space
limitations cannot be repeated here. However, we present a summary
of the results of our ab initio fully relativistic Dirac-Fock SCF as
well as RCI calculations carried out using an extended basis set
(EBS) of 27 valence basis functions for AuH to demonstrate the
importance of the relativistic as well as electron correlation
effects in diatomics of third-row transition elements.

AB INITIO FULLY RELATIVISTIC DF SCF EBS AND RELATIVISTIC
CONFIGURATION INTERACTION (RCI) CALCULATIONS FOR AuH
The chemical basis set for AuH consists of the $5\bar{d}$, 5d and 6s DFAOs
of the gold atom plus the 1s DFAO of the hydrogen atom. Our DF SCF
calculations (14) show that the 6s-6p hybridization is not a
significant feature of the bonding so that $6\bar{p}$ and 6p DFAOs do not
enter the chemical basis set. However, since 1.0 eV out of the
experimental binding energy of 3.36 eV are lost if the $5\bar{d}$ and 5d

DFAOs and their associated electrons are placed in the core, these two DFAOs enter the chemical basis set. Indeed our RCI calculation in which these ten electrons are placed in the core predicts only a binding energy of 0.663 eV, which is about 20% of the experimental value. The computation of an accurate wavefunction with the $5\bar{d}$ and 5d DFAOs in the valence requires the addition of several AFs on both the Au and H atoms. The energy of AuH predicted at various internuclear distances by both the molecular orbital and configuration interaction methods was calculated using a large basis set including AFs. The bond length and fundamental vibration frequency were then predicted from these energies. A comparison of relativistic and exactly corresponding non-relativistic calculations (the latter performed simply by increasing the value of the velocity of light used) shows that relativity **significantly** contracts the bond length (R_e) and **substantially** increases the vibration frequency (ω_e). Thus the relativistic predictions of 2.993 au and 2102 cm^{-1} for R_e and ω_e, respectively, are to be compared with the non-relativistic values of 3.431 au and 1745 cm^{-1}; the latter are in serious disagreement with the experimental values (**18**) of 2.879 au and 2305 cm^{-1}, respectively. The predictions using the improved wavefunctions resulting from RCI calculations are 2.963 au and 2102 cm^{-1}, respectively, for R_e and ω_e.

Our results show that the quantitative predictions of the molecular properties of AuH are substantially changed by relativity. Furthermore, even the **qualitative features** of the bonding, revealed by RMO calculations with chemical basis sets, cannot be understood using non-relativistic theory. This is most readily demonstrated using localized valence RMOs calculated from the RMOs obtained from a RIP computation. The three localized AuH RMOs, two of which become σ orbitals with α spin and the third reducing to a π orbital with β spin in the non-relativistic limit, show that a gold 5d-6s hybrid orbital is responsible for the bonding through its interaction with the hydrogen 1s DFAO. This important hybridization is, however, found to be **absent** in the non-relativistic calculation which **erroneously** predicts that the bond is formed solely from the interaction between the gold 6s and the hydrogen 1s DFAOs. The non-relativistic calculation also **seriously underestimates** the degree of 5d-6s hybridization in one of the nonbonding orbitals. The greater relativistic importance of 5d-6s hybridization arises because the **indirect** relativistic effect **destabilizes** the $5\bar{d}$ and 5d DFAOs whilst the direct relativistic effect stabilizes the 6s DFAO. The combination of these two effects **reduces** significantly the 5d-6s energy gap thus leading to the greater 5d-6s hybridization in the relativistic case. Moreover, the NR wavefunction predicts greater polarity for AuH.

Our above-mentioned conclusions are contrary to the results reported from previous less rigorous calculations; e.g., Hay et al. (**11**) concluded from an effective core potential (ECP) calculation using an approximate treatment of relativity that the 5d^{10} core does not appear to play a dominant role in the chemical bond in AuH; this conclusion is incorrect in view of our result that 5d^{10} electrons cannot be left in the core in AuH. Our extended basis set (EBS), DF SCF calculation predicts a value of 1.682 eV for the D_e of AuH and our relativistic configuration interaction calculation with the

above-mentioned EBS DF SCF wavefunction as the reference and
including all single- and pair-type double excitations predicts
$2.014 \leq D_e \leq 2.376$ eV, whereas the calculation of Ziegler et al.
(12) with a perturbation treatment of relativity and statistical
exchange predicts a much larger value for D_e of 2.95 eV. Moreover,
although the R_e value of 2.929 au calculated by Ziegler et al.
(12) and that of 2.8497 au predicted by the ECP calculation of Hay
et al. (11) is almost in agreement with the experimental value (18)
of R_e = 2.8794 au; our ab initio fully relativistic Dirac-Fock SCF
EBS calculation predicts a value of 2.993 au for R_e. Our results
predicting a very minor role of 6s-6p hybridization in AuH and other
large number of gold compounds (Malli and Pyper, unpublished),
should be contrasted with the claim of Mingos (10) from non-
relativistic extended Hückel theory (EHT) calculations which predict
an important role for such hybridization in gold chemistry.
Finally, the statement of Pyykkö (19) that "At the time of writing
(July 1987) it still seems fair to say that the only DF-LCAO
calculations giving some chemical insight were those of Lee and
McLean (20) on AgH and AuH..." is totally incorrect in view of our
ab initio fully relativistic (four-component) Dirac-Fock SCF
extended basis set (EBS) calculations for AuH as well as our first
relativistic configuration interaction (RCI) calculations with the
DF SCF EBS calculation as reference which includes all single- and
pair-type double excitations. Our Ab Initio DF SCF calculations
have clearly shown that relativity contracts the bond length by
~0.45 au, substantially increases the fundamental vibrational
frequency (ω_e), and doubles the binding energy (D_e) predicted by
using a single determinant wavefunction. Moreover, there is very
significant 5d-6s hybridization whereas the 6s-6p hybridization
plays a very minor role.

The effect of relativity on the bonding in this molecule is so
large that even the qualitative features of the bonding cannot be
correctly described by non-relativistic theory which: (i) fails to
predict any 5d-6s hybridization in the localized bonding orbital;
(ii) seriously underestimates 5d-6s hybridization in one of the nonbonding orbitals (NBOs); (iii) predicts incorrectly that one of the
orbitals with $m_j = 1/2$ is entirely π in character and has pure β
spin; and (iv) erroneously predicts that the bond (in AuH) is formed
solely from the interaction between the gold 6s and the hydrogen 1s
atomic orbitals because the non-relativistic molecular orbital (MO)
wavefunction constructed from these two atomic orbitals predicts the
AuH molecule to be unbound by 0.19 eV.

Moreover, the inclusion of electron correlation effects via our
RCI calculations increases the D_e by ~0.7 eV. Our results for AuH
are presented in Tables I and II.

RELATIVISTIC EFFECTS IN BONDING AND DIPOLE MOMENTS OF HgH^+, TℓH,
PbH^+ and BiH.
Ramos, Pyper and Malli (21) have recently discussed (using the RIP)
the relativistic effects in bonding and dipole moments for these

Table I.   Total SCF and CI Energies Calculated with Various
Basis Sets for Gold Hydride (au)[a]

| | | Relativistic | | | |
| | | -(E_{SCF} + 19040) Basis | | -(E_{CI} + 19040) Basis | |
R	chemical	20 basis	27 basis	chemical	27 basis
2.6294	0.33354	0.36373	0.36525	0.34516	0.39231
2.8794	0.35439	0.37661	0.37850	0.36657	0.40395
3.1294	0.35842	0.37594	0.37824	0.37228	0.40252
3.3794	0.35219	0.36752	0.36987	0.36974	0.39380

| | | Non-Relativistic[b] | | |
| | | -(E_{SCF} + 17865) Basis | | -(E_{CI} + 17865) Basis |
R	chemical	20 basis	27 basis	chemical
2.8794	0.89479	0.91290	0.9147	0.91027
3.1294	0.91182	0.92765		0.92966
3.3794	0.91740	0.93325		0.93854
3.6294	0.91623	0.93129		0.94116
3.8794	0.91123			0.94052

[a]Malli and Pyper (14), reproduced with permission. [b]Basis set
same as the relativistic basis set, except that non-relativistic
calculations were performed using the appropriate increased value of
c (velocity of light).

Table II.   Relativistic and Non-Relativistic Predictions of
Energetic Properties of AuH using Various Basis Sets[a]

| | SCF Results | | | | | |
| | Relativistic Basis | | | NR Basis | | Expt[b] |
	chemical	20 basis	27 basis	chemical	20 basis	
D_e (eV)	1.108	1.625	1.682	0.434	0.864	3.36
R_e (au)	3.083	2.982	2.993	3.445	3.431	2.8794
ω_e (cm^{-1})	2155	2103	2102	1556	1745	2305

| | CI Results | | | |
| | Relativistic Basis | | NR Basis | Expt[b] |
	chemical	27 basis	chemical	
D_e (eV)	1.482	c	1.076	3.36
R_e (au)	3.152	2.963	3.680	2.8794
ω_e (cm^{-1})	1781	2102	1070	2305

[a]Malli and Pyper (14), reproduced with permission.   [b]Huber and
Herzberg (18).   [c]2.014 ≤ D_e ≤ 2.376 eV.

diatomics and a summary of their major conclusions is presented
later in this paper.

AB INITIO FULLY RELATIVISTIC DF SCF CALCULATIONS FOR DIATOMICS OF THIRD-ROW TRANSITION ELEMENTS AND ACTINIDES

Ab initio fully relativistic (4-component) DF SCF calculations using
chemical basis sets have been carried out for a large number of
diatomics involving atoms of sixth and seventh row heavy elements Pt
(Z = 78) to E117; e.g., PtOs, AuH, AuAℓ, AuCℓ, Au$_2$; HgH$^+$, HgLi$^+$,
HgBe^{2+}, HgO, HgS, HgCℓ^+, Hg$_2^{2+}$; TℓH, TℓI; PbH$^+$, PbO, PbS, PbSe,
PbTe; BiH, Bi$_2$; ThPt; ThO, UO, UTh, UPt, U$_2$, Pu$_2$; HE113, HE115,
E113E115, E117Cℓ, etc.

We present below only the results of our fully relativistic DF
SCF calculations using chemical basis (CB) sets for ThO and UO as
representatives of the diatomics involving 6d and 5f atomic spinors
(AS) in bonding since we have discussed above a few diatomics of the
third-row transition elements.

The physics and chemistry of the actinide elements have been
investigated rigorously during the last decade (22-26). However,
since very significant relativistic as well as electron correlation
effects are expected for systems involving these elements, ab initio
fully relativistic (4-component) DF SCF calculations even for the
diatomics of these elements have not been reported so far. An
extensive study of the atomic ground and various excited state
electronic configurations has revealed that the 5f, 6d, 7s and 7p
atomic spinors (DFAOs) of the atoms (and ions) of the actinides have
comparable orbital energies, and hence the large number of
electronic terms arising from the valence configurations lie very
close to each other. Therefore, a multi-configuration Dirac-Fock
(MCDF) treatment involving many configuration state functions (CSF)
is mandatory for describing accurately the ground (and excited)
state terms of the atoms (ions) of these elements. Moreover, in
order to ascertain the relativistic effects correctly, the MCDF
calculations for these systems must be performed judiciously so as
to ensure that these would lead to the correct non-relativistic
limit (NRL) results for a very large value of the velocity of light;
this is achieved in the atomic MCDF program of Grant et al (27).

DF SCF CALCULATIONS FOR ThO

The diatomic ThO has been detected in the vapor phase over a
mixture of Th and ThO$_2$ at high temperatures and a D$_0^o$ of ~9.00 eV,
an ionizational potential of ~ 6.0 eV and a R$_e$ value of 1.8403
angstrom are estimated for ThO (18, 28-29). Ab initio DF SCF
calculations in which each molecular spinor (MS) is expressed as a
linear combination of (4-component) atomic spinors (LCAS) (30) as
well as the corresponding non-relativistic limit (NRL) calculations
were performed for the ground state of ThO at five internuclear
separations viz; 3.077, 3.477, 3.877, 4.277 and 4.677 au.

The Oxford MCDF program of Grant et al (27) was used to
generate a 9 CSF wavefunction (WF) and the corresponding NRL WF for
the neutral Th atom with the ground state electronic configuration
[Rn] 6d^{2}7s^2 where [Rn] stands for the ground state atomic electronic
configuration of the radon atom (Z = 86).

Similarly, a 5 CSF wavefunction was generated for the ground
state of the oxygen atom. In these calculations, the inner 78
electrons of the Th atom and the two (1s)^2 electrons of the O atom

were treated as 'core' and thus the valence basis set for ThO was
taken to consist of 6s, 6p, 7s and 6d DFAOs of the Th atom and the
2s and 2p DFAOs of the atom including thereby a total of 18
electrons in the valence DF SCF wavefunction. Thus, 9 doubly
occupied valence relativistic molecular orbitals (RMO) or molecular
spinors (MS) were constructed via the SCF method as linear
combination of the (numerical) atomic spinors (LCAS), while the
(core) wavefunction for ThO was taken as the appropriate
antisymmetrized products of the 'cores' of the Th and O atoms as
detailed elsewhere(14). The 9 calculated eigenvalues are -2.134,
-1.439, -1.246, -1.055, -0.938, -0.437, -0.428, -0.420 and -0.220 au.
The calculated total relativistic DF SCF and the corresponding non-
relativistic limit (NRL) energies for ThO as well as the predicted
dissociation energies (D_e) at various internuclear separations are
collected in Table III.

Table III. Calculated total Relativistic (E_{DF}) and
Non-Relativistic (E_{NR}) energies (in au)[a] and
dissociation energies (D_e) in eV for ThO at
various internuclear distances (R)

R (au)[b]	Relativistic		Non-Relativistic	
	$(E_{DF}+26599)$	D_e (eV)[c]	$(E_{NR}+24434)$	D_e (eV)
3.077	-0.43588	-6.5693	-0.29442	-3.8069
3.477	-0.65440	-0.6230	-0.38474	-1.3491
3.877	-0.69607	0.5108	-0.50089	1.8114
4.277	-0.66415	-0.3576	-0.46127	-0.7332
4.677	-0.62864	-1.3239	-0.41880	-0.4223

a) 1 au = 27.211 eV b) The experimental R = 3.477 (au) for ThO
(ref. 18), where 1 au = 0.529171 angstrom.
c) The experimental D_e = 9.0 eV for ThO (ref. 18). A positive
(negative) value indicates the molecule to be bound (unbound) with
respect to two atoms.

It can be seen that although both the relativistic and the NRL
calculations predict the molecule to be bound at R = 3.877 au; the
predicted NRL D_e is about 3.5 times larger than the relativistic
prediction. However, the calculated relativistic total molecular
energy at R = 3.877 au is <u>lower</u> by 2165.19 au! (58916eV!). We
shall use the four-component functions defined through the equations
(3.5a - 3.5d) of Malli and Pyper (14) in order to express the
calculated valence RMOs in terms of the large components of the
$|\ell_m m_s>$ functions, where $|p_m\ m_s>$, $|d_m m_s>$ and $|f_m m_s>$ functions have

p, d and f - like angular functions with m_ℓ quantum number m and
$m_s = \pm 1/2$ (the z - component of the spin angular momentum correspond
to the two-component spin functions $|\alpha>$ and $|\beta>$). Therefore, the
valence RMOs can be expressed in terms of the $|p\sigma\alpha>$, $|p\pi\beta>$, $|d\sigma\alpha>$,

$|d\pi\beta>$, $|f\sigma\alpha>$, $|f\sigma\beta>$ functions, etc, which can give insight into the bonding characteristics of the valence RMOs of the molecular species in question. The valence RHOMO of ThO designated as $|9e1/2>$ (using the double group theoretical notation for the additional irreducible representation (AIR) of the heteronuclear diatomic) at R = 3.877 au has the following form:

$$|9e1/2> \simeq -0.936|7s\sigma\alpha1/2> -0.350|6d\sigma\alpha1/2> -0.111|2s\sigma1/2>$$

It consists of a 7s-6d σ-hybrid on the Th atom with a very small bonding contribution from the $|2s\sigma>$ and $|6d\sigma>$ orbitals of the O and Th atoms. The valence RMO $|8e3/2>$ lying just below the valence RHOMO, however is a bonding combination of the $|2p\pi\alpha>$ and $|6d\pi\alpha>$ valence orbitals of the O and Th atoms, viz;

$$|8e3/2> \simeq -0.815|2p\pi\alpha3/2>-0.417|6d\pi\alpha3/2>+0.095|6p\pi\alpha3/2>.$$

The next three calculated valence RMOs (lying below the $8|e3/2>$ RMO) are of the following forms:

$$|7e1/2> \simeq 0.813|2p\pi\beta1/2>+0.418|6d\pi\beta1/2>,$$

$$|6e1/2> \simeq -0.899|2p\sigma\alpha1/2>+0.363|6d\sigma\alpha1/2>-0.253|6p\sigma\alpha1/2>$$
$$-0.215|2s\sigma\alpha1/2>-0.123|6s\sigma\alpha1/2>,$$

$$|5e1/2> \simeq -0.734|6p\sigma\alpha1/2>+0.629|2s\sigma\alpha1/2> -0.379|6p\pi\beta1/2>$$

These valence RMOs indicate substantial bonding involving the $2p\pi$ and $6d\pi$, $2p\sigma$ and $6d\sigma$ and $2s\sigma$ and $6d\sigma$ valence orbitals of the O and Th atoms, respectively. Moreover, the calculated ionization potential (IP) (using the Koopman's theorem) of 6.14 eV is in excellent agreement with the experimentally reported IP of ~6.00 eV (18). The calculated D_e is 0.51 eV (at R=3.877 au) whereas experimentally, the claimed D_e lies between 8.7 and 9.00 eV (18). A single determinantal DF SCF wavefunction is not expected to yield the correct value of D_e due to the neglect of electron correlation effects. The NRL calculations predict D_e of 1.81 and 0.73 eV at R = 3.877 and 4.277 au, respectively. However, the total NRL molecular energy is about 2165 au (58912 eV!) higher than the corresponding DF SCF energy at these internuclear separations. The valence NRL orbital energies at R = 3.877 au for ThO are calculated to be -1.531, -1.140, -0.948,-0.887, -0.837, -0.326, -0.315, -0.299 and -0.165 au for the 9 valence MOs; these values lie much higher than the correponding relativistic orbital energies given above. There is a difference of 0.6 au (16.33 eV) between the relativistic and the NRL calculated orbital energy for the lowest lying valence orbitals whereas the corresponding difference for the HOMOs is 1.50 eV. Thus the IP of 4.49 eV predicted by the NRL calculation is too low by ~1.50 eV compared to the experimental IP of 6.00 eV. This can be easily understood as the calculated valence NRL HOMO (at R = 3.877 au) consists of an almost pure $|6d\delta>$ orbital on the Th atom, whereas in contrast the calculated RHOMO (given above) consists of a hybrid of $|7s\sigma>$ (-0.936) and $|6d\sigma>$ (-0.350) orbitals of the Th atom with the coefficients given in parenthesis. Since the 7s DFAO is

stabilized (due to the direct relativistic effect) and the 6d DFAO
is destabilized (due to the indirect relativistic effect), the
differences in the orbital energies for the orbitals mentioned above
can be easily understood. The lowest lying valence RMO $|1e1/2>$ is
the pure $|6s\sigma>$ valence orbital of the Th atom; whereas the
corresponding NRL MO is calculated to be a hybrid of $|6s\sigma>$ (-0.969)
and $|2s\sigma>$ (-0.124) orbitals. The stabilization of the 6s DFAO of
the Th atom (due to the direct relativistic effect) explains the
difference in the orbital energies for the $|1e1/2>$ MO.

Hence it can be stated that the predicted IPs, bonding energies
and the bonding characteristics predicted for ThO using the
relativistic and the NRL molecular orbital theories differ
considerably and that there are very significant relativistic
effects due to the participation of the 6d and 6p DFAO's of the Th
atom in the bonding of the ThO diatomic.

DF SCF CALCULATIONS FOR UO

The species UO is claimed to exist at high temperatures
(>2000°C) in the gas phase with a dissociation energy of 7.8 eV and
an ionization potential of 5.6 eV. An IR spectrum of $U^{16}O$ in argon
matrix at 15°K has led to ω_e = 825.0 cm^{-1} and $\omega_e x_e$ = 2.5 cm^{-1} ([18]).
The electromic structure of a diatomic involving an actinide (U)
with 5f electrons in the valence atomic spinors is of paramount
importance in order to investigate the role of 5f orbitals in the
actinide-oxygen bond. Moreover, the explanation for the difference
in geometry between the ThO$_2$ (bent) and the isoelectronic (UO$_2$)$^{2+}$
(linear) may be found in terms of the roles of the 6d and 5f
orbitals of the Th and U atoms in the ThO and UO bonds. This
prompted us to investigate the electronic structure of the diatomic
UO in which relativity is expected to be very significant.

For the UO molecule, the calculations were performed at R =
3.05, 3.55, 4.05 and 4.55 au using a valence basis set of 6p, 7s,
5f and 6d DFAOs of the U atom and 2s and 2p DFAOs of the O atom;
i.e., a total of 18 electrons were included in the valence
relativistic molecular wavefunction with the following atomic
configurations, viz U: [core] $6p^6 5f^3 6d7s^2$ where the [core] contains
80 electrons and the $1s^2$ electrons of the O atom were also kept in
its core. The lowest total molecular energy of -28145.039469 au
was found at R = 3.55 au; however, the UO molecule was still
unbound by -0.061739 au (-1.68 eV) while at R = 3.05, 4.05 and
4.55 au it was found to be unbound by -0.14393, -0.11971 and
-0.17596 au, respectively. It should be mentioned that a single
determinant DF SCF wavefunction fails to predict binding for the UO
diatomic. The corresponding non-relativistic limit calculation at
R = 3.55 au predicts the UO molecule unbound by only -0.973 eV;
however, the total non-relativistic energy for the UO molecule is
about 2346 hartrees (1 hartree = 27.211 eV) above the DF SCF total
molecular energy at R = 3.55 au. Thus, although the NRL
wavefunction for UO predicts a D_e almost twice as much as the
relativistic calculation, it should not be regarded as better than
the DF SCF calculation.

The relativistic highest occupied molecular orbital (RHOMO)
$9|e1/2>$ consists of a $7s\sigma$-$6d\sigma$ hybrid on U with small contributions
from $|5f\pi\beta>$ and $|5f\sigma\alpha>$ U DFAOs. Using the relations (3.5a-3.5d) of
Malli and Pyper ([14]), all the valence RMOs of UO can also be written

in terms of the $|\ell_m m_s>$ functions. The RHOMO $9|e1/2>$ of UO has the
following form (where 'e' denotes the two-dimensional additional
irreducible representation of the RMO and 1/2,3/2,5/2 corresponds to
ω = 1/2,3/2,5/2), where ω indicates the total angular momentum of
the one-electron RMO), viz,

$$9|e1/2> = -0.873|7s\sigma\alpha1/2> -0.398|6d\sigma\alpha1/2> +0.183|5f\pi\beta1/2>$$
$$-0.176|5f\sigma\alpha1/2> +0.129|6d\pi\beta1/2> -0.093|2s\alpha1/2>$$
$$-0.069|2p\sigma\alpha1/2>$$

It is an almost nonbonding RMO since it contains very small
contributions from the $|2s\sigma\alpha>$ (-0.093) and $|2p\sigma\alpha>$ (-0.069) DFAOs of
the O atom, where the coefficients are given in parentheses.
However, the RMO $8|e3/2>$ lying just below the RHOMO has the
following form which clearly shows that $|5f\delta>$ and $|6d\pi>$ DFAOs of U
are definitely involved in the bonding of UO, viz,

$$8|e3/2> \simeq 0.727|2p\pi\alpha3/2> -0.465|5f\delta\beta3/2> +0.344|6d\pi\alpha3/2>$$
$$-0.111|6p\pi\alpha3/2>$$

The valence RMOs lying below the $8|e3/2>$ RMO also contain very
significant contributions from the 5f and 6d DFAOs of the U atom
and, in fact, the $5|e3/2>$ RMO (with orbital energy of -0.502 au)
has the coefficients of -0.74, +0.49 and -0.38 for the $|5f\delta\beta>$,
$|5f\pi\alpha>$ and $|2p\pi\alpha>$ DFAOs, respectively. These results clearly
demonstrate that the 5f atomic spinors (DFAOs) and their associated
electrons are very significant for bonding in actinides (which
contain 5f DFAO in their ground state electronic configuration)
except that in ThO, the 6d DFAOs are involved significantly in
bonding. Both these results can only be understood due to the
indirect relativistic effect which destabilizes the (electrons in
the) 6d and 5f DFAOs so that these DFAOs are pushed up in energy and
act as valence atomic spinors. The orbital energies of the valence
RMOs also indicate substantial roles for the 5f and 6d DFAOs in
bonding for the UO and ThO diatomics, respectively.
We give below the expressions for the $7|e1/2>$ and $|5e3/2>$ RMOs
of UO which also clearly indicate π-bonding arising due to the
interaction of the 2pπ DFAOs of O and the 6dπ and/or 5fπ DFAOs of U
atom, viz,

$$7|e1/2> \simeq 0.788|2p\pi\beta1/2> +0.346|6d\pi\beta1/2> -0.225|5f\pi\beta1/2>$$
$$-0.098|6p\pi\beta1/2>,$$

$$5|e3/2> \simeq -0.74|5f\delta\beta3/2> +0.49|5f\pi\alpha3/2> -0.38|2p\pi\alpha3/2>.$$

Although, the 6pπ DFAO was not found to be involved significantly in
these valence RMOs (except a very minor contribution to $8|e1/2>$ as
noted above); the 6pσ DFAO of U contributes as much as the $|6d\sigma>$ and
$|5f\sigma>$ DFAOs contribute to the 6|e1/2) RMO which has the following
form:

$$6|e1/2> \simeq -0.87|2p\sigma\alpha1/2> +0.30|6d\sigma\alpha1/2> -0.29|6p\sigma\alpha1/2>$$
$$-0.29|5f\sigma\alpha1/2> -0.23|2s\sigma\alpha1/2>$$

The calculated relativistic orbital eigenvalues of the $5|e3/2>$, $6|e1/2>$, $7|e1/2>$, $8|e3/2>$ RMOs and the RHOMO $9|e1/2>$ of UO are -0.5017, -0.4551, -0.4444, -0.4281 and -0.2266 au, respectively; and using Koopman's theorem, the calculated ionization potential of 6.17 eV for UO is in excellent agreement with the corresponding experimental value of 6 ±0.5 eV (<u>18</u>). The calculated total molecular energies, etc., at various internuclear separations for UO are collected in Table IV. It can be seen that the UO molecule is calculated to be unbound at all these internuclear separations.

Table IV. Calculated total Relativistic (E_{DF}) and Non-Relativistic (E_{NR}) energies in au (1au = 27.211 eV) and dissociation energies (in eV) for UO at various internuclear seperations (R) in au (1au = 0.529171 angstrom)

R (au)	RELATIVISTIC		NON-RELATIVISTIC	
	(E_{DF}+28144)	D_e (eV)[a]	(E_{NR}+25379)	D_e (eV)
3.05	-0.95728	-3.9165	-0.02107	-5.0213
3.447	-1.0394	-1.6800	-0.14147	-1.7451
3.877	-0.98150	-3.2573	-0.12528	-2.1855
4.277	-0.92535	-4.7852	-0.07869	-3.5142
4.677	-	-	-0.06620	-3.7930

a) The experimental D_e is reported to be 7.8 eV (<u>18</u>). A positive (negative) value indicates the molecule to be bound (unbound) with respect to the two atoms.

Since the lowest total molecular energy was calculated at R = 3.55 au, the corresponding NRL calculation was also performed at R = 3.55 au in order to gain insight into the major differences in bonding arising due to relativity. It was found that at R = 3.55 au, the NRL calculation predicts for the UO molecule a D_e of -0.97 eV which is about 0.70 eV greater than that predicted by the corresponding DF SCF calculation; however, the total NRL molecular energy is about 2406 hartrees <u>above</u> the DF SCF total molecular energy at R = 3.55 au. Moreover, the NRHOMO $9|e1/2>$ predicts the ionization potential of 4.42 eV, which is about 1.6 eV lower than the corresponding DF SCF value of 6.14 eV which, however, agrees excellently with the experimental value. The $9|e1/2>$ NRHOMO of UO has the following expression, <u>viz</u>,

$$9|e1/2>NR \approx -0.539|5f\pi\beta1/2> -0.483|5f\sigma\alpha1/2> -0.449|6d\sigma\alpha1/2>$$
$$-0.348|7s\sigma\alpha1/2> -0.288|6d\pi\beta1/2> -0.213|2p\sigma\alpha1/2>$$
$$+0.144|6p\sigma\alpha1/2>$$

The NRHOMO has a much larger contribution from the $5f\pi$, $5f\sigma$, $6d\sigma$, $6d\pi$ DFAOs of the U atom and the $2p\sigma$ DFAO of the O atom as compared to the RHOMO; however, the contribution of the $7s\sigma$ DFAO of the U atom to the NRHOMO is much smaller. Moreover, whereas the NRMO

8|e3/2> consists mostly (with a coefficient of 0.90) of |5fδ3/2>
DFAO (with contributions from |5fπ> (0.338) and |2pπ> (0.20) DFAO),
the RMO 8|e1/2> has maximum contributions from the |2pπ> (0.727),
|5fδ> (-0.465) and |6dπ> (0.344) DFAOs of the U atom. Similarly,
the 7|e1/2> and 6|e1/2> RMOs and NRMOs differ substantially; e.g.,
the NR 7|e1/2> consists mostly of 2pσ (0.746), 6fσ (0.435), 6pσ
(0.381), 6dσ (-0.288) (and |5fπβ1/2>) DFAOs of U; while, the
corresponding RMO as discussed above is a π-type MO with major
contributions from the 2pπ (0.788), 6dπ (0.346) and 5fπ (-0.225)
DFAOs of the U atom.

    The orbital eigenvalues of the RMO (the NRMOs) 8|e3/2> and
7|e1/2> are -0.4281 (-0.2642) and -0.4444 (-0.2989) <u>a.u.</u>
respectively, and it is clear that the NRL orbital eigenvalues are
about 4.4 eV <u>lower</u> than the corresponding DF SCF orbital energies
for the 8|e3/2> and 7|e1/2> MOs.

    Our results therefore <u>clearly demonstrate</u> that there are <u>marked
qualitative</u> as well as <u>quantitative differences</u> between the
predictions of the NRL and DF SCF calculations for the nature of
bonding, total energies, orbital energies, dissociation energies
etc., for the diatomics involving actinides due to very significant
relativistic effects in such systems.

RELATIVISTIC EFFECTS FOR DIPOLE MOMENTS OF DIATOMICS OF HEAVY
ELEMENTS

The RIP has been adapted by Ramos (<u>21</u>) to evaluate dipole moments
for the diatomic species AuH, HgH⁺, TℓH, TℓI, PbH⁺, PbTe and BiH,
using ab initio relativistic as well as non-relativistic limit (NRL)
chemical basis set wavefunctions (WF) calculated at the experimental
internuclear separation (R_e) of each species except PbH⁺ for which a
value of 3.5884 au was used for the internuclear distance. In
addition, dipole moments calculated from the relativistic and NRL
WFs, obtained using extended basis sets with 20 (EB20) and 27 (EB27)
basis functions, CB set augmented by a 6p Slater-type orbital (STO)
and CB set augmented by a 6p DFAO, have been reported for AuH (<u>21</u>).
Moreover, dipole moment curves were calculated for AuH using the
relativistic CB set, extended basis set (EB20 and EB27) and
configuration interaction (CI) wavefunctions reported by Malli and
Pyper (<u>14</u>). Unfortunately, at present, experimental dipole moments
are not available for AuH, HgH⁺, TℓH, PbH⁺ and BiH; however, the
predicted dipole moments (with the CB set relativistic
wavefunctions) of 1.9078 au and 1.2655 au agree very well with
the experimental values (<u>18</u>) of 1.8137 au and 1.0623 au for TℓI
and PbTe, respectively, where a positive dipole moment for the
species AB indicates its polarity as A⁺B⁻.

    It turns out that the dipole moment calculated for AuH (at R_e =
2.8794 au) using the relativistic wavefunction is <u>smaller</u> (by
about 40% to 50%) than that predicted by the corresponding NRL
wavefunction, depending upon the basis set used in the calculation
of the wavefunction. However, the predicted dipole moments (using
the relativistic chemical basis set wavefunctions) of 0.976, 0.323
and 0.371 au <u>differ considerably</u> from the values of 1.372, -0.120
and 0.019 au, predicted by the NRL wavefunctions, for AuH, TℓH and
BiH, respectively. In the case of TℓH, although the dipole moment
calculated with the relativistic (CB set) wavefunction <u>predicts</u> the
expected polarity <u>viz</u> Tℓ⁺H⁻; the value of dipole moment (-0.12 au)

obtained from the corresponding NRL WF indicates the <u>opposite</u> <u>polarity</u> for this molecule <u>viz</u> $T\ell^- H^+$. A comparison of the dipole moments (μ), calculated for a set of internuclear separations for AuH, indicates that the μ values predicted from relativistic CB set (μ_{CB}), EB27 set (μ_{EB27}) and RCI (μ_{RCI}) WFs decrease in the order: $\mu_{CB} > \mu_{EB27} > \mu_{RCI}$, except that at R = 3.3794 au, $\mu_{EB27} > \mu_{CB}$ whereas at R_e = 2.8794 au, the μ_{CB} and μ_{RCI} are 0.967 and 0.846 au, respectively, indicating thereby that a <u>lesser</u> polarity (μ_{CB}), EB27 set (μ_{EB27}) and RCI (μ_{RCI}) WFs decrease in the order: $\mu_{CB} > \mu_{EB27} > \mu_{RCI}$, except that at R = 3.3794 au, $\mu_{EB27} > \mu_{CB}$ whereas at R_e = 2.8794 au, the μ_{CB} and μ_{RCI} are 0.967 and 0.846 au, respectively, indicating thereby that a <u>lesser</u> polarity is predicted by the RCI wavefunctions than that predicted <u>both</u> by the non-relativistic limit and the relativistic chemical basis set wavefunctions. This is the <u>first</u> study of the effect of relativity on dipole moments of diatomic systems involving heavy atoms or ions using DF SCF LCAS MS calculations, and it can be concluded from our results (<u>21</u>) presented in Tables V-VI that relativistic and electron correlation effects are fairly significant for dipole moments of such systems.

Table V. Dipole Moments (μ) for LiH, $T\ell$I and PbTe Calculated by Using Chemical Basis Wavefunctions[a]

A-B[b]	μ(au)[c]	EXP[d]
LiH	2.575 (2.367)[e]	2.314
$T\ell$I	1.908	1.814
PbTe	1.266	1.062

[a]Reproduced with permission from Ref. 21. [b]All values indicate A^+B^- polarity. [c]1 au = 2.542 D. [d]Experimental values from reference (<u>18</u>). [e]Using extended basis function of Malli and Pyper (<u>14</u>).

CONCLUDING REMARKS
We have conclusively shown from ab initio fully relativistic DF SCF LCAS MS calculations that the <u>qualitative</u> as well as the <u>quantitative</u> features of electronic structure and bonding in diatomics involving heavy and very heavy atoms (Z ≥ 90) cannot be properly understood using the traditional non-relativistic theory based on the Schrödinger equation. In addition, it has been shown that, for the sixth row elements, the $5\bar{d}$ and 5d DFAOs participate in the chemistry of gold and mercury compounds, whereas they belong to the core in heavier elements. Moreover, it is safe to state that the $6\bar{p}$ and 6p DFAOs are not involved in gold chemistry (in contrast to the semi-empirical non-relativistic theory predictions), but they are significant for the chemistry of heavier elements.

Table VI. AuH Relativistic Dipole Moment (μ) Curves,[a] and
Non-Relativistic Dipole Moment[b]
Values at R_e,[c] in au[d]

	Wavefunctions					
R(au)	CB	CB + 6pSL [e]	CB + 6pDF [f]	EB20	EB27	RCI
2.6294	0.968	-	-	0.905	0.829	0.738
2.8794[g]	0.976	1.065	1.271	0.984	0.901	0.802
	(1.373)	(1.443)	(1.925)	(1.434)	(1.363)	
3.1294	0.967	-	-	1.046	0.956	0.846
3.3794	0.968	-	-	1.132	1.045	0.909

[a]Calculated by using the wavefunctions from reference (14);
Chemical Basis (CB), extended basis EB27 including polarization
functions, EB20 (same as EB27 but without 5f' polarization
functions), and relativistic configuration interaction (RCI)
wavefunction. [b]Non-relativistic values are given in parentheses.
[c]Reproduced with permission from Ref. 21. [d]1 au = 2.542 D. All
values indicate Au$^+$H$^-$ polarity. [e]CB plus 6p' Slater-type
polarization function with exponent ζ = 2.75 centered on gold. [f]CB
plus 6p DFAO as obtained from the MCDF calculation for the gold atom
(see reference 14). [g]Experimental R_e (reference 18).

We also conclude from our ab initio DF SCF calculations that
the 5d, 6d and 5f DFAOs (and their associated electrons) are
definitely involved (due to relativistic effects) in the electronic
structure and bonding of the diatomics of the heavy third-row
transition elements and actinides, and they present the formidable dual
challenge to quantum chemists of the accurate calculation of
the relativistic and electron correlation effects for such systems.
Furthermore, relativistic effects have been shown to be fairly
significant in non-energetic properties, e.g., dipole moment, and it
is hoped that the accurate prediction of non-energetic properties
would supplement the criterion for the quality of a relativistic
wavefunction in future. Thus, the knotty bottlenecks of ab initio
fully relativistic DF SCF calculations have been broken, and it is

gratifying that the computational machinery is currently at hand
for performing reliable ab initio fully relativistic DF SCF
calculations for diatomics containing heavy and very heavy atoms.
It is hoped that with the availability of faster supercomputers
ab initio (all-electron) fully relativistic calculations for
polyatomics containing heavy atoms will become feasible in the
near future.

ACKNOWLEDGMENTS
I sincerely thank Professors Dennis Salahub and Mike Zerner for
inviting me to this symposium. This work has been made possible due
to the cooperation and the enthusiasm of my colleagues and coworkers
over many years; in particular, I would like to acknowledge my debt
to Dr. N.C. Pyper, Dr. R. Arratia-Perez, Messrs A.F. Ramos and D.
Yu, for their contributions to the research reported in this paper.
My thanks also go to the operations staff of our Computing Services
for their cordial cooperation. The Natural Sciences and Engineering
Research Council of Canada (NSERC) is thanked for their continuous
financial support through grant no. A3598.

LITERATURE CITED
1. Schrödinger, E. Ann. Physik. 1926, 81, 109.
2. Klein, O. Z. Physik. 1926, 37, 895.
3. Gordon, W. Z. Physik. 1926, 40, 117.
4. Dirac, P. A. M. Proc. Roy. Soc. Lond. 1928, A117, 610.
5. Burrau, O. Kgl. Danske. Videnskab. Mat. Fys. 1927, 7, 14.
6. Dirac, P. A. M. Proc. Roy. Soc. Lond. 1928, A123, 714-33.
7. Mayers, D. F. Proc. Roy. Soc. Lond. 1957, A241, 93.
8. Swirles, B. Proc. Roy. Soc. Lond. 1935, A152, 625-49.
9. Boyd, R. G.; Larson, A. C.; Waber, J. T. Phys. Rev. 1963,
    129, 1629-30.
10. Mingos, D. M. P. Phil. Trans. Roy. Soc. Lond. 1982, A308, 75-
    83.
11. Hay, P. J.; Wadt, W. R.; Kahn, L. R.; Bobrowicz, F. W.
    J. Chem. Phys. 1978, 69, 984.
12. Ziegler, T.; Snijders, J. G.; Baerends, E. J. J. Chem. Phys.
    1981, 74, 1271.
13. Jiang, Y.; Alarez, S.; Hoffmann, R. Inorg. Chem. 1985, 24,
    749-57.
14. Malli, G. L.; Pyper, N. C. Proc. Roy. Soc. Lond. 1986, A407, 377-
    404.
15. Grant, I. P.; Mckenzie, B. J.; Norrington, P. H.; Mayers, D.
    F.; Pyper, N. C. Comput. Phys. Commun. 1980, 21, 207.
16. Brown, G. E.; Ravenhall, D. G. Proc. Roy. Soc. Lond. 1951,
    A208, 552-9.
17. Sucher, J. Phys. Rev. 1980, A22, 348-62.
18. Huber, K. P.; Herzberg, G. Molecular Spectra and Molecular
    Structure IV. Constants of Diatomic Molecules; Van Nostrand
    Reinhold, New York, 1979.
19. Pyykkö, P. Chem. Rev. 1988, 88, 563.
20. Lee, Y. S.; McLean, A. D. J. Chem. Phys. 1982, 76, 735.

21. Ramos, A. F.; Pyper, N. C.; Malli, G. L. Phys. Rev. 1988, A 38
    2729-2739.
22. Katz, J. J.; Seaborg, G. T.; Morss, L. R. The Chemistry of the
    Actinide Elements. Chapman and Hall: London, 1986.
23. Oetting, F. L.; Rand, M. H.; Ackermann, R. J. The Chemical
    Thermodynamics of Actinide Elements and Compounds Part 1;
    International Atomic Energy Agency: Vienna, 1976.
24. Oetting, F. L.; Fuger, J. The Chemical Thermodynamics of
    Actinide Elements and Compounds Part 2; International Atomic
    Energy Agency: Vienna, 1976.
25. Erdos, P.; Robinson, J. M. The Physics of Actinide Compounds;
    Plenum Press: New York, 1983.
26. Handbook on the Physics and Chemistry of the Actinides Vols. 1-
    5; Freeman, A. J.; Lander, G. H. Eds.; North Holland:
    Amsterdam, 1987.
27. Grant, I. P.; McKenzie, B. J.; Norrington, P. H.; Mayers, D.
    F.; Pyper, N. C. Computer Phys. Commun. 1980, 21, 218.
28. Ackermann, R. J.; Rauh, E. G. Higher Temp. Sci. 1973, 5, 463;
    J. Chem. Phys. 1974, 60, 2266.
29. Hildenbrand, D. L.; Murad, E. J. Chem. Phys. 1974, 61, 1232.
30. Malli, G. L.; Oreg, J. J. Chem. Phys. 1975, 63, 830-841.

RECEIVED March 21, 1989

Chapter 22

Relativistic Effective Potentials in Quantum Monte Carlo Studies

Phillip A. Christiansen

Department of Chemistry, Clarkson University, Potsdam, NY 13676

An overview of quantum Monte Carlo electronic structure
studies in the context of recent effective potential
implementations is given. New results for three
electron systems are presented. As long as care is
taken in the selection of trial wavefunctions, and
appropriate frozen core corrections are included,
agreement with experiment is excellent (errors less
than 0.1 eV). This approach offers promise as a means of
avoiding the excessive configuration expansions that
have plagued more conventional transition metal studies.

In the last ten years considerable effort has gone into the study
of small metal clusters. Several reviews on the subject have
appeared in the literature. Volume 156 of Surface Science, Volume 86
of Chemical Reviews and a portion of Volume 91 (especially No. 10) of
the Journal of Physical Chemistry are devoted to this topic. The
small transition metal clusters are particularly intriguing as a
result of their unique structures (multiple d bonding, etc.) and as
possible models for catalytic processes. Furthermore, as can be seen
from the compendium by Huber and Herzberg (1) and also from the
encyclopedic reviews by Weltner and Van Zee (2) and more recently by
Morse (3) and by Salahub (4), relatively little is known about the
detailed structures of even the simplest clusters (diatomics) of the
elements beyond the first transition row. The field would appear to
be wide open for computational chemists. For clusters of only a few
atoms one would expect rigorous electronic structure studies (SCF
plus large CI, etc.) to yield useful imformation regarding molecular
geometries, dissociation energies, vibrational frequencies, etc.
Unfortunately, in contrast to recent light element work, early
transition element studies proved somewhat disappointing. The
chromium diatom (5-9) is probably the best known example. However a
more disturbing case is Sc_2, the simplest transition metal diatomic.
The Sc_2 dissociation energy (1.65 eV) is known from the mass
spectrometric work of Verhaegen et al. (10) although there may be
some error due to the use of rather imprecise molecular partition
functions (11,12). (A value of 1.22 eV was originally given but it

0097–6156/89/0394–0309$06.00/0

has since been reported in various references ([1,3]) as 1.65.)
Resonance Raman ([13]), ESR ([14]) and MCD ([15]) matrix isolation studies
suggest a $^5\Sigma$ ground state with a fundamental vibrational frequency of
about $239cm^{-1}$, which is consistent with the $^5\Sigma_u$ assignment given by
the best ab initio calculation. Nevertheless, in their fairly
extensive study, Walch and Bauschlicher ([12]) were able to account for
only a fraction of the experimentally determined dissociation energy.
As with diatomic chromium the difficulty involves correlation in a
multiple d bonded system. These discouraging results have prompted
Morse ([3]) to suggest that for transition element problems the density
functional approaches ([4,17-19]) might be more appropriate.
    Of course the transition metal electron correlation problems do
not necessarily begin at the molecular level. Ab initio studies ([20-24]) typically show errors in atomic excitation energies of about 0.3
eV or more for transitions involving the outer s and d electrons.
    The errors seen in the above examples are of course the result
of the necessary incompleteness of orbital and configuration basis
sets. The power of these expansion approaches is that if one works
hard enough (uses a sufficiently complete, or at least appropriate,
basis) one should get the right answer. The recent extensive
transition metal hydride studies indicate the possibilities ([25-30]).
Nevertheless, heavy atom electron correlation involving d and even f
subshells is such an enormous problem that every alternative should
be explored.

Effective Potential Quantum Monte Carlo

As configuration expansions approach the multi-million range,
alternatives such as quantum Monte Carlo (QMC) techniques begin to
appear attractive. A useful overview of QMC has been given by
Ceperly and Alder ([31]). Pioneering work in this field was done in
the mid 70's by Anderson ([32,33]) as well as by Kalos and coworkers
([34,35]). This has been followed by considerable development work as
well as molecular and atomic applications ([36-56]). The advantage of
QMC is that it does not depend on the exhaustive configuration and
orbital basis set expansions that have plagued conventional studies.
As one moves down the periodic table to the transition elements with
occupied d shells and to the Lanthanides and Actinides with f shells,
the QMC advantage becomes more apparent. Unfortunately, to a
considerable extent, what one gains in the elimination of infinite
basis set expansions, one looses to statistical sampling error. And
furthermore the sampling error increases rapidly as a function of the
nuclear charge. Doll ([57]), Ceperly ([58]), and most recently Hammond
et al. ([59]) have given arguments indicating that the QMC computer
requirements increase with about the sixth power of the nuclear
charge. As a result QMC has, to date, offered little competition for
conventional calculations, and we are aware of no all-electron QMC
studies involving elements beyond the first row.
    Although all-electron heavy element QMC studies are at the
present time out of the question, we have recently shown that by
replacing the core electrons (and the corresponding fraction of the
nuclear charge) with an appropriate relativistic effective potential
(REP) the QMC domain can be quite readily extended to the lower
portion of the periodic table ([60-62]). To our knowledge, reference
([60]) is the first QMC study involving an element from below the first

row and is also the first to include relativity. This work was followed closely by a study by Hammond et al. (59) who used an almost identical approach in alkali and alkaline earth atomic and molecular studies.

As pointed out in the review by Ceperly and Alder (31) the diffusion interpretation of the Schroedinger equation has an extensive history. The diffusion analogy becomes apparent if one writes the time–dependent Schroedinger Equation (one electron for simplicity) in terms of imaginary time, t,

$$\frac{d\psi}{dt} = \frac{1}{2}\nabla^2\psi - (V - E_T)\psi \qquad (1)$$

ψ then corresponds to a concentration and the equation is simulated by a combination of random particle movement (first term on the right) as well as particle birth and death according to the first-order rate constant, $V-E_T$ (32,33). The arbitrary reference energy, E_T, can be adjusted to maintain normalization. However a far more efficient approach involves importance sampling (34,36,37,41,42). By defining the function $f=\psi\psi_T$, the product of ψ with a time-independent trial wavefunction, ψ_T, Equation 1 becomes

$$\frac{df}{dt} = \frac{1}{2}\nabla^2 f - \nabla \cdot (f\nabla\ln\psi_T) - (\frac{H\psi_T}{\psi_T} - E_T)f. \qquad (2)$$

The middle term on the right adds a drift velocity ($\nabla\psi_T/\psi_T$) to the simulation which greatly reduces sampling in regions of low electron density. In addition the nodes in ψ_T (resulting from either orbital nodes or antisymmetry) can be used to define sampling region boundaries, an assumption of the fixed node approximation (32,33). Electronic energies are ultimately obtained from averages of the local energies, $H\psi_T/\psi_T$. Detailed descriptions of algorithms based on Equation 2 can be found in the literature (31,39,42,52).

For QMC simulations involving atomic or molecular systems with more than a small number of electrons, potential sources of difficulty are fairly obvious (58-60). In regions of high electron density (such as near nuclei) one sees the corresponding high density of singularities in the hamiltonian resulting from the two-electron electrostatic interaction. At the same time, wavefunction antisymmetry causes a high nodal density. The dense nodal structure forces one to employ short time steps, thereby greatly increasing the computational requirements. And although the effects of electron-nucleus and two-electron singularities in the potential can be controlled to considerable extent using pair-correlation functions [see reference (37) for instance], unless ψ_T is a good approximation to ψ, correlation error will become overwhelmingly apparent in the local energies, leading to large statistical errors in the average. The more densely packed the electrons become (this will be most serious in the core region) the more acute the difficulties will be. Furthermore, with the exception of the work by Vrbik et al. (63) it is not clear how relativity (essential for heavy element studies) would be included in all-electron QMC work.

In this context the advantages in the use of effective potentials are quite clear. The potentials eliminate the high electron density (and associated nodes) near the nuclei, thereby reducing sampling in a region in which both the wavefunction and the potential are rapidly varying. This eliminates a difficult fraction of the multi-electron potential from the wave equation and at the same time makes the use of much longer time steps appropriate. Perhaps equally important, relativistic effective potentials allow one to introduce relativity in a particularly convenient form.

The key to the use of conventional semi-local REPs in QMC involves the transformation of the REP to local form. In reference (60) Hurley et al. proposed the many-electron local potential, V^{REP},

$$V^{REP} = U^{REP}\psi_T/\psi_T. \tag{3}$$

U^{REP} is the conventional effective potential and ψ_T is the trial wavefunction (at least the determinant portion) from Equation 2. To carry out REP-QMC calculations one simply adds V^{REP} to the valence-electron hamiltonian, H, in the importance sampling algorithm. In light of the local energy expression, $H\psi_T/\psi_T$, in Equation 2, this definition of the local potential is rather obvious and has also been used in the work by Hammond et al. (59) but without relativity.

Equation 3 obviously adds approximations. These include the usual effective potential assumptions (frozen core, etc.) in addition to the localization shown in the equation. However in one sense it is a trade-off in that the local potential effectively eliminates the fixed node approximation in the core region.

Atomic Studies   In Table I electron affinities for Li, Na and K computed using Equation 3 with either relativistic (60) or nonrelativistic (59) effective potentials are compared with the respective experimental values (64-66). Only in the relativistic Li calculation do we see a significant discrepancy, and even then the error is well below 0.1 eV. In all of these calculations single determinant trial wavefunctions were employed. While this is no approximation for the one-electron neutral atoms we might see minor problems for the anions, and Li could be a case in point.

Table I.   Alkali Electron Affinities (in eV) obtained
from Relativistic and Nonrelativistic Effective
Potential QMC Simulations

Atom	reference	effective potential QMC	Expt.
Li	60	0.56(2)	0.62
Li	59	0.61(2)	
Na	"	0.56(2)	0.55
K	60	0.52(1)	0.50

Conventional shape-consistent effective potentials (67-70), whether relativistic or not, are typically formulated as expansions of local potentials, $U_\ell(r)$, multiplied by angular projection operators. The expansions are truncated after the lowest angular function not contained in the core. The last (residual) term in the expansion typically represents little more than the simple coulombic interaction between a valence electron and the core (electrons and corresponding fraction of the nuclear charge) and is predominantly attractive. The lower ℓ terms, on the other hand, include strongly repulsive "Pauli" contributions. This is illustrated in Figure 1 where we have plotted the two terms in the Li REP (68). Note that the difference between the two curves goes rapidly to zero for large values of r. A single determinant s^2 trial wavefunction for Li would result in a V^{REP} that included only the repulsive curve. However the inclusion of s to p promotions in the trial wavefunction would introduce a small contribution from the attractive term and would tend to lower the electronic energy. Fortunately, for Li the anion electron density is relatively diffuse and the correction quite small.

In some cases however serious errors can result from the use of such a simple trial wavefunction in Equation 3. The terms in the Be REP are quite similar to those of Li, but the ground state electron distribution is considerably more compact and the correlation correction from the p^2 configuration far more important. In Table II we have listed SCF and REP-QMC energies for the lowest ^1S, ^3P, and ^3D states of Be along with experimental values (71) for comparison. Numbers in square brackets include core polarization corrections (72).

For the Be ^3P and ^3D excited states the correlation corrections are relatively small and as can be seen from the table, the single determinant approximation in Equation 3 is quite good. In contrast the ^1S state is too high by 0.3 eV. However, as Equation 3 would suggest, a simple two-configuration ($2s^2 + 2p^2$) wavefunction brings in the attractive "p" contribution and we get the value listed in column four. Curiously, in an all-electron QMC study Harrison et al. (74) observed a similar difficulty with the Be ground state. They found that a single determinant trial wavefunction gave the energy about 0.3 eV too high due to the fixed node approximation. The use of a multiconfiguration trial wavefunction eliminated the error.

Table II. Effective Potential QMC Energies (in eV)
for various states of Be and Mg

State	reference	SCF	Single	Multi.	Expt.
Be$^+$ ^2S		0.0	0.0	0.0	0.0
Be ^3D	61	-1.54	-1.63(2)	--------	-1.63
Be ^3P	"	-6.39	-6.56(2)	--------	-6.59
Be ^1S	"	-8.06	-9.04(3)	-9.32(1)	-9.32
"	"			[-9.34(1)]	
Mg$^+$ ^2S		0.0	0.0	0.0	0.0
Mg ^1S	62	-6.59	-7.58(1)	-7.55(1)	-7.65
"	"			[-7.66(1)]	
Mg "	59		-7.64(3)		
Mg "	73		-7.57(3)		

In the effective potential approximation Mg is isoelectronic with Be. But, as can be seen in Figure 2, the Mg REP is composed of three terms (s, p and d) with the s and p both repulsive. As a result, even though the correlation correction is almost as large as in Be the multi-determinant correction resulting from Equation 3 is only a tenth as big (see Table II). The discrepancy between values from references (62) and (59) is due to large statistical or extrapolation error. Note that unlike Be one cannot make comparisons with experimental results without first taking core-valence

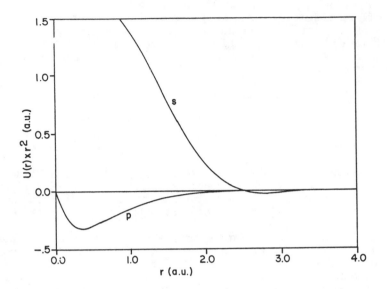

Figure 1. Radial plots of the s and p terms of a Li shape consistent effective potential. (Data are from ref. 68.)

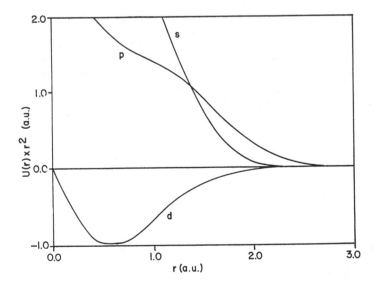

Figure 2. Radial plots of the s, p, and d terms of a Mg shape consistent effective potential. (Data are from ref. 70.)

correlation (72) into account. For the heavier alkalis and alkaline
earths such corrections can amount to several tenths of an eV.
    The effective potential QMC studies discussed above have all
employed more or less conventional shape consistent relativistic
(60,61) or nonrelativistic (59,62) effective potentials localized
according to Equation 3. Yoshida and Iguchi (73) on the other hand
have recently published Mg, Ca and Sr studies employing model
potentials of the type developed by Huzinaga et al. (75). By
comparison with the nodeless orbitals in the preceeding studies the
model potential approach employs representations of normal Hartree-
Fock valence orbitals. In reference (73) the potential was just the
coulombic interaction between normal core and valence orbitals which
is essentially equivalent to the residual term in the shape
consistent REPs. The advantage to this approach is that the
potential is already in simple local form and there is no need for
Equation 3. The dissadvantage is that the trial wavefunctions
include numerous nodes in the core region (which forces one to employ
shorter time steps) and also may include a sizable amplitude near the
nucleus which for Gaussian basis sets might require the use of
nuclear cusp functions. In the QMC simulation this approach almost
looks more like a frozen core study rather than effective potential.
An additional (not nearly so well understood) complication is that in
the simple shielded nucleus potential a trial wavefunction formed
from the valence Hartree-Fock orbitals looks like an excited state.
One must therefore choose the trial wavefunction carefully to ensure
orthogonality to the false lower energy solutions.
    The above studies all involved only one and two-electron
systems. And with the exception of the Be high spin excited states
(61) none required the use of "Fermi statistics" (wavefunction
antisymmetry) in the Monte Carlo simulations. This is of course a
prerequisite for multi-electron systems. We have recently carried
out REP-QMC simulations on some three-electron systems. Aluminum is
probably the simplest. In Table III we show energies for two states
of Al and also for Al$^+$.

Table III. Comparison of SCF, REP-QMC and Experimental
Energies (in eV) for various states of Al.

State	SCF	Single	Multi.	Expt.
Al$^+$ ^{1}S	0.0	0.00(2)	0.00(1)	0.0
Al ^{4}P	-3.11	-2.42(3)	[-2.42(2)]	-2.38
Al ^{2}P	-5.48	-5.92(4)	-5.92(2)	-5.98

    For the Al ground state and also for the cation we carried out
simulations using both single and multiple determinant trial
wavefunctions. (The brackets indicate a single determinant trial
function for the ^{4}P state but multiple for the ^{1}S.) As opposed to
either Be or Mg we see no significant adjustment in the energies
resulting from the use of the more accurate trial functions. We do
however see a reduction in the statistical error. The agreement with
experiment is excellent except for the 0.06 eV systematic error in
the ground state. This could be due to a still inadequate trial
wavefunction in Equation 3, but based on Muller's Mg work (72) we
also suspect core polarization. This remains to be determined.
    As indicated in the introduction a major motivation in the

development of REP–QMC involves applications to transition element problems. Although to our knowledge there are as yet no such applications in the literature we have recently been running preliminary three–electron simulations involving the ^{2}D ground and ^{4}F excited states of Sc and Y. Our results so far have been somewhat disappointing. For both elements we see excitation energy errors of about 0.5(2) eV which are two to three times what we would have anticipated for the three–electron approximation. The poor results could be due to the single determinant trial functions but it might also be the result of core polarization or some other aspect of the frozen–core approximation. We may be forced to use 11–electron REPs to achieve adequate results. In this event, a recently developed "frozen core" approach (76) could prove invaluable in combination with the 11–electron REP.

Molecules   To date the only molecular effective potential QMC results that we are aware of are those of Hammond et al. (59). The difficulty in extending this work to diatomic or polyatomic systems involves the efficient evaluation of the local potential, v^{REP}, in conjunction with the projection operators in the potential and basis functions centered on different nuclei. In the NaH and Na$_2$ work of reference (59) the angular projection integrations were carried out by more or less conventional means. In preliminary work however we have found a simple alternative useful. Since the radial functions in U^{REP} decay very rapidly with increasing r, the product of the $U_\ell(r)$ with a function centered on another nucleus can be approximated quite accurately by short one–center expansions. Conveniently, the angular part of the expansion need not go beyond the highest ℓ quantum number in the core. For first row atoms (Li, Be, C, etc.) only a small number of s functions are required. K or Mg studies would require only s and p functions. In preliminary studies we have found that for K$_2$ the use of s functions alone results in an error of only about a tenth of an eV due to the neglect of p functions in the one–center expansion.

For linear molecules a particularly efficient scheme (in terms of the QMC sampling) might be to tabulate the product of the U^{REP} with each orbital on a coarse two–dimensional grid. In this sense one can see that an interesting approach to diatomic REP–QMC could be based on numerically determined SCF or MCSCF trial wavefunctions (77,78).

Discussion

Quantum Monte Carlo techniques have considerable potential for application to problems involving open d or f shells where the treatment of electron correlation has proven particularly difficult. However if QMC is to be a viable alternative one must be able to limit the simulations to small numbers of electrons and in addition relativeity must be included. Relativistic effective potentials offer one avenue (at the present time the only avenue) for achieving these conditions. However, as we have indicated, REPs do introduce complications.

Because of Equation 3 one must be somewhat more careful in the selection of trial wavefunctions and multiconfiguration algorithms are essential. Unfortunately the additional configurations

necessarily increase the computer requirements. For the Be ground
state for instance the addition of the p^2 configuration just about
doubles the amount of processing per QMC time step. On the other
hand as one can see from the Tables the sampling error is typically
reduced by more than a factor of two. Noting that the sampling error
varies inversely with the square root of the number of samples we see
that the use of multiconfiguration trial wavefunctions actually
reduces the overall computer requirements (for a given error level)
by more than a factor of two. This is consistent with the arguments
given in references (58,59) and with the results of Moskowitz et al.
(48,49).

Though not discussed above, in all the studies mentioned the
trial wavefunctions included pair correlation functions, J_{ij}, as
prescribed by Reynolds et al. (42). Moskowitz et al. (48,49) have
shown that the product of a relatively simple multiconfiguration
wavefunction with pair correlation functions can provide a rather
accurate approximation to the exact wavefunction. In our
calculations and in those of Hammond et al. (59) the many-electron
local potential, V^{REP}, has been obtained by allowing the REP to
operate only on the determinantal portions of the trial wavefunction.
The effects of the pair correlation functions have been ignored. As
pointed out in (61) the effects of the pair correlation functions on
the local potentials could be included by means of zeta-function
expansions (79). However in our multiconfiguration calculations the
J_{ij} were parametrized to correct for short range difficulties only,
(We assumed that the configuration expansions properly accounted for
long-range and near-degeneracy effects), and we would therefore
expect the J_{ij} to have only a negligible effect on the V_{REP}.

All of the effective potential QMC studies that we are aware of
have employed relatively simple fixed-node diffusion Monte Carlo
algorithms. This is not to suggest that these are preferable, but
rather easy to program. One should not underestimate the advantages
of Green's Function [see references (50,51) for instance] or other
more recently developed approaches (80).

Essentially all of the effective potential QMC work in the
literature to date has been, or could be, carried out on small mini
or microcomputers. For instance, using QMC algorithms similar to
those of references (39) and (52), our Be, Mg and Al (two and three
electron) studies required around 50 to 200 MicroVAX hours for single
atomic state energies with standard errors of about 0.01 eV. However
work on larger systems will obviously require more powerful computing
machinery. Although we would not anticipate dramatic increases in
speed due to vectorization, (at least not for problems involving
smaller numbers of electrons) QMC algorithms are almost trivially
adaptable to "massively parallel" computing environments (81-83). By
simply distributing configurations (particles) evenly among
processors one should be able to obtain near peak parallel
efficiency. The only complication we forsee would be due to the
occasional renormalization required by the particle multiplication
and destruction events. This would necessitate the communication of
particle coordinates, etc. between processors to maintain an even
configuration distribution. Fortunately the time required to
transfer configurations between processors will be proportional to
the number of electrons per configuration whereas the processing time
(per time step) is proportional to the number of electrons squared or

cubed (42). As a result one would expect inefficiencies due to processor imbalance to become less important for bigger problems. Furthermore for "hypercube" machines (81-83) with a thousand or fewer processors the transfer of configurations could probably be limited to adjacent nodes, thereby minimizing the transfer time.

Massively parallel (multiple instruction, multiple data) computers with tens or hundreds of processors are not readily accessible to the majority of quantum chemists at the present time. However the cost of currently available hypercube machines with tens of processors (each with about the power of a VAX) is comparable to that of superminis but with up to a hundred times the power. For applications of the type discussed above the performance of a machine with as few as 32 or 64 processors would be comparable to (or perhaps even exceed) that of a single processor supercomputer. Although computer requirements currently limit QMC applications (even with effective potentials) the proliferation of inexpensive massively parallel machines could conceivably make the application of relativistic effective potentials with QMC quite competitive with more conventional electronic structure techniques.

Our REP-QMC work (60-62) along with the studies by Hammond et al. (59) provide evidence that with the proper precautions the combination of relativistic effective potentials with quantum Monte Carlo procedures may provide an alternative for obtaining accurate electronic structure information. The possible elimination of excessive basis set and configuration expansions for the transition and heavier elements is especially appealing. And at the same time the transition to parallel computing is particularly simple. The possibility of carrying out definitive calculations on small transition metal clusters should not be dismissed offhand. Although we would not expect QMC (or REP-QMC) to replace conventional approaches (for one thing an accurate conventional trial wavefunction appears to be an essential prerequisite) the electron correlation problem for elements containing occupied d and f subshells is potentially so enormous that all possible avenues should be thoroughly researched.

Acknowledgments

This work has been supported in part by the National Science Foundation under Grant No. CHE-8214665 and also by the Research Corporation.

Literature Cited

1.   K.P. Huber and G. Herzberg, Molecular Spectra and Molecular Structure IV. Constants of Diatomic Molecules, Van Nostrand Reinhold Company: New York, 1979.
2.   W. Weltner and R.J. Van Zee, Ann. Rev. Phys. Chem. 1984, 35, 291.
3.   M.D. Morse, Chem. Rev. 1986, 86, 1049.
4.   D.R. Salahub, Adv. Chem. Phys. 1987, 69, 447.
5.   M.M. Goodgame and W.A. Goddard, III, J. Phys. Chem. 1981, 86, 215.
6.   S.P. Walch, C.W. Bauschlicher, Jr., B.O. Roos and C.J. Nelin, Chem. Phys. Lett. 1983, 103, 175.

7.   G.P. Das and R.L. Jaffe, Chem. Phys. Lett. 1984, 109, 206.
8.   A.D. McLean and B. Liu, Chem. Phys. Lett. 1983, 101, 144.
9.   M.M. Goodgame and W.A. Goddard, III, Phys. Rev. Lett. 1985, 54, 661.
10. G. Verhaegen, S. Smoes and J. Drowart, J. Chem. Phys. 1964, 40, 239.
11. G. Das, Chem. Phys. Lett. 1982, 86, 482.
12. S.P. Walch and C.W. Bauschlicher, Jr., J. Chem. Phys. 1983, 79, 3590.
13. M. Moskovits, D.P. DiLella and W. Limm, J. Chem. Phys. 1984, 80, 626.
14. L.B. Knight, Jr., R.J. Van Zee and W. Weltner, Jr., Chem. Phys. Lett. 1983, 94, 296.
15. L.B. Knight, Jr., R.W. Woodward, R.J. Van Zee and W. Weltner, Jr., J. Chem. Phys. 1983, 79, 5820.
16. R.J. Singer and R. Grinter, Chem. Phys. 1987, 113, 99.
17. N.A. Baykara, B.N. McMaster and D.R. Salahub, Mol. Phys. 1984, 52, 891.
18. B. Delley, A.J. Freeman and D.E. Ellis, Phys. Rev. Lett. 1983, 50, 488.
19. J. Bernholc and N.A.W. Holzwarth, Phys. Rev. Lett. 1983, 50, 1451.
20. C.W. Bauschlicher, Jr., S.P. Walch and H. Partridge, J. Chem. Phys. 1982, 76, 1033.
21. K.K. Sunil and K.D. Jordan, J. Chem. Phys. 1985, 82, 873.
22. C.M. Rohlfing and R.L. Martin, Chem. Phys. Lett. 1985, 115, 104.
23. S.R. Langhoff and C.W. Bauschlicher, Jr., J. Chem. Phys. 1986, 84, 4485.
24. C.W. Bauschlicher, Jr., J. Chem. Phys. 1987, 86, 5591.
25. D.P. Chong, S.R. Langhoff, C.W. Bauschlicher, Jr., S.P. Walch and H. Partridge, J. Chem. Phys. 1986, 85, 2850.
26. S.R. Langhoff, L.G.M. Pettersson, C.W. Bauschlicher, Jr. and H. Partridge, J. Chem. Phys. 1987, 86, 268.
27. L.G.M. Pettersson, C.W. Bauschlicher, Jr., S.R. Langhoff and H. Partridge, J. Chem. Phys. 1987, 87, 481.
28. A.E. Alvarado-Swaisgood, J. Allison and J.F. Harrison, J. Phys. Chem. 1985, 89, 2517.
29. J.B. Schilling, W.A. Goddard III and J.L. Beauchamp, J. Am. Chem. Soc. 1986, 108, 582.
30. C.W. Bauschlicher, Jr. and S.P. Walch, J. Chem. Phys. 1982, 76, 4560.
31. D.M. Ceperly amd B.J. Alder, Science 1986, 231, 555.
32. J.B. Anderson, J. Chem. Phys. 1975, 63, 1499.
33. J.B. Anderson, J. Chem. Phys. 1976, 65, 4121.
34. M.H. Kalos, D. Levesque and L. Verlet, Phys. Rev. A 1974, 9, 2178.
35. D.M. Ceperly, G.V. Chester and M.H. Kalos, Phys. Rev. B 1977, 16, 3081.
36. J. B. Anderson, J. Chem. Phys. 1980, 73, 3897.
37. F. Mentch and J.B. Anderson, J. Chem. Phys. 1981, 74, 6307.
38. F. Mentch and J.B. Anderson, J. Chem. Phys. 1984, 80, 2675.
39. J.B. Anderson, J. Chem. Phys. 1985, 82, 2662.
40. D.R. Garmer and J.B. Anderson, J. Chem. Phys. 1987, 86, 4025.
41. D.M. Ceperley and B.J. Alder, Phys. Rev. Lett. 1980, 45, 566.

42. P.J. Reynolds, D.M. Ceperley, B.J. Alder and W.A. Lester, Jr., J. Chem. Phys. 1982, 77, 5593.
43. P.J. Reynolds, M. Dupuis and W.A. Lester, Jr., J. Chem. Phys. 1983, 82, 1983.
44. R.N. Barnett, P.J. Reynolds and W.A. Lester, Jr., J. Chem. Phys. 1985, 82, 2700.
45. R.N. Barnett, P.J. Reynolds and W.A. Lester, Jr., J. Chem. Phys. 1986, 84, 4992.
46. R.N. Barnett, P.J. Reynolds and W.A. Lester, Jr., J. Chem. Phys. 1987, 91, 2004.
47. P.J. Reynolds, R.N. Barnett, B.L. Hammond and W.A. Lester, Jr., J. Stat. Phys. 1986, 43, 1017.
48. J.W. Moskowitz, K.E. Schmidt, M.A. Lee and M.H. Kalos, J. Chem. Phys. 1982, 76, 1064.
49. J.W. Moskowitz, K.E. Schmidt, M.A. Lee and M.H. Kalos, J. Chem. Phys. 1982, 77, 349.
50. D.W. Skinner, J.W. Moskowitz, M.A. Lee, P.A. Whitlock and K.E. Schmidt, J. Chem. Phys. 1985, 83, 4668.
51. K.E. Schmidt and J.W. Moskowitz, J. Stat. Phys. 1986, 43, 1027.
52. J. Vrbik and S.M. Rothstein, J. Comput. Phys. 1986, 63, 130.
53. J. Vrbik, J. Phys. A 1985, 18, 1327.
54. J. Vrbik and S.M. Rothstein. Int. J. Quant. Chem. 1986, 29, 461.
55. S.M. Rothstein, N. Patil and J. Vrbik, J. Comput. Chem. 1987, 8, 412.
56. S.M. Rothstein and J. Vrbik, J. Comput. Phys. 1988, 74, 127.
57. J.D. Doll, Chem. Phys. Lett. 1981, 81, 335.
58. D.M. Ceperley, J. Stat. Phys. 1986, 43, 815.
59. B.L. Hammond, P.J. Reynolds and W.A. Lester, Jr., J. Chem. Phys. 1987, 87, 1130.
60. M.M. Hurley and P.A. Christiansen, J. Chem. Phys. 1987, 86, 1069.
61. P.A. Christiansen, J. Chem. Phys. 1988, 88, 4867.
62. P.A. Christiansen and L.A. LaJohn, Chem. Phys. Lett. 1988, 146, 162.
63. J. Vrbik, M.F. DePasquale and S.M. Rothstein, J. Chem. Phys. 1988, 88, 3784.
64. D. Feldmann, Z. Phys. A 1976, 277, 19.
65. R.D. Mead, P.A. Schulz and W.C. Lineberger, Phys. Rev. A.
66. J. Slater, F.H. Read, S.E. Novick and W.C. Lineberger, Phys. Rev. A 1978, 17, 201.
67. P.A. Christiansen, Y.S. Lee and K.S. Pitzer, J. Chem. Phys. 1979, 71, 4445.
68. L.F. Pacios and P.A. Christiansen, J. Chem. Phys. 1985, 82, 2664.
69. P.J. Hay and W.R. Wadt, J. Chem. Phys. 1985, 82, 270.
70. W.J. Stevens, H. Basch and M. Krauss, J. Chem. Phys. 1984, 81, 6026.
71. C.E. Moore, Atomic Energy Levels, Natl. Bur. Stand. Circ. 467, Vols. I and II, U.S. Government Print Office: Washington, DC, 1949, 1952.
72. W. Muller, J. Flesch and W. Meyer, J. Chem. Phys. 1984, 80, 3297.
73. T. Yoshida and K. Iguchi, J. Chem. Phys. 1988, 88, 1032.
74. R.J. Harrison and N.C. Handy, Chem. Phys. Lett. 1985, 113, 257.

75. S. Huzinaga, L. Seijo, Z. Barandiaran and M. Klobukowski, J. Chem. Phys. 1987, 86, 2132.
76. W.A. Lester, Jr., private communication.
77. E.A. McCullough, Jr., J. Chem. Phys. 1975, 62, 3991.
78. P. Pyykko, G.H.F. Diercksen, F. Muller-Plathe and L. Laaksonen, Chem. Phys. Lett. 1987, 134, 575.
79. M.P. Barnett, Methods Comput. Phys. 1963, 2, 95.
80. J.B. Anderson, J. Chem. Phys. 1987, 86, 2839.
81. R.A. Whiteside, J.S. Binkley, M.E. Colvin and H.F. Schaefer III, J. Chem. Phys. 1987, 86, 2185.
82. Science News 1988, 133.
83. J.L. Gustafson, G.R. Montry and R.E. Benner, SIAM J. Sci. and Stat. Computing 1988, 9.

RECEIVED February 16, 1989

Chapter 23

Relativistic Effects on Compounds Containing Heavy Elements

The Influence of Kinetic Energy on Chemical Bonds

T. Ziegler[1], J. G. Snijders[2], and E. J. Baerends[2]

[1]Department of Chemistry, University of Calgary, Calgary, Alberta T2N
1N4, Canada
[2]Department of Theoretical Chemistry, The Free University, Amsterdam,
Netherlands

It is shown that relativity will reduce the
kinetic energy of the electrons in a number
of compounds containing heavy elements.
The reduction of the kinetic energy leads to
bond stabilization and bond contraction
and influences significantly the chemistry
of third row transition metals .

Valence electrons in atoms and molecules have a
finite (albeit small) probability of being close to the
nuclei and they can as a consequence acquire high
instantaneous velocities.In fact,the velocities for the
valence electrons can approach that of light as they pass
in close proximity to heavier nuclei with Z >72.It is for
this reason not too surprising that relativistic effects
become of importance for the chemical properties of
compounds containing 5d-block elements in the third
transition series or 5f-block elements in the actinide
series.
        We shall here discuss how relativistic effects
,related to the high instantaneous velocities of electrons
near heavy nuclei, will influence the chemical bond
involving 5d- and 5f-elements.In particular,we shall

0097–6156/89/0394–0322$06.00/0
© 1989 American Chemical Society

demonstrate that relativity in many cases will strengthen the bonds and contract the bond distances by a reduction in the kinetic energy of the electrons.It will further be illustrated how relativistic effects influence the periodic trends within a triad of transition metals.

The field of relativistic quantum chemistry has been reviewed by Pitzer (1) and by Pyykko (2-3). The results presented here are based on the relativistic Hartree-Fock-Slater method due to Snijders and Baerends (4),augmented in some cases by the recently proposed density functional method by Becke (5).Our analysis will build on previous studies (6) due to Ziegler,Snijders and Baerends.

Relativistic Theories

It is possible to extend Dirac's (7) relativistic theory for the hydrogen atom to n-electron systems by neglecting retardation as well as certain magnetic effects.Dirac's Hamiltonian for a many electron system can be written as

$$\hat{H}^D - \frac{1}{2} \sum_{i=1}^{i=n} \hat{h}_D(i) + \sum_{i=1}^{i=n} V_N(\vec{r_i}) + \frac{1}{2} \sum_{i \neq j} 1/|\vec{r_i} - \vec{r_j}| \tag{1}$$

where

$$\hat{h}_D(i) = c\vec{\alpha}(i).\hat{p}(i) + c^2\beta \tag{2}$$

and $V_N(\vec{r_i})$ is the electron-nuclear attraction potential.

In Eq.(1) $\alpha_j^o(i)$, $j=1,2,3$ and β are the 4x4 Dirac matrices, $\hat{p}(i)$ is the momentum operator,and c the velocity of light.The corresponding Dirac wave equation reads

$$H^D\Psi^D = E^D\Psi^D \tag{3} ,$$

where the wave function Ψ^D can be expressed as a linear combination of Slater determinants constructed from four-component Dirac-spinors $\{a^i\}$.

The Dirac wave equation is somewhat cumbersome to solve due to the presence of four,in general complex, components. Foldy and Wouthuysen (8) have fortunately proposed a transformation which allows one to approximate the four-component Dirac equation, Eq. (3), by a two-component Schrödinger equation in the familiar 2x2 Pauli representation to any given order in the fine structure constant α. The Hamiltonian $\hat{H}^D$ of Eq. (1) takes after the Foldy-Wouthuysen transformation the form of an infinite sum of terms in increasing orders of α^2

$$\hat{H}^{FW} = \hat{H}^0 + \alpha^2 \hat{H}_1 + \alpha^4 \hat{H}_2 + O(\alpha^6) \qquad (4),$$

where $\hat{H}^0$ is the non-relativistic Hamiltonian. The eigenfunctions Ψ^{FW} of $\hat{H}^{FW}$ can be written as a linear combination of Slater determinants constructed from spin orbitals as in the non-relativistic case.

The method of Snijders and Baerends (4) retains terms in Eq. (4) up to $\alpha^2 \hat{H}_1$, where $\hat{H}_1$ is given by

$$\hat{H}_1 = \hat{H}_{MV} + \hat{H}_{Darw} + \hat{H}_{SO} \qquad (5).$$

The so-called mass-velocity term $\hat{H}_{MV}$, which represents the first order (in α^2) relativistic correction to the non-relativistic kinetic energy operator

$$\hat{T}^{NR} = -\frac{1}{2} \sum_i \nabla_i^2 = \frac{1}{2} \sum_i \hat{p}^2(i) \qquad (6)$$

is given by

$$\hat{H}_{MV} = -\frac{1}{8} \sum_i \nabla_i^4 = -\frac{1}{8} \sum_i \hat{p}^4(i) \qquad (7a)$$

whereas the Darwin term, from the Zitterbewegung of the electrons (9), after neglecting some numerically insignificant two-electron operators (6a), takes on the form

$$\hat{H}_{Darw} = \frac{1}{8} \sum_i \nabla_i^2 (V_N(\vec{r_i})) \qquad (7b)$$

The spin-orbit operator $\hat{H}_{SO}$ does not contribute to the total energy of the closed shell molecules considered in the following and need not thus be specified here.

The relativistic wave function, and related total energy, is obtained by first order perturbation theory, with the non-relativistic wave function as zero-order solution.

Relativistic Calculations on Metal Dimers and Metal Hydrides .Relativistic Bond Contraction and Relativistic Bond Stabilization

We present in Table I results from calculations on bond energies ,bond distances and vibrational frequencies for the simple MH hydrides of the coinage triad M=Cu,Ag,and Au as well as the isoelectronic series MH^+ ,with M=Zn,Cd,and Hg.Table I contains experimental data(10) as well as results from non-relativistic (11) and relativistic (4) Hartree-Fock-Slater(HFS) calculations. Results from a similar set of calculations on the metal-dimers M_2 (M=Cu,Ag,and Au) as well as the dications M_2^{+2} (M=Zn,Cd,and Hg) are presented in Table II.

It follows from Tables I and II that relativistic effects have a sizable influence on bond energies,bond distances,as well as vibrational frequencies for the compounds containing the heavier 5d-member (Au or Hg) within the triad.

In the non-relativistic case bond energies follows the wrong order(compared with experiment) of first row>second row>third-row.Relativistic corrections,which for Au and Hg stabilize the bonds by some 30 Kcal mol^{-1} ,provide on the other hand the correct ordering of third row>first row>second row.

Bond distances are, in the non-relativistic limit, calculated to follow the order first row<second row<third row ,whereas in fact experiment finds that homologous second and third row compounds have similar bond distances.However,relativistic effects ,which for Au and Hg contract the bonds by 0.3 Å or more,bring bond distances of second and third row metals into close proximity to each other.

Relativistic effects are finally seen to increase vibrational frequencies of the third row compounds by up to 100%.The relativistic frequencies are further in better accord with experiment than the non-relativistic values.Results similar to ours have been found in studies based on other methods (12) .

Table I. Calculations of bond distances (R_{AB}), dissociation energies $(-\Delta E_{AB})$, and vibrational frequencies (ω_{AB}) from relativistic HFS-calculations and non-relativistic HFS calculations (figures in parentheses) on metal hydrides

| Compound | R_{AB} (Å) | | $-\Delta E_{AB}$ (Kcal mol^{-1}) | | ω_{AB} (cm^{-1}) | |
	HFS[b]	Exp.[a]	HFS[b].	Exp.[a]	HFS.	Exp.[a]
CuH	1.50(1.51)	1.46	61(59)	66 ±2	1905(1884)	1940
AgH	1.61(1.71)	1.61	47(39)	53 ±2	1709(1605)	1760
AuH	1.55(1.78)	1.52	68(37)	74 ±3	2241(1704)	2305
ZnH$^+$	1.58(1.58)	1.52	58(58)	65 ±9	1810(1803)	1916
CdH$^+$	1.74(1.78)	1.68	48(46)	48 ±9	1669(1665)	1775
HgH$^+$	1.64(1.88)	1.59	62(41)	53±10	2156(1267)	2034

[a]Ref. 10.
[b]With an exchange factor, α_{ex}, of 0.7

Table II. Calculations of bond distances (R_{AB}), dissociation energies $(-\Delta E_{AB})$, and vibrational frequencies (ω_{AB}) from relativistic HFS-calculations and non-relativistic HFS calculations (figures in parentheses) on metal dimers

| Compound | R_{AB} (Å) | | $-\Delta E_{AB}$ (Kcal mol^{-1}) | | ω_{AB} (cm^{-1}) | |
	HFS[c].	Exp.	HFS[c].	Exp.	HFS[c].	Exp.
Cu_2	2.24(2.26)	2.22	53(51)	45 ±2	274(268)	266
Ag_2	2.52(2.67)	-	47(40)	37 ±2	203(184)	192
Au_2	2.44(2.90)	2.47	58(27)	52 ±2	201(93)	191
Zn_2^{2+}	2.40(2.42)	-	-30(-30)[b]	-	187(183)	-
Cd_2^{2+}	2.73(2.84)	-	-34(-39)[b]	-	160(141)	-
Hg_2^{2+}	2.62(3.12)	-	-11(-46)[b]	-	182(107)	-

[a]Ref. 10. [b]The Di-cations Zn_2^{2+}, Cd_2^{2+}, and Hg_2^{2+} are all unstable in vacuum (negative dissociations energies). They will, however, be stabilized in a medium by counterions
[c]With an exchamge factor, α_{ex}, of 0.7

Relativistic Reduction of the Kinetic Energy.

How does relativity do it? That is, how are relativistic effects able to contract and stabilize the chemical bonds? Rather detailed numerical analyses(6a) have shown that the crucial term in the first order Hamiltonian of Eq.(5) is H_{MV}, representing the first order relativistic correction to the kinetic energy .We shall for this reason first discuss how the kinetic energy of a particle is modified in going from the non-relativistic to the relativistic limit.

The kinetic energy T^R of a particle with the rest mass m_0 is given by(9)

$$T^R = \sqrt{p^2c^2 + m_0^2c^4} \quad -m_0c^2 \qquad (8),$$

in classical relativistic kinematics ,where p is the momentum.Further, the relation to the non-relativistic limit can be established by expanding T^R in powers of $\alpha^2 = (1/c)^2$ as

$$\tilde{T}^R = \frac{1}{2m_0} p^2 - \frac{\alpha^2}{8m_0} p^4 + O(\alpha^4) \qquad (9),$$

where the first term in Eq.(9) represents the classical non-relativistic kinetic energy T^{NR} and the second term the first order relativistic correction.Note that the kinetic energy operator in the first order relativistic Hamiltonian can be obtained from Eq.(8) by replacing p with the operator p, Eq.(6) and Eq.(7a).

We provide in Figure 1 a plot of $T^R, \tilde{T}^R$, and T^{NR} as a function of $x=|p|/m_0c$. It follows from Figure 1 that $T^R, \tilde{T}^R$, and T^{NR} for small values of x ($x< .1$) are similar in size.However, as x and $|p|$ increase T^R (and $\tilde{T}^R$) drops below T^{NR}.That is, relativity will <u>reduce</u> the kinetic energy for larger values of $|p|$.In fact, the relativistic reduction in kinetic energy(compared to T^{NR}) is seen to increase with $|p|$.

We shall now tie the relativistic reduction of the kinetic energy(for large $|p|$) to the bond contraction and bond stabilization in molecules of heavy elements.However, we must, before we can do that ,recall a few facts about potential energy surfaces.To this end consider first ,in the non-relativistic limit, a diatomic molecule AB of the total non-relativistic energy E_{AB}^{o}

,formed from A and $_\circ$ B with the total combined non-relativistic energy E°_{A+B} .The total energies of AB ,as well as those of the constituting atoms ,can further be decomposed as

$$E^\circ = T^\circ + V^\circ_{Ne} + V^\circ_{ee} = T^\circ + V^\circ \quad (10),$$

into contributions from the non-relativistic kinetic energy (T°),the non-relativistic electron-nuclei attraction energy V°_{Ne} ,as well as the non-relativistic electron-electron repulsion energy V°_{ee} .The two terms, V°_{ee} and V°_{Ne} ,can in addition be combined into the total Coulomb energy V°.We give in Figure 2 a plot of $\Delta E^\circ_{AB} = E^\circ_{AB} - E^\circ_{A+B}$ as a function of the interatomic distance, R_{AB},decomposed into contributions from the kinetic energy and the Coulomb energy.The kinetic energy ,T°_{AB} ,of the molecule might well be smaller than the sum of the kinetic energies of the two constituting atoms,T°_{A+B},at large interatomic distances ,as it is shown in Figure 2.However,at the equilibrium distance ,$R_{AB_\circ}$,we have that T°_{AB} is larger than T°_{A+B} and is given by $T^\circ_{AB} = T^\circ_{A+B} - \Delta E^\circ_{AB}$ according to the virial theorem,where $-\Delta E^\circ_{AB}$ is the (positive) bond energy for AB.The Coulomb energy , V°_{AB}, of the molecule might on the other hand at large interatomic distances be above the Coulomb energies of the constituent atoms ,V°_{A+B}.However,at the equilibrium distance R°_{AB} the molecular Coulomb energy ,V°_{AB}, is below the Coulomb energy of the constituting atoms and given by $V^\circ_{AB} = V^\circ_{A+B} + 2\Delta E^\circ_{AB}$ according to the virial theorem.

What really is of importance in this somewhat lengthy discussion of a text book subject(13) ,is that at distances smaller than R°_{AB} the kinetic energy ,T_{AB} ,will rise faster in energy than V°_{AB} will decrease,resulting in a minimum for the non-relativistic energy,E°_{AB}, of the diatomic molecule AB.

We can now turn to a discussion of how relativistic effects will modify the molecular energies, E_{AB} (or ΔE_{AB}), as well as their functional dependence on the interatomic distance ,R_{AB}.There are two non-zero relativistic corrections from the first order Hamiltonian, H_1 ,of Eq.(5).One is the first order correction ,T^1 ,to the kinetic energy from the mass-velocity term , $\hat{H}_{MV}$,and

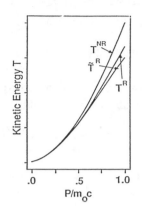

Figure 1. Non-relativistic kinetic energy ,T^{NR},the first order relativistic kinetic energy ,$\tilde{T}^{R}$,and the total kinetic energy ,T^{R}, as a function of the absolute momentum $|p|$

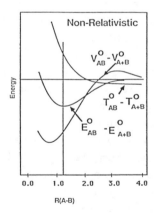

Figure 2. Non-relativistic bond energy $-\Delta E_{AB}^{o}$ as a function of the interatomic distance ,R_{AB}, in the diatomic molecule ,AB. The total non-relativistic energy of AB,E_{AB}^{o},is decomposed into kinetic energy,T_{AB}^{o}, and Coulomb energy ,V_{AB}^{o}. The corresponding combined energies for the constituting atoms are given by E_{A+B}^{o},T_{A+B}^{o},and V_{A+B}^{o},respectively.

the other a correction ,V^1, ₌to the nuclear electron-attraction from the Darwin term H_{Darw}.Thus,the total energy to first order can be written as

$$E^R = E^0 + T^1 + V^1 + O(\alpha^4) \qquad (11)$$

The correction ,T^1, is ,with respect to a Slater-determinantal wave function $\Psi = |\mu_1\mu_2\mu_3...\mu_n|$,given by

$$T^1 = -\frac{\alpha^2}{8} \sum_i <i|\nabla_i^4|i> = -\frac{\alpha^2}{8} \sum_i \int \nabla_1^2\mu_i(1)\nabla_1^2\mu_i(1)\,d\tau_1 \qquad (12)$$

It follows from Eq.(12) that T^1 constitutes a negative definite correction.That is ,T^1 will <u>reduce</u> the kinetic energy just as in the classical case illustrated in Figure 1.The influence of T^1 on the molecular energy surface is illustrated in Figure 3.We observe that the relativistic reduction in the kinetic energy ,T^1,is enhanced as we shorten the bond distance ,R_{AB}.The term T^1 will ,as a consequence ,stabilize and contract the chemical bond.The dependency of T^1 on R_{AB} can readily be understood from Figure 1 when we recall that T^0 and $<p^2>$ increase as we shorten R_{AB}.The increase in T^0 has two sources,the contraction of the valence AOs toward the nuclei at distances shorter than R_e and the core-valence orthogonality requirement.Both effects cause increaced wavefunction and density gradients.There is a concommittant increase in the average momentum,i.e. an expansion of the momentum density.This gives rise to an increase in $<p^4>$ and corresponding decrease in T^1.There is therefore an increasing reduction in the relativistic kinetic energy(compared to T^0) as the interatomic distance is reduced.

The second relativistic corection ,V^1, comes from the Darwin term H_{Darw} .It is positive definite since(9)

$$\nabla_i^2 (V_N(\vec{r_i})) = \sum_g 4\pi\delta(|\vec{r_i}-\vec{Rg})|)Z_g \qquad (13),$$

and constitutes a reduction in the nuclear-electron attraction energy (see Figure 3) which will weaken the bond.It does not seem to be possible to relate T_{AB}^1 and V_{AB}^1 by a virial theorem in the same way as T_{AB}^o is related to V_{AB}^o at the non-relativistic equilibrium dist∍nce

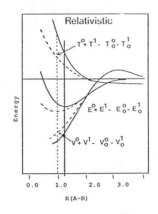

Figure 3. Influence of T^1_{AB} and V^1_{AB} on the energy and equilibrium distance in AB. Solid lines represent non-relativistic curves of Figure 2.

R_{AB}^{o}.However,numerically, T_{AB}^{1} prevails over V_{AB}^{1} with the result that the combined relativistic corrections will increase the bond energy in AB and contract the bond distance(Table I,Table II and Figure 3).

It should be pointed out that Schwarz (20),using double perturbation theory,has demonstrated that it is possible to rationalize the relativistic bond length contraction in terms of the attractive Hellmann-Feynman force due to the relativistic change in electron density.In such an approach it would be necessary to analyze and get a physical picture of the relavant density changes

We shall now turn to a general discusion of the trends in bond energies involving a triad of transition metals,with special emphasis on the role played by relativistic effects.

Periodic Trends within a Triad of Transition Metals

a. Metal-Metal Bonds

Valence electrons in s-orbitals have a relatively large probability for being near the nuclei and it is thus not surprising that relativistic effects are substantial in Au_2 and Hg_2^{2+} ,where the bonding primarily involves the two 6s orbitals. Relativistic effects seem also to be substantial in the hextuple bonded dimers W_2,Mo_2,and Cr_2 (14),where one of the σ-bonds involve s-orbitals(15).Thus ,we calculate (14) relativity to stabilize the W-W bond by 25 Kcal mol^{-1} and change the non-relativistic order of stability ,$Mo_2>W_2>Cr_2$,to $W_2>Mo_2>Cr_2$.One might also speculate that relativistic effects increase the cohesive energy in bulk metal of third row elements.Third row metals have in general larger cohesive energies in bulk than their lighter congeners within a triad.

Density functional calculations (4) on the strengths of multiple metal-metal bonds in binuclear complexes such as $M_2Cl_4(PR_3)_4$ and M_2X_6 ,where the bonding primarily involves nd-orbitals rather than (n+1)s orbitals,reveal (13) on the other hand only modest contributions from relativistic corrections (6-10 Kcal mol^{-1}) for the 5d homologues.The calculated order of stability for the M-M bonds in these systems ,given by third row>second row>first row ,is ,instead caused by optimal d-d overlaps for the 5d orbitals ,combined with less repulsive interactions involving core orbitals of the third row metals compared with the first row metals.It still remains to be seen to what degree relativity influences the M-M bond strengths in metal cluster and polynuclear metal complexes.

b. Metal-ligand σ-bonds.

Single metal-ligand bonds involving primarily σ-type orbitals on both the metal center and the ligand are quite common.Small size examples would be the two hydrides AuH and HgH^+.However,the two species AuH and HgH^+ are somewhat atypical in that the σ-interaction involves the 6s-orbital on the metals rather than 5d .The involvement of 6s is largely responsible for the substantial relativistic contribution to the M-H bond energies in AuH and HgH^+ (Table I).

Metal-ligand σ-bonding involves in most cases d_σ-type metal orbitals rather than the (n+1)s orbital,and we have found ,from a number of studies(16) ,that the relativistic corrections to metal-ligand σ-bond strengths are modest(2-10 Kcal mol^{-1}) in cases where d_σ rather than (n+1)s is involved.Our studies include M-R bond energies (R=H,CH_3) in $RM(CO)_5$ (M=Mn,Tc,Re) (16a,16b),$RM(CO)_4$ (M=Co,Rh,Ir) (15a,15b),Cp_2MR(M=Sc,Y,La,V,Mn,Tc,Re)(15c),$CpFe(CO)_2R$,CpNi COR (15c),and $RMCl_3$ (M=Ti,Zr,Hf)(15d),as well as the M-L bond energies in $LMCl_3$ (M=Ti,Zr,Hf) and $LCo(CO)_4$ (16d) with L = SiH_3,OH,OCH_3,SH,NH_2, PH_2 ,and CN.

The order of stability for the M-L σ-bonds was in all the studies (16) found to be third row>second row>first row for a homologous series of compounds .An analysis revealed further that this trend is caused by an increase in the overlap between d_σ and the σ-ligand orbital as we decend the triad.

c. Synergic Metal-Ligand Bonds.

A number of ligands such as CO,O_2,C_2H_2,C_2H_4,C_6H_6,and Cp^-(cyclopentadienyl) bind to metal centers as π-acceptors.That is,they have empty π-orbitals capable of accepting electron density from occupied metal based d-orbitals ,as it is shown for CO in **1a** .The ligands have in addition occupied σ-type or π-type orbitals with the ability to donate density to empty metal-orbitals(**1b**).

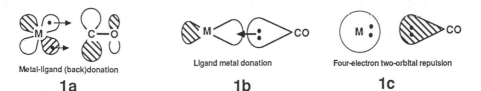

Metal-ligand (back)donation   Ligand metal donation   Four-electron two-orbital repulsion

1a   **1b**   **1c**

A number of density functional calculations on metal-carbonyls(17a),complexes of O_2,C_2H_2, and C_2H_4 (17b) as well as X_2,CX,CX_2,H_2CX (X=O,S,Se,Te) (17c) and C_6H_6,Cp^- (17d) have shown that the strength of synergic metal-ligand bonds primarily is due to the (back)donation of charge from the metal center to the ligands(**1a** in the case of metal carbonyls) .The same conclusion has also been reached in studies based on other methods ,see (17) for references.

In an attempt to quantify the synergic effect it was found(17e) that if the σ-bonding and π-back-bonding are allowed to operate simultaneously they do indeed significantly reinforce each other.The orbital interaction contribution to the bond energy increases by 50 % compared to the sum of the σ-bonding and π-back-bonding when each acts alone.

Our investigations (17) indicate further that the strengths in the synergic metal-ligand bonds ,for a homologous series ,follow the order first row>second row~third row in the non-relativistic limit ,as shown in Figure 4 in the case of the metal carbonyls $M(CO)_6$ (M=Cr,Mo,W) and $M(CO)_5$ (M=Fe,Ru,Os).This trend, which is set by the (back)donation from metal to ligand , stems from the fact that 3d-orbitals are of higher energy than their 4d and 5d counterparts.The 3d orbitals are as a consequence the better donor orbitals as they are closer in energy to the π-acceptor orbitals on the ligands(see (17a) and (17b) for a qualifier of this rationalization).

Relativistic effects are found (17) to change the order of stability to first row>third row>second row by stabilizing the bonds involving 5d-elements of the third transition series ,as illustrated for the metal-carbonyls in Figure 4. A similar relativistic order of stability is also illustrated for complexes of C_2H_2, C_2H_4, and O_2 in Figure 5.The relativistic corrections do not constitute a change(enhancement) in the metal to ligand (back)donation .Instead,the relativistic corrections are related to the ligand σ-type or π-type donor orbitals.These orbitals are not only involved in favorable donations to empty metal orbitals (**1b**) but also in destabilizing two-orbital four-electron interactions with occupied metal orbitals.These interactions ,which are illustrated for the carbonyls in **1c**,give rise to a substantial increase in the kinetic energy.Relativistic effects will,as discussed ,reduce the kinetic energy in compounds of third row metals,in particular for the type of two-orbital four electron interactions involving the core-like 5s and 5p orbitals.

Relativistic Corrections to Bond Energies in Compounds of Actinides.

We shall finally probe the degree to which relativistic effects might be of importance for bond energies in compounds involving actinides ,by representing results from calculations on $RMCl_3$ with M=Th,U and R=H,CH_3.

Table III. Calculated Bond Energies[18] (Kcal mol^{-1}) in $RMCl_3$ (M=Th,U) and (R=CH_3,H)

	D(M-R)^{NR}	D(M-R)^R	[a]Exp.
$HThCl_3$	30.1	76.0	~ 80
CH_3THCl_3	35.8	79.8	~ 80
$HUCl_3$	10.5	70.1	76
CH_3UCl_3	16.8	72.2	72

[a]D(M-R) bond energies from CpM(Cl)R of Ref. 19

It follows from Table III that relativistic corrections to the actinide-R bond energies are necessary in order to reproduce experimental results.Non-relativistic bond energies are seen to be too small by some 50-60 Kcal mol^{-1}.The importance of relativistic

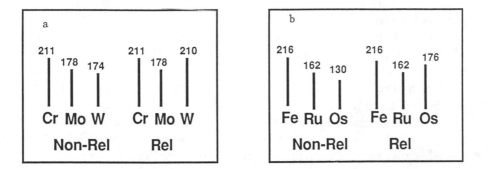

Figure 4. (**a**) Relativistic and non-relativistic averaged intrinsic bond energies (17a) in M(CO)$_6$ (m=Cr,Mo,W) . (**b**) intrinsic bond energies (17a) in M(CO)$_5$ (m=Fe,Ru,Os). Energies in kJ mol^{-1}.

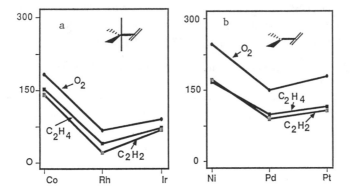

Figure 5. (**a**) Relativistic bond energies (17b) in (PH$_3$)$_2$MX ,with M=Ni,Pd, Pt and X= C$_2$H$_2$,C$_2$H$_4$,O$_2$. (**a**) Relativistic bond energies (17b) in (PH$_3$)$_4$MX ,with M=Co,Rh,Ir and X=C$_2$H$_2$,C$_2$H$_4$,O$_2$.Energies in kJ mol^{-1}.

corrections for the chemistry of actinides is currently under investigation (18,21).

Acknowledgment

All calculations were carried out at the Cyber-205 installations in Calgary(ACS) and Amsterdam(SARA).This investigation was supported by the Natural Science and Engineering Research Council of Canada(NSERC) as well as the Netherlands Organization for the Advancement of Pure Research (ZWO).

Literature Cited

1. Pitzer, K.S. Acc.Chem.Res. 1979,12,271 .
2. Pyykkö,P.;Desclaux,J.P.Acc.Chem.Res 1979,12,276.
3. Pyykkö,P.Chem.Rev. ,in press.
4   (a) Snijders,J.G.;Baerends,J.G.Mol.Phys.1978,36,1789.
    (b) Snijders,J.G.;Baerends,J.G.;Ros,P.Mol.Phys.1979,38, 1909.
5. Becke,A. J.Chem.Phys. 1986,84,4524.
6   (a) Ziegler,T.;Snijders,J.G.;Baerends,E.J.J.Chem.Phys. 1981,74,1271.
    (b) Ziegler,T.;Snijders,J.G.;Baerends,E.J. Chem.Phys.Lett. 1980,75,1.
7. Dirac,P.A.M. Proc.R.Soc.London Ser.A 1928 ,117,610.
8. Foldy, L.L.;Wouthuysen,S.A. Phys.Rev. 1950,78,29.
9. Moss,R.E. Advanced Molecular Quantum Mechanics ,Chapman and Hall,London,1973.
10. Krasnov,K.S.;Timoshinin,V.S.;Danilova,T.G.;Khandozhko ,S.V.Handbook of Molecular Constants of Inorganic Compounds ,Jerusalem 1970.
11. Baerends,E.J.;Ellis,D.E.;Ros,P.Chem.Phys. 1973,2,71.
12. (a) Lee,Y.S.;Ermler,W.C.;Pitzer,K.S.;McLean,A.D. J.Chem.Phys. 1979,70,293.
    (b) Hay,P.J.;Wadt,W.R;Kahn,L.R.;Bobrowicz,F.W. J.Chem.Phys. 1978,69,984.
13. Pilar F.L. Elementary Quantum Chemistry ,McGraw Hill,New York 1968.
14. Ziegler,T.;Tschinke,V.;Becke,A. Polyhedron 1987,6,685.
15. Baykara,N.A.;McMaster,B.N.;Salahub,D.R. Mol.Phys.1984,52 891.
16. (a) Ziegler,T. Organometallics 1985,4,5675.

(b) Ziegler,T.;Tschinke, V.;Becke,A. <u>J.Am.Chem.Soc.</u>
1987,<u>109</u>,1351.

(c) Ziegler,T.;Cheng,W.;Baerends,E.J.;Ravenek,W.
<u>Inorg.Chem.</u>,1988,accepted.

(d) Ziegler,T.;Tschinke,V.;Versluis,L; Baerends,E.J.;
Ravenek,W.<u>Polyhedron</u>,in press.

17   (a) Ziegler,T.;Tschinke,V.;Ursenbach,C. <u>J.Am.Chem.Soc.</u>
1987,<u>109</u>,4825.

(b) Ziegler,T. <u>Inorg.Chem</u>. 1985,<u>24</u>,1547.

(c) Ziegler,T. <u>Inorg.Chem</u>. 1986,<u>25</u>,2721.

(d) Ziegler,T.;Cheng,W. ,unpublished work.

(e) Baerends,E.J.;A.Rozendaal <u>NATO ASI</u> **1986**,<u>C176</u>,159.

18.   Ziegler,T.;Baerends,E.J.;Snijders,J.G.;Ravenek,W.
<u>J.Chem.Phys</u>,submitted.These calculations are based on
a quasi-relativistic approach in which the valence
density is determined variationally ,rather than by
first order perturbation theory.

19.   Bruno,J.W.;Stecher,H.A.;Morss,L.R.;Sonnenberg,D.C.
;Marks,T.J. <u>J.Am Chem.Soc.</u> 1986,<u>108</u>,7275.

20.   Schwarz,W.H.F.;Chu,S.Y.;Mark,F. <u>Mol.Phys.</u> 1983,<u>50</u>,603

21.   (a) Boerrigter,P.M.;Baerends,E.J.;Snijders,J.G.
<u>Chem.Phys.</u> 1988,<u>122</u>,357.

(b) Boerrigter,P.M.,thesis,Vrije Universiteit,<u>1987</u> .

RECEIVED October 24, 1988

Chapter 24

Ground-State Properties of Heme Complexes in Model Compounds and Intact Proteins

Frank U. Axe[1], Lek Chantranupong[1], Ahmad Waleh[1], Jack Collins[2], and Gilda H. Loew[2]

[1]The Rockefeller University, 701 Welch Road, Suite 213, Palo Alto, CA 94304
[2]SRI International, 333 Ravenswood Avenue, Menlo Park, CA 94025

The ground and low-lying spin states of ferric porphyrin (heme) complexes found in model compounds and intact proteins have been studied using an INDO-RHF-SCF method parameterized to include transition metals. The results for the model compounds using known crystal structure geometries are consistent with and help explain the origin of their observed electromagnetic properties. These studies demonstrate the ability of this method to determine with a high degree of reliability the relative energies of the manifold of spin states of ferric heme complexes. The same method has been used to address unresolved questions regarding the resting states of four heme proteins, cytochrome P450 which is a monofunctional oxidase, cytochrome c peroxidase (CCP) and catalase (CAT) both with peroxidase oxidizing activity, and metmyoglobin (MMB) which lacks significant peroxidase oxidizing activity. The characterization of the P450$_{cam}$ resting state leads to a possible explanation of the origin of its low-spin (S = 1/2) form. This result helps resolve the apparent contradiction between the presence of water as an axial ligand as determined by the x-ray structure and the absence of hyperfine splittings in ESR spectra with 65% ^{17}O enriched water. Comparisons of the resting state calculations of CCP and MMB provides an understanding of the origins of the differences in observed electromagnetic properties for MMB and CCP in spite of the similarity of their active sites. Differences in function between MMB and CCP could not, however, be understood from properties of the active site alone. Changing the imidazole ligand to an imidazolate in CCP makes its active site more similar to CAT. Thus, the anionic form of the imidazole ligand of CCP, thought to be partially formed by H-bonding to a nearby Asp residue, could account at least in part for the similar activities of CCP and CAT as oxidizing enzymes.

0097–6156/89/0394–0339$06.00/0
© 1989 American Chemical Society

Heme proteins all share a common active site or prosthetic group consisting of an iron-porphyrin (heme) unit, a nearly planar entity embedded in the globular protein and connected to it by one or at most two nearby amino acids which serve as axial ligands. This important family of proteins performs three basic biological functions (1-5): reversible oxygen transport (globins), electron transfer (cytochromes), and metabolic oxidation of small organic molecules and peroxides (peroxidases, catalases, and cytochrome P450s). In all heme proteins, the biological function is centered on the heme unit and primarily on the iron itself (1-5). Thus the oxidation and spin state of the iron, the nature of the axial ligands, and the protein environment of the heme unit serve as subtle modulators of biological behavior. The heme group is also the principal origin of spectroscopic features of these proteins. Both electronic spectra (6-9) and ground-state electromagnetic properties such as quadrupole splittings in Mössbauer resonance spectra (10-14), anisotropic g values and hyperfine splittings in electron and nuclear spin resonance spectra (15-23) and temperature-dependent magnetic moments (24-29) originate almost entirely on the heme unit. Consequently, a large field dedicated to the study of model heme complexes has emerged in an effort to understand the effect of changes in the heme unit itself on these observed properties. These studies are useful in understanding the properties of intact heme proteins since isolated heme complexes have electromagnetic properties very similar to heme units embedded in proteins. Some of these model heme systems have also been shown to mimic the biological activity of intact heme proteins. For instance, model oxo-iron compounds have been found to epoxidize olefins much like the cytochrome P450s (34, 35). The relative simplicity of model heme complexes makes it possible to study the important role of the axial ligands (30-33) in modulating electronic structure and geometries without the effect of the nearby amino acid residues present in the proteins. The insights gained from such studies can help to separately assess the relative importance of the heme unit itself and of its protein environment on the function of intact heme proteins.

Up to now most quantum mechanical studies of the ground and excited states of model heme complexes have focused primarily on diamagnetic systems (36), with less frequent treatment of heme systems with unpaired spins (37-42). With the inclusion of a restricted Hartree-Fock treatment (37, 38) within an INDO formalism parameterized for transition metals (39, 40, 42), it is now possible to calculate the relative energies of different spin states of ferric heme complexes in an evenhanded fashion at a semiempirical level.

In the work reported here we have used this method in two types of studies. The first study is a systematic investigation of the effect of changes in geometry and ligand type on the relative energies of the low-lying spin states and observable properties of eight model ferric heme complexes. This study also represents a test of the capabilities of the INDO-RHF method to characterize the lowest lying doublet (S = 1/2), quartet (S = 3/2) and sextet (S = 5/2) spin states of these model ferric heme complexes. The eight complexes chosen all have known crystal structures and include those with varying axial ligands and high-, intermediate-, and low-spin ground states as inferred from magnetic susceptibility measurements ($\underline{43-50}$) and g values in electron spin resonance spectra ($\underline{47-48}$). They also have known quadrupole splittings (ΔE_Q) from Mössbauer resonance spectra ($\underline{44}$, $\underline{45}$, $\underline{49}$, $\underline{51-53}$), a quantity which we directly calculate.

In the second type of study, using insights gained from the model compound studies, the active site of the resting state (i.e., state of the enzyme when not involved in its biochemical cycle) of four heme proteins, cytochrome P450$_{cam}$, metmyoglobin (MMB), cytochrome c peroxidase (CCP), and catalase (CAT) have been characterized. These four proteins belong to different classes of heme proteins. P450, CCP, and CAT are oxidative metabolizing enzymes thought to share a similar highly oxidized biologically active state, and MMB is the oxidized form of an oxygen transport protein with little or no peroxidase or monofunctional oxidase activity ($\underline{1}$). Each of these proteins have ferric resting states which have been characterized by x-ray structure determinations ($\underline{54-57}$). Paradoxically, while a number of long-standing questions have been resolved by these structure determinations, new ones are emerging. This study addresses five such specific questions.

The first two questions involve properties of the resting state of cytochrome P450$_{cam}$, the only P450 with a known structure ($\underline{54}$). The camphor-free resting state is mostly in a low-spin (S = 1/2) form while the camphor-bound state is a high-spin (S = 5/2) ferric complex. The x-ray structure ($\underline{58}$) reveals, that as previously deduced, the camphor-bound state is 5-coordinated with a cysteine residue as the single axial ligand and the iron significantly out of the porphyrin plane. The camphor-free state retains the cysteine ligand, but surprisingly, a water, and not a second amino acid as previously thought ($\underline{5}$), binds to the iron in the distal ligand binding site. There is also evidence that this water is part of an H-bonded network involving four more water molecules.

With the insight gained from the x-ray structure, two puzzling aspects of the camphor-free resting state have emerged. One is the origin of the low-spin form deduced from observed electromagnetic properties ($\underline{5}$). This result is surprising since other ferric heme proteins with H_2O as a sixth ligand such as MMB have primarily high-spin (S = 5/2) ground states. The other question raised is: If water is an axial ligand, as reported in the x-ray structure, why is no broadening of the ESR spectra from hyperfine interactions observed in 65% enriched ^{17}O H_2O (H. Beinert, private communication), as it is in MMB ($\underline{59}$). Since the magnitude of the hyperfine splitting depends directly on the amount of unpaired spin density on the water oxygen atom, the possibility that the negative results obtained for P450 could be a consequence of reduced spin density on the water has been investigated.

All the remaining questions focus on comparisons of CCP, MMB, and CAT. The first question addressed is: Can the differences in the active site of CCP, MMB, and CAT account for the differences in their observed electromagnetic properties? MMB and CCP ($\underline{55}$, $\underline{56}$) have been found to have the same heme unit, a ferric protoporphyrin-IX with a water and an imidazole as axial ligands. CAT has a single

phenolate group from a nearby tyrosine residue as an axial ligand
(57). In addition to the crystal structures, the nature of the
resting states of CCP, MMB, and CAT have been probed by Mössbauer
(11-13), temperature-dependent magnetic susceptibility (27-29) and
electron paramagnetic resonance (21-23) experiments. The observed
properties of both CAT and MMB have been interpreted in terms of a
definite high-spin resting state, while the properties of CCP have
been interpreted in terms of a thermal distribution of high- and low-
spin states (11, 29). The determination of the relative energies of
the low-lying sextet (S = 5/2), quartet (S = 3/2) and doublet (S =
1/2) states of the active sites of these proteins should lead to a
better understanding of the origin of these properties.

The final two questions raised are the extent to which the
active site itself controls the function of CCP, CAT, and MMB.
Specifically, we have asked: To what extent can similarities in
function between CAT and CCP be understood in light of their different active sites? Finally, we have asked: To what extent can
differences in the function of MMB and CCP be understood in terms of
their active site characteristics alone? Knowledge of the extent to
which the active site can account for function should help to understand the relative importance of the protein environment around the
heme in determining the function of each protein.

Methods

All calculations were carried out within the approximation of
intermediate neglect of differential overlap (37-42) (INDO-RHF-SCF)
which includes parameterization for transition metals. A restricted
open-shell formalism, developed by Zerner et al. (37,38), was
employed to prevent spin contamination and to make the quantitative
evaluation of the relative spin state energies possible. This method
has been used successfully to study simple transition metal complexes
like $[FeCl_4]^-$ (42), $[CuCl_4]^{2-}$ (42), and ferrocene (41) as well as
larger and more complicated systems like model oxyheme (61) and
carbonylheme (61) and model oxyhorseradish peroxidase (62) complexes.

Energies of the lowest lying sextet, quartet and doublet states
were calculated for each of the heme units studied. The geometries
of the complexes were taken from crystal structures and simplified to
unsubstituted porphyrins. The orientations of the porphyrin
macrocycles were such that the pyrrole nitrogens were on the x- and
y-axes. The choice of the lowest energy configurations for each
state was as follows:

Doublet state: $(d_{xz} \text{ and } d_{yz})^3$
Quartet state: $(d_{xz})^1(d_{yz})^1(d_{z^2})^1(d_{xy})^2$
Sextet state: $(d_{xz})^1(d_{yz})^1(d_{xy})^1(d_{x^2-y^2})^1(d_{z^2})^1$

The sextet state configuration is unique. The choice of the
lowest energy configuration for the doublet and quartet states was
confirmed by comparisons of the relative energies of various quartet
and doublet configurations, obtained by assignment of unpaired
electron(s) to different iron d orbitals, in some representative
complexes. It is also corroborated by a recent detailed study of

hemin using the INDO/RHF method by Edwards, Weiner and Zerner (J. Phys. Chem., in press).

The quadrupole splitting (ΔE_Q) observed in Mössbauer resonance of heme compounds was calculated from the INDO-RHF eigenvectors. This quantity was determined by first calculating the nine components (V_{ij}) of the electric field gradient tensor, using the appropriate one-electron operator, and considering only the contribution of the iron from all its filled orbitals. The 3×3 electric field gradient tensor was then diagonalized and the principal values ordered: $|V_{ii}|$ > $|V_{jj}|$ > $|V_{kk}|$. These values were then used in the expression:

$$\Delta E_Q = 8(1 - R)Qq[1 + \eta^2/3]^{\frac{1}{2}} . \qquad (1)$$

where $q = V_{ii}$, $\eta = (V_{kk}-V_{jj})/V_{ii}$ ($0 \le \eta < 1$), (1-R) = Sternheimer Shielding constant, and Q = nuclear quadrupole moment. The sign of ΔE_Q is the sign of the largest component V_{ii}. Values of Q and (1-R) used in these calculations are 0.187 and 0.68, respectively.

Results and Discussion

Model Heme Complexes. Presented in Table I are the calculated relative energy differences in kcal/mol for the sextet, quartet, and doublet states of the model ferric heme complexes included in the present study. Also included in Table I are the calculated quadrupole splittings (ΔE_Q) for the relevant spin state, along with the experimentally observed values of ΔE_Q and the measured effective magnetic moments.

The results clearly demonstrate that the ground spin state calculated for each model complex agrees with the one inferred from measured effective magnetic moments. Moreover the energy separations between these ground states and the other two spin states are clearly consistent with observable electromagnetic properties and help explain their origins.

The observed effective magnetic moments (43-46) for the 5- and 6-coordinated complexes found to have sextet ground states are all in the range of 5.9-6 μ_b typical of high-spin complexes. The calculated ΔE_Qs for the high-spin state of these complexes are also in good agreement with the experimental values known for three of them (44-46, 50).

Both 5- and 6-coordinated high-spin complexes have significant spin density on the porphyrin ring, 60% of which is on the pyrrole nitrogens. This should be manifest in hyperfine splittings observable in ESR or ENDOR experiments. The unpaired spin density on the axial ligands is much less than on the porphyrin ring and greater on anionic than neutral ligands.

The calculated results for the 5-coordinated high-spin complexes indicate that it is definitively more stable than the doublet and quartet states by ~30 kcal/mol and ~12 kcal/mol, respectively. The 6-coordinated high-spin complexes exhibit a significant reduction of the energy separation between the sextet and the quartet states,

Table I. Relative Energies of Different Spin States
of Model Ferric-Heme Complexes

	High Spin				Intermediate Spin		Low Spin	
L_1	Cl^-	$(S\phi pNO_2)^-$	$(NCS)^-$	TMSO	ClO_4^-	3-Clpyridine	CN^-	N_3^-
L_2	--	--	pyridine	TMSO		3-Clpyridine	CN^-	pyridine
ΔE (kcal/mol)								
$S = 5/2$	0	0	0	0	2	3	0	0
$S = 3/2$	12	12	5	5	0	0	22	12
$S = 1/2$	33	28	18	18	25	17	19	12
ΔE_Q (mm/sec)								
Calc.	0.44	1.01	0.12	1.12	3.20	3.14	2.31	2.19
Exp.	0.46	0.76	--	1.22	3.5	2.7	0.35	2.25[a]
μ_{eff} Exp. (μ_B)	5.92	5.90	5.9	6.05	4.5-5.3	3.7-4.7	2.26	2.09

[a]Experimental value of azide complex of MMB and CCP (Reference 53, page 3)

attributable mainly to the presence of the second axial ligand which increases the tendency of the iron atom to move into the plane of the porphyrin.

In both 5- and 6-coordinated complexes the energy of the doublet state is predicted to be too high to play a significant role in determining the observed electromagnetic properties. However, because of the smaller separation of the sextet and quartet states in the 6-coordinated high-spin compounds compared with the 5-coordinated compounds, spin mixing ($\lambda \hat{L} \cdot \hat{S}$) between them should be enhanced. Therefore larger zero field splittings and more aniso-tropic g values (63) in the ESR spectra should be observed.

Intermediate-spin (63) heme complexes are rare and two complexes inferred to have quartet ground states have been included in our studies. As shown in Table I, the predicted ground state of each complex is a S = 3/2 state in agreement with the spin state assign-ment deduced from observed properties. The calculated relative energy of the doublet spin state is ~ 20 kcal/mol above the quartet spin state, while the sextet states of [Fe(TPP)(ClO$_4$)] and [Fe(OEP)(3-Clpy)$_2$]$^+$ are only ~ 1.8 and 2.8 kcal/mol above their respective quartet state. These results strongly suggest that observable properties can best be understood in terms of significant spin-orbit coupling of these two low-lying states, together with the possibility of a thermal equilibrium of such spin-mixed states.

Effective magnetic moment measurements of [Fe(TPP)(ClO$_4$)] (47) have yielded values in the range of 4.5-5.3 μ_b at 77-300°K. This temperature dependence and range of values is consistent with contri-butions from sextet and quartet states with the quartet lower in energy. ESR data (47) for this complex yielded values for $g_\perp$ and $g_\parallel$ of 4.75 and 2.03, respectively. These results are atypical for a high-spin complex and lend further support to the conclusion that the ground state is a S = 3/2 or a 3/2,5/2 mixture with predominant S = 3/2 character.

Magnetic susceptibility measurements (48) for [Fe(OEP)(3-Clpy)$_2$]$^+$, the other intermediate complex studied, yield a range of μ_{eff} between 3.7-4.7 μ_b for the temperature range of 77-294°K. This range of magnetic moments is also consistent with an intermediate-spin or spin-mixed ground state. The EPR spectrum (48) for this complex yielded values for $g_\perp$ and $g_\parallel$ of 4.92 and 1.97 respectively which are similar to values obtained for [Fe(TPP)(ClO$_4$)].

The ΔE_Q values calculated for the quartet state of these complexes also agree very well with the experimentally observed values (Table I) for the same complexes. All these results taken together are highly suggestive that the S = 3/2 spin state is the principal contributor to the ground electronic state of these complexes.

For the dicyano and azide-pyridine complexes, our calculated results indicate that in each case a S = 1/2 spin state is the lowest energy state with the quartet and sextet states much higher in energy. The observed effective magnetic moments (Table I) of both the dicyano (49) and the azide-pyridine (50) complexes indicate that these two complexes are indeed essentially low-spin at all temperatures.

The only experimental ΔE_Q measured for either of the two low-spin complexes is for the dicyano complex. Our calculated value of 2.31 mm/sec for the S = 1/2 spin state is in poor agreement with the experimental value (53) of 0.35 mm/sec. However, the reported experimental ΔE_Q seems to be anomolously low for what is considered to be a low-spin complex. There is apparently no experimental ΔE_Q measured for the [Fe(TPP)(N$_3$)(py)] complex. The calculated value of 2.19 mm/sec for the doublet state is, however, in good agreement with the experimental values (53) of 2.45 and 2.25 mm/sec for CCP-N$_3$ and MMB-N$_3$, respectively, which differ only by one axial ligand being an imidazole rather than pyridine. This provides further evidence that [Fe(TPP)(N$_3$)(py)] has a doublet ground state.

An important conclusion from this study of model compounds is the additional evidence obtained for the key role of the S = 3/2 spin state in the chemistry of ferric heme complexes. There are no complexes for which S = 5/2 and 1/2 spin states are close enough to interact without an even greater contribution of the S = 3/2 spin state. Thus, the widely used assumption (64) of high-spin/low-spin thermal contributions to explain observable properties of heme complexes appears to be incorrect. Explanations involving high-spin/intermediate-spin interaction are much more plausible, since small energy separations between these states were found.

In general, the relative spin state energies calculated for all the model heme complexes studied are consistent with and help explain their observed electromagnetic properties. Thus the INDO-RHF method used appears to be sensitive to the effect of the varying axial ligands and predicts the correct energy order of spin states produced by each of them.

The ability of the method to predict the patterns of spin state behavior in these model complexes lends credence to the use made of it in the second part of these studies, to further characterize the heme units in the resting state of four heme proteins.

Comparative Studies of Resting State Active Sites of Four Heme Proteins. In this second type of study reported, we have used the x-ray structure of the active site of four heme proteins: cytochrome P450$_{cam}$ (54), CCP (55), MMB (56), and CAT (57) simplified to the ferric porphyrin complexes, shown in Table II, to calculate the relative energies and electron and spin distributions in their low-lying sextet, quartet and doublet states.

As shown in Table III, a high-spin ground state is definitively obtained for camphor-bound P450$_{cam}$ in which the single axial ligand is a mercaptide. For camphor-free P450$_{cam}$, with water and mercaptide as axial ligands at their x-ray structure values, the sextet state is still the lowest energy, but the energy separation to the low-spin (S = 1/2) state is greatly diminished. In the x-ray structure deter-mination of the resting state, the value of the Fe-water distance was

Table II. Ligand Distances (Å, X-Ray) Used for Oxidized
Resting State of Four Model Heme Proteins

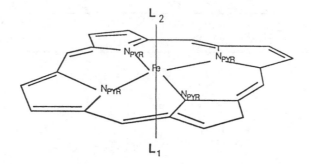

	[MMB]	[CCP]	[CAT]	[P450]
L_2	H_2O	H_2O	--	H_2O
L_1	Imidazole	Imidazole	Phenolate	$(SCH_3)^-$
Fe-O	1.90	2.40	--	2.24
Fe-L	2.02	1.93	1.76	2.32
Z_{Fe}[a]	0.25	0.13	0.13	0.24

[a]Extent of out-of-planarity of the iron atom from the
mean porphyrin ring plane.

constrained and the exact position of the water was not explicitly
optimized.   Thus we have considered the consequences of a shorter
iron-water distance (2.0 Å) and movement of the iron into the heme
plane.   In this geometry, a low-spin state is calculated to
predominate. An alternative origin of the stabilization of the low-
spin state comes from the possibility that some anionic character is
imparted to the water ligand by its postulated interaction with the
network of H-bonded water molecules, seen in the x-ray structure
(54).   This effect was simulated by using an OH^- as an axial ligand
with an Fe-O bond length of 1.75 Å. As seen in Table III, in this
model of the resting state, the low-spin (S = 1/2) state is favored
by 16 kcal/mol over the high-spin state.   While this is an extreme
model for the effect of H-bonding, it does demonstrate that partial
anionic character of the water ligand could account for the
predominant low-spin ground state observed.

Table IV gives the spin densities calculated on the water oxygen
for P450$_{cam}$ and for comparison, in MMB. Experimental values of μ_{eff}
and our calculated results (Table V) indicate MMB is in a high-spin

Table III. Origin of the Low-Spin Form
of the Resting State of Cytochrome $P450_{cam}$

	Models for Resting State			Substrate Bound State
L_1	$(SCH_3)^{-a}$	$(SCH_3)^{-b}$	$(SCH_3)^{-c}$	$(SCH_3)^{-d}$
L_2	H_2O	H_2O	$(OH)^-$	--
ΔE (kcal/mol)				
S = 1/2	6	0	0	15
S = 5/2	0	1.5	16	0

[a] Geometry from x-ray structure as shown in Table II (Ref. 54)

[b] X-ray structure with Fe-Water distance of 2.00 Å and iron moved into the porphyrin plane

[c] X-ray structure with Fe-OH⁻ distance at 1.75 Å as it is in model Fe-O_2 complexes.

[d] Geometry form x-ray structure as in Reference 58.

ground state. In its crystal structure geometry, the water oxygen of the high-spin ferric MMB is calculated to have 0.057e or about 1.1% of the total spin. For this protein, a barely detectible amount of broadening of the g=2 signal was observed in the ESR spectra in the presence of [17]O enriched H_2O (59). By contrast, in both the high-spin state and the low-spin state of P450, in its crystal structure geometry, the spin density on the axial water ligand is much lower than in the corresponding state of MMB. Allowing the Fe-O distance of the water ligand to decrease, or simulating its ionic character by an OH⁻, both of which favor a low-spin ground state, somewhat increases the spin density on the oxygen. However it remains at most about 1/6 that of MMB. Since the broadening in MMB was barely detectible, no measurable broadening of the ESR spectra in [17]O enriched water would be expected for either low-spin model of $P450_{cam}$ currently proposed here. These results then account for the absence of such broadening in a manner consistent with the presence of water as an axial ligand in the resting state of $P450_{cam}$ as observed in its x-ray structure.

Turning now to comparisons of CCP, MMB, and CAT, the relative energies of the germane doublet, quartet, and sextet spin states have been calculated using the same INDO-RHF-SCF method as for the model complexes and the results are presented in Tables V and VI.

The geometries for the resting states of CCP, MMB, and CAT used here were taken directly from their respective x-ray crystal

Table IV. Calculated Spin Densities of Oxygen[a]
of H_2O Ligands in P450$_{cam}$ and MMB[a]

	P450			MMB
	H_2O		OH^-	
S = 5/2	0.01	(0.03)[b]	0.16	0.057
S = 3/2	--	(0.03)	0.01	0.030
S = 1/2	0.0005	(0.004)	0.01	0.004

[a]Values underlined are for lowest lying state.
[b]Values in parenthesis calculated for H_2O
at 2.00 Å and Fe in the porphyrin plane.

Table V. Effect of Geometry on Resting States
of CCP, MMB, and CAT

	CCP		MMB		CAT	
	X-Ray	D_{4h}	X-Ray	D_{4h}	X-Ray	D_{4h}
ΔE (kcal/mol)						
S = 5/2	0 (0)[a]	0	0 (0)[b]	0	0	0
S = 3/2	-1.2 (-2.3)	-1.1	2.7 (5.6)	1.5	2.1	2.8
S = 1/2	8.0 (13.0)	7.7	7.5 (11.4)	5.7	5.8	6.3
ΔE$_Q$ (mm/sec)						
Calc.	3.20		0.76		0.72	
Exp.	--		1.33		0.84	
μ$_{eff}$ Exp. (μ$_b$)	4.86		6.00		5.92	

[a]Values in parenthesis without distal water
[b]Values in parenthesis with iron moved 0.1 Å further out of mean
plane of porphyrin ring.

Table VI. Effect of Ionization State
of Axial Ligand in CCP and Catalase

	CCP		Catalase	
L_1	H_2O	H_2O	--	--
L_2	Im	Im$^-$	Phenolate	Phenol
ΔE (kcal/mol)				
S = 5/2	1.2	0	0	4.7
S = 3/2	0.0	2.4	2.1	0
S = 1/2	9.2	9.5	5.8	12.2
Net Charge				
Fe	1.25[a]	1.35[b]	1.37[b]	1.28[a]
L_1	0.11	0.10	--	--
L_2	0.18	-0.63	-0.55	0.14
Porph.	-0.55	-0.83	-0.82	-0.42
Spin				
Fe	2.79	4.26[b]	4.24	2.74
L_1	0.02	0.02	--	--
L_2	0.07	0.16	0.25	0.11
Porph.	0.12	0.56	0.51	0.15

[a]For S = 3/2 spin state
[b]For S = 5/2 spin state

coordinates (55-57). However, in order to examine various geometric
effects on the spin states of each heme unit, calculations were also
carried out at several step-wise regularized geometries, starting
from the crystal geometry of each protein. The effects of porphyrin
ruffling and doming were examined by regularizing the porphyrin
crystal geometry to D_{4h} symmetry for CCP, MMB, and CAT, while leaving
the axial ligands at the same geometry as in their crystal structure.
Further differences in the geometries of CCP and MMB were examined by
removal of the axial water in CCP and by increasing the out of plane
distance of the iron in MMB by 0.1 Å.

    The calculated relative energies of CCP (Table V) indicate that
the S = 3/2 state is the lowest energy spin state in CCP, with the
S = 5/2 and S = 1/2 spin states being ~1 kcal/mol and ~9 kcal/mol
higher in energy. Furthermore, the energy ordering and separation of
the spin states are rather insensitive to regularization of the
porphyrin and axial ligand geometries. The predominance of the
quartet state in CCP appears to be due to a combination of near
planarity of the iron and a weak axial ligand. The Fe-water distance
in CCP at 2.4 Å is considerably longer than that in MMB. Indeed,
calculations at the crystal geometry in which the water ligand is

removed results in the enhanced stability of the quartet state of the complex in spite of the 5-coordinate nature of the iron (Table V).

For MMB (Table V), the sextet state is the lowest energy state for the neutral imidazole ligand with the quartet and doublet states ~3 and ~8 kcal/mol higher in energy. In contrast to CCP where regularization of the porphyrin ring to D_{4h} symmetry had little effect; for MMB it lowers the energy of both the quartet and doublet states relative to the sextet. These results suggest that the enhanced doming of the porphyrin ring observed in MMB relative to CCP is a factor in stabilizing the sextet spin state in MMB. However, since the sextet state is still lowest in energy even when the porphyrin is regularized to D_{4h} symmetry, the enhanced out-of-plane distance of the iron must be the main contributor to the stabilization of the sextet state. This effect is verified by the further stabilization of the sextet relative to the quartet state when the iron atom in MMB is moved by an additional 0.1 Å out of the mean porphyrin plane (Table V).

The calculated relative spin state energies for CAT (Table V) at the crystal geometry shows that the the sextet state is the most stable state with the S = 3/2 and S = 1/2 states, respectively, ~2 kcal/mol and ~6 kcal/mol higher in energy. The close energetic proximity of the S = 3/2 spin state is a result of the small displacement of the iron atom from the pyrrole nitrogen plane. Changing the highly ruffled porphyrin macrocycle of CAT to one of pure D_{4h} symmetry leads only to a very small stabilization of the sextet spin state of ~0.5 kcal/mol relative to the quartet and doublet spin states. Thus, the highly irregular porphyrin macrocycle in the crystal structure has very little effect upon the relative spin state orderings in this system.

In addition to geometry variations, the effects of hydrogen bonding and the resulting ionicity of the proximal ligands in CCP and CAT were simulated by deprotonation of the imidazole N_δ in CCP and protonation of the tyrosine oxygen in CAT. Deprotonation to form an Im⁻ ligand in CCP reverses the order of the sextet and quartet state energies (Table VI). Since this is an extreme model for the partial proton transfer that could occur as a result of the imidazole binding to a nearby aspartate residue in CCP, the partial anionic nature could result in near degeneracy of the quartet and sextet states.

Both the qualitative and quantitative results obtained for the active sites of the three proteins provide an improved basis for understanding the observed electromagnetic properties of the resting states of CCP, MMB, and CAT. An important aspect of the present results is that for these protein active sites, the sextet and quartet states are close in energy and the doublet state is significantly higher. Thus, the dominant contributions to observed properties in these proteins are expected to come from the S = 5/2 and S = 3/2 spin states, which can mix by spin-orbit coupling (63) as well as be in thermal equilibrium. These results provide a consistent explanation of the electromagnetic properties of the resting states of CCP, MMB, and CAT. The alternative explanation, a thermal equilibrium between sextet and doublet states, without a contribution from the quartet state, does not seem possible in these proteins.

These results are particularly important for a correct qualitative understanding of the the observed properties of CCP. For CCP the experimentally observed magnetic susceptibility ($\underline{29}$) and Mössbauer spectra ($\underline{11}$) have been interpreted in terms of a thermal mixing between high- and low-spin states, ignoring any contribution from the intermediate-spin state. This explanation is contrary to our findings of $E_{5/2} \sim E_{3/2} << E_{1/2}$. Measured values of μ_{eff} ($\underline{29}$) for CCP that are in the range of 3.7 to 4.0 μ_b over a temperature range of 77-250°K can more correctly be understood in terms of a thermal contribution from heavily spin-mixed sextet and quartet spin states. These results also strongly indicate that a re-analysis of the Mössbauer resonance spectra of CCP ($\underline{11}$) as a mixture of quartet and sextet states would also be more appropriate.

The observed properties of MMB are also best understood in the light of our calculated sextet ground state and close lying quartet state. For example, magnetic susceptibility measurements ($\underline{27}$) of MMB indicate that it is primarily a high-spin complex at low temperature ($\underline{27}$) but that μ_{eff} decreases from pure high-spin values as the temperature increases. Mössbauer spectra ($\underline{12}$) of MMB gave a value of ΔE_Q = 1.33 mm/sec. This value also appears to reflect a contribution from both the sextet state with a calculated value of ΔE_Q = 0.76 mm/sec and from the quartet state with ΔE_Q = 2.72.

The calculated spin state ordering in CAT with a sextet ground state and a low-lying quartet state is also consistent with and helps explain the observed μ_{eff} values ($\underline{28}$) between 5.9 and 4.45 μ_b in the temperature range of 4.2-293°K. The calculated ΔE_Q of 0.72 mm/sec for the sextet spin state of CAT (Table V) is in good agreement with the experimental value of 0.84 mm/sec from Mössbauer spectra ($\underline{12}$).

These studies of the active sites of four heme proteins appear to allow a consistent explanation of their observed electromagnetic properties and confirm the central role of the heme unit itself in determining these properties.

The question remains as to the extent to which the heme unit itself controls function. In this context, it is of interest to examine whether comparisons of the resting state description of CCP and CAT can help us understand why both function in a similar manner as oxidizing enzymes. The resting states of these two enzymes have a number of common features. Both are functionally 5-coordinated. While a water is found as a second axial ligand in CCP, it is too far away to have a significant effect on the heme unit. In fact, as shown in Table V, with or without the H_2O ligand, the relative order of spin states is the same for CCP. A second common feature is that the extent of anionic character in the endogenous axial ligand modulates active site properties in a similar manner. As shown in Table VI, protonation of the phenolate oxygen leads to a quartet ground state in CAT just as the neutral imidazole ligand does for CCP. Moreover, when a proton is removed from the imidazole of CCP, a sextet state is favored as it is for the phenolate resting state of CAT. The calculated net charges, and spin distributions for CCP (Im⁻), shown in Table VI, are also very similar to those calculated for CAT in both the S = 5/2 and 3/2 spin states. In the Im⁻ form of CCP, there is electron donation (0.15e) from the proximal ligand to the metal-porphyrin unit and an increased amount (0.1e) of spin density on this ligand which was also seen in the model systems.

The similarities between CCP (Im⁻) and CAT tend to reinforce the belief that the anionic character of the proximal ligand in these two proteins promotes peroxidase activity or at least stabilizes the active form of these proteins. A recent study ([65]) of a model ferric heme system in which internally hydrogen-bonded imidazoles were shown to increase peroxidase activity also supports this hypothesis.

The results obtained in this study however cannot account for differences in function between MMB and CCP. The Im⁻ form of CCP also closely resembles MMB: Both have sextet ground states with low-lying quartets and both have similar electron and spin distributions. Thus, for these proteins, the amino acid environment around the heme unit, rather than the heme itself appears to be a predominant factor in controlling function. These results then, by ruling out a major role of the heme unit itself, lend support to the proposed ([1]) key role of polar residues surrounding the distal ligand binding pocket, present in CCP and absent in MMB, in contributing to peroxidase function. Full protein studies of CCP are underway to further explore this possiblity (Collins, Axe, and Loew).

Acknowledgments

Financial support of the National Science Foundation grant PCM8410244 (for FUA and LC) and NIH Grant GM27943 (for GHL and JC) is gratefully acknowledged. We also thank the National Science Foundation for a supplement to the grant for generous use of the San Diego and Pittsburgh Supercomputer Centers. The helpful guidance and support of the staff at these centers is also gratefully acknowledged.

Literature Cited

1. (a) Poulos, T. L. Adv. Inorg. Biochem. 1988, 7, 1-36. (b) Poulos, T. L.; Finzel, B. C. In Peptide and Protein Reviews, Hearn, M. T. W., Ed.; Marcel Dekker: New York, 1984; Vol. 4, p 115.
2. Dunford, H. B.; Stillman, J. S. Coord. Chem. Rev. 1976, 19, 187.
3. Schonbaum, G. R.; Chance, B. In The Enzymes, 3rd ed., Boyer, P. D., Ed.; Academic Press: New York, 1976; pp 295-332.
4. Hewson, W. D.; Hager, L. P. In The Porphyrins, Dolphin, D., Ed.; Academic Press: New York, 1979; Vol. 7, pp 295-332.
5. Dawson, J. H.; Eble, K. S. Advanced Inorg. Bioinorg. Mech. 1986, 4, 1.
6. Gouterman, M. In The Porphyrins, Dolphin, D., Ed.; Academic Press: New York, 1978; Vol. 3, p 1.

7. Eaton, W. A.; Hofrichter, J. Methods in Enzymology 1981 76, 175.
8. Makinen, M. W.; Eaton, W. A. Nature (London) 1974, 247, 62.
9. Makinen, M. W.; Churg, A. K.; Shen, Y.; Hill, S. C.; In Electron Transport and Oxygen Utilization, Proc. of the International Symposium on Interaction between Iron and Proteins in Oxygen and Electron Transport, Ho, C. et al., Eds.; Elsevier North Holland: New York, 1980; p 101.
10. Moss, T. H.; Ehrenberg, A.; Bearden, A. Biochemistry 1969, 8, 4159.
11. Lang, G.; Asakura, T.; Yonetani, T. J. Phys. Chem. 1969, 2, 2246.
12. Maeda, Y.; Trautwein, A.; Gonser, U.; Yoshida, K.; Kikuchi, K.; Homma, T.; Ogura, Y. Biochim. Biophys. Acta 1973, 303, 230.
13. Maeda, Y.; Morita, Y.; Yoshida, K. J. Biochem. 1971, 70, 509.
14. Sharrock, M.; Debrunner, P. G.; Schulz, C.; Lipscomb, J. D.; Marshall, V.; Gunsalus, I.C. Biochim. Biophys. Acta 1976, 420, 8-26.
15. Collman, J. P.; Sorrell, T. N.; Hoffman, B. M. J. Am. Chem. Soc. 1975, 97, 913.
16. Harada, N.; Omra, T. J. Biochem. (Tokyo) 1980, 87, 1539.
17. Lang, G. Atomic Energy Research Establishment Report No. 6171, 1969.
18. Lang, G.; Asakura, T.; Yonetani, T. Atomic Energy Research Establishment Report No. 6170, 1969.
19. Wittenberg, B. A.; Kampa, L.; Wittenberg, J. B.; Blumberg, W. E.; Peisach, J. J. Biol. Chem. 1968, 243, 1863.
20. Tamura, M.; Hori, H. Biochim. Biophys. Acta 1972, 284, 20.
21. Torii, K.; Ogura, Y. J. Biochem. 1968, 64, 171.
22. Bennett, J. E.; Gibson, J. F., Ingram, D. J. E. Proc. Roy. Soc. London, Ser. A. 1957, 240 67.
23. Hori, H.; Yonetani, T., J. Biol. Chem. 1984, 260, 349.
24. Tamura, M. Biochim. Biophys. Acta 1971, 243, 239.
25. Tamura, M. Biochim. Biophys. Acta 1971, 243, 249.
26. Yoshida, K.; Iizukas, T.; Ogura, Y. J. Biochem. 1970, 68, 849.
27. Tasaki, A.; Otsuka, J.; Kotani, M. Biochim. Biophys. Acta 1967, 140, 284.
28. Torii, K.; Ogura, Y.; Otsuka, J.; Tasaki, A. J. Biochem. 1969, 66, 791.
29. Iizuka, T.; Kotani, M.; Yonetani, T. J. Biol. Chem. 1971, 246, 4731.
30. Scheidt, W. R.; Reed, C. A. Chem. Rev. 1981, 81, 543.
31. Williams, R.J.P. Fed. Proc., Fed. Am. Soc. Exp. Biol. 1961, 20, 5.
32. Hoard, J. L.; Hamor, M. J.; Hamor, T. A.; Caughey, W. S. J. Am. Chem. Soc. 1965, 87, 2312.
33. Hoard, J. L. Science 1971, 174, 1295.
34. Groves, J. T.; Hanshalter, R. C.; Nakamura, M.; Nemo, T. E.; Evans, B. J. J. Am. Chem. Soc. 1981, 103, 2284.
35. Groves, J. T.; Watanabe, Y. J. Am. Chem. Soc. 1986, 108, 507.
36. Loew, G. H. In Iron Porphyrins, Part I; Lever, A.B.P., Gray, B., Eds.; Addison-Wesley, 1983; pp 1-87.
37. Edwards, W. D.; Weiner, B.; Zerner, M. C. J. Am. Chem. Soc. 1986, 108, 2196.
38. Edwards, W. D.; Zerner, M. C. Theor. Chim. Acta 1987, 72, 347.

39. Ridley, J.; Zerner, M. Theor. Chim. Acta 1973, 32, 111.
40. Ridley, J. E.; Zerner, M. C. Theor. Chim. Acta 1976, 42, 223.
41. Zerner, M. C.; Loew, G. H.; Kirchner, R. F.; Muller-Westerhoff, U. T. J. Am. Chem. Soc. 1980, 102, 589.
42. Bacon, A. D.; Zerner, M. C. Theor. Chim. Acta 1979, 53, 21.
43. Maricondi, C.; Straub, D. K.; Epstein, L. M. J. Am. Chem. Soc. 1972, 94, 4157.
44. Tang, S. C.; Koch, S.; Papefthymiou, G. C.; Foner, S.; Frankel, R. B.; Ibers, J. A.; Holm, R. H. J. Am. Chem. Soc 1976, 98, 2414.
45. Mashiko, T.; Kastner, M. E.; Spartalin, K.; Scheidt, W. R.; Reed, C. A. J. Am. Chem. Soc. 1978, 100, 6354.
46. Scheidt, W. R.; Lee, Y. J.; Geiger, D. K.; Taylor, K.; Hatano, K. J. Am. Chem. Soc. 1982, 104, 3367.
47. Reed, C. A.; Mashiko, T.; Bently, S. P.; Kastner, M. E.; Scheidt, W. R.; Spartalian, K.; Lang, G. J. Am. Chem. Soc. 1979, 101, 2948.
48. Scheidt, W. R.; Geiger, D. K.; Hayes, R. G.; Lang, G. J. Am. Chem. Soc. 1983. 105, 2625.
49. Scheidt, W. R.; Haller, K. J.; Hatano, K. J. Am. Chem. Soc. 1980, 102, 3017.
50. Adams, K. M.; Rasmussen, P. G.; Scheidt, W. R.; Hatano, D. Inorg. Chem. 1979, 18, 1892.
51. Maricondi, C.; Swift, W.; Straub, D. K. J. Am. Chem. Soc. 1969, 91, 5205.
52. Spartalian, K.; Lang, G.; Reed, C. A. J. Chem. Phys. 1979, 71, 1832.
53. Rhynard, D.; Lang, G.; Spartalian, K. J. Chem. Phys. 1979, 71, 3715.
54. Poulus, T. L.; Finzel, B.C.; Howard, A. J. Biochemistry 1986, 25, 5314.
55. Finzel, B. C.; Poulos, T. L.; Kraut, J. J. Biol. Chem. 1984, 259, 13027.
56. Takano, T. J. Mol. Biol. 1977, 110, 537.
57. Murthy, M. R. N.; Reid, T. J.; Sicignano, A.; Tanaka, N.; Rossmann, M. G. J. Mol. Biol. 1981, 152, 465.
58. Poulos, T. L.; Finzel, B. C.; Ginnsalus, I. C.; Wagner, G. C.; Krant, J. J. Biol. Chem. 1985, 260, 16122.
59. Vuk-Pavlovic', S.; Siderer, Y. Biochem. Biphys. Res. Comm. 1977, 79, 885.
60. Herman, Z. S.; Loew, G. H. J. Am. Chem. Soc. 1980, 102, 1815.
61. Herman, Z. S.; Loew, G. H.; Rohmer, M. M. Int. J. Quant. Chem. QBS 1980, 7, 137.
62. Loew, G. H.; Herman, Z. S. J. Am. Chem. Soc. 1980, 102, 6173.
63. (a) Harris, G. Theor. Chem. Acta 1968, 10, 119.
    (b) Harris, G. Theor. Chem. Acta 1968, 10, 155.
64. Griffith, J. S. Proc. Roy. Soc. London Ser. A 1956, 235 23.
65. Traylor, T. C.; Popovitz-Biro, R. J. Am. Chem. Soc. 1988, 110, 239.

RECEIVED October 24, 1988

Chapter 25

Atom-Transfer Reactivity of Binuclear d^8 Complexes

Photochemical and Electrocatalytic Reactions

David C. Smith and Harry B. Gray

Arthur Amos Noyes Laboratory, California Institute of Technology, Pasadena, CA 91125

The long-lived $^3(d\sigma^*p\sigma)$ state of binuclear d^8 complexes undergoes a variety of reactions. One prominent reaction, photooxidative addition of halocarbons, apparently proceeds by halogen atom transfer rather than outer-sphere electron transfer. Excited-state hydrogen atom transfer occurs in reactions between several binuclear d^8 complexes and a number of organic and organometallic substrates. Specific results for $Pt_2(P_2O_5H_2)_4^{4-}$ and $Ir_2(TMB)_4^{2+}$ (TMB = 2,5-diisocyano-2,5-dimethyl-hexane) are discussed. Production of a hole in the dσ^* orbital is believed to be an important factor in these photochemical atom-transfer reactions. Electrochemical generation of such a hole produces a highly reactive intermediate that can undergo atom abstraction, thereby yielding net oxidation of an organic substrate. The net reaction is electrocatalytic in metal complex.

An electronic excited state of a metal complex is both a stronger reductant and a stronger oxidant than the ground state. Therefore, complexes with relatively long-lived excited states can participate in intermolecular electron-transfer reactions that are uphill for the corresponding ground-state species. Such excited-state electron transfer reactions often play key roles in multistep schemes for the conversion of light to chemical energy ($\underline{1}$).

While electron-transfer processes are common in inorganic photochemistry, excited-state atom transfer is limited to a small class of inorganic complexes. For UO_2^{2+}, the diradical excited state ($\cdot$U-O$\cdot$) is active in alcohol oxidation ($\underline{2}$). The primary photoprocess is hydrogen atom abstraction by the oxygen-centered radical. Photoaddition to a metal center via atom transfer has been observed for binuclear metal complexes such as $Re_2(CO)_{10}$ ($\underline{3}$-$\underline{5}$). The primary photoprocess is metal-metal bond homolysis. The photogenerated metal radical undergoes thermal atom-abstraction reactions. Until recently, atom transfer to a metal-localized excited state had not been observed.

Atom transfer to a metal complex is facilitated if localized electron or hole generation occurs at one or more open coordination sites. Binuclear d^8 complexes have been found to undergo photochemical atom transfer to one

0097–6156/89/0394–0356$06.00/0

of the metal centers (Roundhill, D. M.; Gray, H. B.; Che, C.-M. Accts. Chem. Res., in press, 6-8). These complexes possess open coordination sites in addition to an electronic structure that localizes the electron (hole) necessary for atom transfer to the metal center.

Much effort has been directed toward elucidating the electronic structure of binuclear d^8 metal complexes (9-14). Several years ago, a simple molecular-orbital model was presented to explain the electronic spectroscopic features (15). Starting from a monomer orbital scheme, two square-planar units are brought together in a face-to-face orientation (Figure 1). The orbitals perpendicular to the molecular plane, d_{z2} and p_z, interact strongly, yielding $d\sigma, \sigma^*$ and $p\sigma, \sigma^*$ orbitals. Perturbational mixing of orbitals of the same symmetry will stabilize the lower set and destabilize the upper one; stabilization of the filled lower set was viewed as a source of intermonomer binding. This scheme has been used to assign spectral transitions and to explain many of the photophysical properties. Recently, detailed spectroscopic studies have suggested that a valence-bond interpretation may be more appropriate in describing certain excited states (Smith, D. C.; Miskowski, V. M.; Gray, H. B., in preparation). Adoption of a valence-bond model, however does not alter the interpretation of the thermal chemistry, photophysics, photochemistry, or the conceptual picture of the electronic structure-to-chemistry relationship.

The lowest energy optical transition for the binuclear systems is $d\sigma^* \rightarrow p\sigma$. The excitation results in formation of a formal metal-metal single bond in the excited state. The transition is metal localized and can be viewed as movement of an electron from an orbital localized on the exterior of the M_2 unit (the $d\sigma^*$ orbital) to an orbital localized in the interior of the dimer cage (the $p\sigma$ orbital). The excitation results in hole formation localized on a metal center at an open coordination site (Figure 2). The lifetime of the $^3(d\sigma^*p\sigma)$ excited state, normally in between 100 ns and 10 μs, makes M_2 systems attractive for bimolecular photoprocesses.

Our initial interest in these systems was stimulated by observations of their photochemical electron-transfer reactivity (6, 12). From spectroscopic and electrochemical studies, the $^3(d\sigma^*p\sigma)$ excited state is predicted to be a powerful reductant, with $E°(M_2{}^+/{}^3M_2{}^*)$ estimated to range from -0.8 to -2.0 V vs SSCE in CH_3CN. That this state is a powerful reductant has been confirmed by investigation of the electron-transfer quenching of $^3M_2{}^*$ by a series of pyridinium acceptors with varying reduction potentials (13). For several binuclear complexes, the excited-state reduction potential cannot be calculated accurately due to the irreversibility of the ground-state electrochemistry; but it can be estimated from bimolecular electron-transfer quenching experiments.

For systems that are powerful excited-state reductants, photo-reduction of alkyl halides is observed (6,16). This reaction was initially interpreted to be an outer-sphere electron transfer to form the radical anion, which rapidly decomposes to yield R· and X^-. Subsequent thermal reactions yield the observed products, an $S_{RN}1$ mechanism (Figure 3a). While such a mechanism, $S_{RN}1$, appears plausible for a metal complex with $E°(M_2{}^+/{}^3M_2{}^*) < -1.5$ V (SSCE), it seems unlikely for complexes with $E°(M_2{}^+/{}^3M_2{}^*) > -1.0$ V (SSCE). Reduction potentials for alkyl halides of interest are generally more negative than -1.5 V (SSCE) (17). Alkyl halide photoreduction is observed for binuclear d^8 complexes whose excited-state reduction potentials are more positive than -1.0 V (SSCE) in CH_3CN.

An alternative pathway to outer-sphere electron transfer, which yields similar photoredox products with alkyl halides, is excited-state atom transfer (Figure 3b). Data obtained for $Pt_2(P_2O_5H_2)_4{}^{4-}$ indicate that alkyl

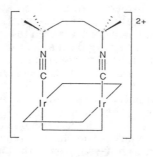

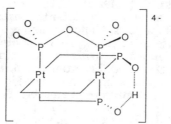

Figure 1.   Structural representations of $[Ir_2(TMB)_4]^{2+}$ (TMB =2,5-diisocyano-2,5-dimethylhexane) and $[Pt_2(P_2O_5H_2)_4]^{4-}$.

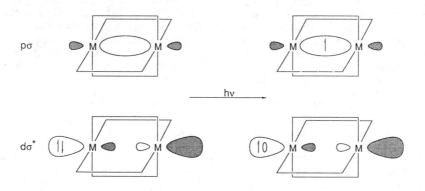

Figure 2.   Pictorial representation of the M_2-localized hole in a $^3(d\sigma^*p\sigma)$ state.

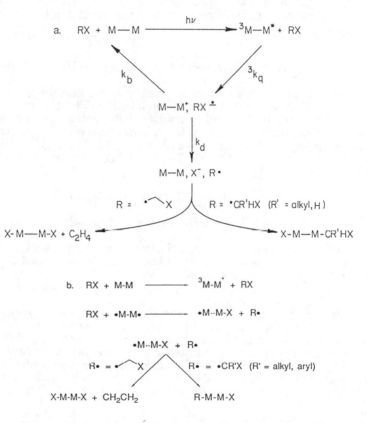

Figure 3.   a. $S_{RN}1$ mechanistic scheme for halocarbon photooxidative addition to binuclear d^8 complexes.   b. Atom-transfer mechanism for halocarbon photooxidative addition.

and aryl halides react with the $^3(d\sigma^*p\sigma)$ excited state via halogen atom transfer (Roundhill, D. M.; Che, C.-M.; Gray, H. B. Accts. Chem. Res., in press). Additional work in this area is underway.

Although the primary photoprocess for alkyl halide photoreduction may not be atom transfer in all cases, $^3(d\sigma^*p\sigma)$ excited-state hydrogen atom transfer has been established as the mechanism of the reactions between several binuclear d^8 complexes and a number of organic and organometallic substrates (Roundhill, D. M.; Che, C.-M.; Gray, H. B. Accts. Chem. Res., in press, 6-8). Initial work in this area focused on $Pt_2(P_2O_4H_2)_4{}^{4-}$, for which the catalytic conversion of isopropanol to acetone (Equation 1) had been first observed (18).

$$Pt_2(P_2O_5H_2)_4{}^{4-} + (CH_3)_3CHOH \xrightarrow{h\nu} Pt_2(P_2O_5H_2)_4{}^{4-} + H_2 + (CH_3)_2CO \quad (1)$$

From detailed studies of this system, it was concluded that the primary photoprocess is abstraction of the α-hydrogen by the $^3Pt_2^*$ to form a mono-hydride species (directly observed by transient absorption spectroscopy for a number of substrates) and the organic radical (Equation 2), with the final photoproduct being Pt_2H_2 and acetone (Equation 3). The Pt_2H_2 complex has been characterized by NMR, UV-Vis, and IR (but has not been successfully isolated) (19).

$$^3Pt_2^* + (CH_3)_2CHOH \rightarrow Pt_2H + (CH_3)_3COH \quad (2)$$

$$Pt_2H + (CH_3)_3COH \rightarrow Pt_2H_2 + (CH_3)_2CO \quad (3)$$

The hydrogen atom-transfer reactivity of the $^3(d\sigma^*p\sigma)$ excited state has been generalized to other d^8-d^8 complexes. Among these metal complexes, $Ir_2(TMB)_4{}^{2+}$ has been studied in detail (Smith, D. C.; Gray, H. B., in preparation). Stern-Volmer quenching rates for $^3Ir_2^*$ are listed in Table I and compared to rates for $^3Pt_2^*$.

Results for $^3Pt_2^*$ have been interpreted as supporting a hydrogen-abstraction pathway in which the H atom transfers via a linear Pt-H-E transition state with negligible charge transfer. The reactivity of $^3Pt_2^*$ has been compared to that of the $n\pi^*$ excited states of ketones with similar triplet energies (7). For $^3Ir_2^*$, a similar interpretation is possible; however, the analogy to the $n\pi^*$ excited states of ketones cannot be made in that the triplet energy for this complex is 30 kcal/mol compared to 60 kcal/mol for $^3Pt_2^*$.

If the photoreaction is truly an atom-transfer process, the observed rate in the absence of steric effects should track the homolytic C-H bond energies of the substrates. In a series of quenchers, the observed rate constant decreases with increasing C-H bond energy (2 > > 3 > 4 > 5) (20). However, this trend is not general. For 1, whose homolytic C-H bond energy is ~10 kcal/mol greater than that of 2, a larger rate constant is observed. However, steric arguments may be sufficient to reconcile the order of rates. Observation of a large kinetic isotope effect is supportive of an atom-transfer process for $^3Ir_2^*$.

In comparing rate constants for $^3Ir_2^*$ and $^3Pt_2^*$, some surprising trends emerge. For small substrates, such as 2, the rate observed for $^3Pt_2^*$ is greater. This is expected, due to the more energetic excited state. However, for 3 and possibly 4, a larger rate is observed for $^3Ir_2^*$. This might be understood from the greater steric demands about the open axial site in $Pt_2(P_2O_5H_2)_4{}^{4-}$ in comparison to $Ir_2(TMB)_4{}^{2+}$.

Table I. Stern-Volmer quenching rate constants for $Ir_2(TMB)_4^{2+}$ and $Pt_2(P_2O_5H_2)_4^{4-}$

Substrate	BDE (kcal/mol)[a]	$^3Ir_2^*$ $(M^{-1}s^{-1})$	$^3Pt_2^*$ $(M^{-1}s^{-1})$
1	82 [b]	8.1×10^5 ($k_H/k_D > 3$)	
2	73 [b]	5.4×10^5	8.2×10^6
3	84.4 [b]	3.2×10^5	10^4
4	85.5 [b]	2.2×10^5	
5	88.0 [b]	9.7×10^4	10^4
6		6.5×10^4	
7	91.0 [b]		5×10^3
8	c		1.9×10^6 ($k_H/k_D = 4.6$)
Et_3SiH	d	e	2.0×10^4
Bu_3SnH	d	e	1.2×10^7

a. C-H bond dissociation energy.
b. Reference 20.
c. Not known.
d. The gas phase (298 K) E-H bond energies of $(CH_3)_3EH$ species have been reported: 90 (Si) and 74 kcal/mol (Sn). Jackson, R.A. J. Organomet. Chem. 1979, 166, 17-19.
e. Reacts upon mixing.

The dihydride of $Ir_2(TMB)_4{}^{2+}$ (Ir_2H_2) has been characterized as the primary photoproduct. For some substrates (3 and 4), very little Ir_2H_2 product is observed. For 1, Ir_2H_2 appears slowly. This behavior might be understood if efficient Ir_2H_2 formation requires a rapid in-cage reaction of Ir_2H with the organic radical, or if the factors that govern the second H atom-transfer reaction are unfavorable with either the substrates or substrate radicals present.

The Ir_2H_2 complex has been isolated and characterized. (Characterization of $[Ir_2(TMB)_4(H)_2](BPh_4)_2 \cdot CH_3C_6H_5$. Calculated: C, 64.3; H, 6.5; N, 6.3. Found: C, 64.8; H, 6.8; N, 6.2. NMR (400 MHz JNM-GX400 FT NMR spectrometer, CD_3CN, 20°C): δ(ppm) -10.6 (singlet, 1H), 1.4 (broad singlet, CH_3), 1.6 (broad singlet, CH_3), 1.8 (broad singlet, CH_2), 6.9 (triplet, 8H), 7.04 (triplet, 16H), 7.33 (multiplet, 16H). IR (Nujol mull, NaCl plates): 1940 cm^{-1} (m), ν(Ir-H); 2160 cm^{-1} (s), ν(CN). UV-Vis (CH_3CN, 25°C): 320 nm (ε = 17000 M^{-1} cm^{-1}). In addition to NMR, UV-Vis, and IR spectra, the complex has been characterized crystallographically. Crystallographic data for $[Ir_2(TMB)_4(H)_2](BPh_4)_2 \cdot CH_3C_6H_5$: M_r = 1774.03; monoclinic, space group $P2_1/c$; a = 10.54 (2) Å; b = 31.02(2) Å; c = 27.05(4) Å; β = 91.57(3)°; V = 8840.7 Å^3; Z = 4; ρ_{calc} = 1.333 g/cm^3; MoKα radiation, λ = 0.71069 Å; Goodness-of-fit = 1.16 for 4692 measured intensities; R = 0.118 for 3915 reflections with I>0; R = 0.045 for 2005 reflections with I>3.0σ(I): Smith, D. C.; Marsh, R. E.; Schaefer, W. P.; Gray, H. B., in preparation.) This is the first example of a binuclear metal complex with a trans-dihydride (H-M-M-H) structure. The metal-metal separation is 2.920(2) Å, approximately 0.3 Å shorter than in the starting d^8 dimer, indicating formation of an Ir-Ir bond. This Ir-Ir separation is ~0.1 Å longer than that in the analogous Ir_2I_2 complex, presumably a result of the two trans-hydride effects. The hydrogen atom positions were not refined; in the final difference Fourier map, electron density was observed ~1.6 Å away from each Ir atom and assigned to the hydrogen positions.

An exciting aspect of atom-transfer reactivity is the possibility of photocatalytic processes. The metal dihydrides formed in the initial photochemical hydrogen atom-transfer reaction may be turned over to produce H_2 and d^8-d^8 complex (M_2) or reacted with a substrate to effect hydrogenation (and again production of M_2).

Isolation and characterization of M_2H_2 afford an opportunity to explore its reactivity. Some of this work has already been reported for Pt_2H_2 (Roundhill, D. M.; Gray, H. B.; Che, C.-M. Accts. Chem. Res., in press). For Ir_2H_2, photolysis results in production of the starting Ir_2 complex. This reaction is not clean and we observe a large amount of decomposition of the iridium material. A potentially more useful reaction has been found. Reaction of Ir_2H_2 with styrene results in a slow, clean conversion to Ir_2 and ethylbenzene. The reaction may be related to the hydrogenation of alkenes by $Co(CN)_5{}^{3-}$, which involves a coordinatively saturated metal hydride that can effect hydrogen atom transfer to give a stable 1-electron reduced transition metal ion (21). Thus, combination of the photochemical hydrogen atom-transfer reaction with the hydrogenation step completes a catalytic cycle in which the Ir_2 complex functions as a photochemical two-hydrogen transfer reagent (Figure 4).

The ^{3}(dσ*pσ) excited state of the d^8-d^8 complexes has been shown to be involved in the photochemical hydrogen atom-transfer reaction. The atom-transfer reactivity of this state is attributed to the presence of a hole in the dσ* orbital, analogous to the ^{3}nπ* state of organic ketones. Interaction of the oxidizing hole with the electron pair of the C-H bond is the presumed pathway.

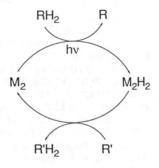

Figure 4. Photochemical two-hydrogen transfer catalytic cycle.

If production of an oxidizing hole in the $d\sigma^*$ orbital is the important factor in the photochemical reaction, then electrochemical generation of such a hole should produce a highly reactive intermediate that would mimic the initial step in the $^3(d\sigma^*p\sigma)$ photoreaction. Several of the binuclear d^8 complexes undergo reversible one-electron oxidations in noncoordinating solvents (22-24). The complex $Rh_2(TMB)_4{}^{2+}$ possesses a quasireversible one-electron oxidation at 0.74 V (Electrochemical measurements for $[Rh_2(TMB)_4](PF_6)_2$: $CH_2Cl_2/TBAPF_6$ (0.1 M), glassy carbon electrode, 25°C, SSCE reference electrode). Electrochemical oxidation of $Rh_2(TMB)_4{}^{2+}$ in the presence of 1,4-cyclohexadiene exhibits an enhanced anodic current with loss of the cathodic wave, behavior indicative of an electrocatalytic process (25). Bulk electrolysis of $Rh_2(TMB)_4{}^{2+}$ in an excess of 1,4-cyclohexadiene results in the formation of benzene and two protons (Equation 4).

$$C_6H_8 \rightarrow C_6H_6 + 2H^+ + 2e^- \ (E° = 0.4 \ V \ NHE) \hspace{2cm} (4)$$

The maximum number of turnovers observed is ten. This is limited by the amount of substrate relative to solvent. The complex $Rh_2(TMB)_4{}^{2+}$ is slowly lost due to a competitive reaction with CH_2Cl_2 to produce $Rh_2(TMB)_4(Cl)_2{}^{2+}$.

Extension of our work on binuclear d^8 species to include electrochemical oxidations points to new directions. We hope to develop systems that possess either long-lived excited states or reversible one-electron oxidations whose oxidizing potentials are increased from those we have observed so far. Increasing the oxidizing power of the oxidized form of the binuclear complex should allow us to activate C-H bonds in the most inert hydrocarbons. One way to achieve highly oxidizing species is to employ fluorinated ligands. The first step will be to examine several prototype photochemical and electrochemical oxidations by oxidizing binuclear complexes. In subsequent research, we intend to elucidate the mechanisms of these reactions, as they should be a promising start in the search for effective thermal catalysts for hydrocarbon oxidations.

Acknowledgments. Our research on binuclear complexes has been supported by the Sun Company and by National Science Foundation Grant CHE84-19828. This is contribution no. 7897 from the Arthur Amos Noyes Laboratory.

Literature Cited

1. Balzani, V.; Bolletta, F.; Gandolfi, M. T.; Maestri, M. T. Top. Curr. Chem. 1978, 75, 1-64.
2. Bergamini, P.; Sostero, S.; Traverso, O. In Fundamental and Technological Aspects of Organo-f-Element Chemistry; Marks, T. J.; Fragala, I. L., Eds.; D. Reidel: Boston, 1985; pp 361-385.
3. Stiegman, A. E.; Tyler, D. R. Comm. Inorg. Chem. 1986, 5, 215-245.
4. Hanckel, J. M.; Lee, K.-W.; Roshman, P.; Brown, T. L. Inorg. Chem. 1986, 25, 1852-1856.
5. Wayland, B. B.; Del Rossi, K. J. J. Organomet. Chem. 1984, 276, C27-C30.
6. Marshall, J. L.; Stiegman, A. E.; Gray, H. B. In Excited States and Reactive Intermediates; Lever, A. B. P., Ed.; ACS Symposium Series 307; Americn Chemical Society: Washington, D.C., 1986; pp 116-176.
7. Vlcek, A., Jr.; Gray, H. B. J. Am. Chem. Soc. 1987, 109, 286-287.

8. Vlcek, A., Jr.; Gray, H. B. Inorg. Chem. 1987, 26, 1997-2001.
9. Che, C.-M.; Butler, L. G.; Gray, H. B. J. Am. Chem. Soc. 1981, 103, 7796-7797.
10. Rice, S. F.; Gray, H. B. J. Am. Chem. Soc. 1983, 105, 4571-4575.
11. Rice, S. F.; Milder, S. J.; Gray, H. B.; Goldbeck, R. A.; Kliger, D. S. Coord. Chem. Rev. 1982, 43, 349-354.
12. Milder, S. J.; Goldbeck, R. A.; Kliger, D. S.; Gray, H. B. J. Am. Chem. Soc. 1980, 102, 6761-6764.
13. Marshall, J. L.; Stobart, S. R.; Gray, H. B. J. Am. Chem. Soc. 1984, 106, 3027-3029.
14. Dallinger, R. F.; Miskowski, V. M.; Gray, H. B.; Woodruff, W. H. J. Am. Chem. Soc. 1981, 103, 1595-1596.
15. Mann, K. R.; Gordon, J. G., II; Gray, H. B. J. Am. Chem. Soc. 1975, 97, 3553-3555.
16. Caspar, J. V.; Gray, H. B. J. Am. Chem. Soc. 1984, 106, 3029-3030.
17. Hawley, M. D. Encyclopedia of Electrochemistry of the Elements. Vol. XIV; Bard, A. J., Ed.; Marcel Dekker: New York, 1980; pp 1-135.
18. Roundhill, D. M. J. Am. Chem. Soc. 1985, 107, 4354-4356.
19. Harvey, E. L.; Stiegman, A. E.; Vlcek, A., Jr.; Gray, H. B. J. Am. Chem. Soc. 1987, 109, 5233-5235.
20. McMillen, D. F.; Golden, D. M. Ann. Rev. Phys. Chem. 1982, 33, 493-532.
21. Kwiateh, J. Catal. Rev. 1967, 1, 37-72.
22. Womack, D. R.; Enlow, P. D.; Woods, C. Inorg. Chem. 1983, 22, 2635-2656.
23. Enlow, P. D.; Woods, C. Inorg. Chem. 1985, 24, 1273-1274.
24. Boyd, D. C.; Matsch, P. A.; Mixa, M. M.; Mann, K. R. Inorg. Chem. 1986, 25, 3331-3333.
25. Bard. A. J.; Faulkner, L. R. Electrochemical Methods: Fundamentals and Applications; John Wiley & Sons; New York, 1980.

RECEIVED March 8, 1989

Chapter 26

Spin Coupling and Electron Delocalization in Mixed-Valence Iron–Sulfur Clusters

Stephen F. Sontum[1], Louis Noodleman[2], and David A. Case[2]

[1]Department of Chemistry, Middlebury College, Middlebury, VT 05753
[2]Department of Molecular Biology, Research Institute of Scripps Clinic, La Jolla, CA 92037

We extend our previous analysis of spin coupling in fully oxidized Fe_3 clusters [L. Noodleman, D.A. Case and A. Aizman, *J. Am. Chem. Soc.* 1988, 110, 1001] to cases where one or two of the iron atoms is reduced. This requires a spin Hamiltonian that contains "resonance" delocalization terms in addition to the usual Heisenberg spin coupling. Energies from broken symmetry $X\alpha$ scattered wave calculations have been fit to simple spin Hamiltonians in order to gain a better understanding of the interactions between the metal ions in these clusters. We find that the Heisenberg coupling constants J are significantly smaller in the reduced than in the oxidized clusters, and that resonance parameters B are negative, with typical values of $|B|/J$ in the range 1.5 to 3.0. Complexation with Zn at one apex of a tetrahedron stabilizes the reduced forms of the Fe_3 complexes, but has little effect on the spin interactions. Prospects for the use of such an analysis in other transition metal complexes are discussed.

Many metalloproteins involved in electron transfer and related catalytic tasks contain polynuclear transition metal complexes. Examples include nitrogenases, cytochrome oxidase, hemerythrin, and various iron-sulfur proteins. All of these exhibit strong magnetic coupling, in which the state energy depends to an important extent on the relative orientation of the metal ion spin vectors. These clusters typically contain high-spin metal ions in mixed-valence oxidation states, and may differ in important ways from "classical" mixed-valence complexes containing low-spin metal ions. Some of the simplest examples are found in synthetic and naturally-occurring iron-sulfur clusters, in which high-spin Fe(II) or Fe(III) ions are bridged by sulfides (1). Although useful qualitative theories of the origins of antiferromagnetic coupling are well known (2), it remains a difficult task to compute magnitudes of "coupling constants," their dependence upon molecular geometry, or even the ground spin state to be expected for a particular cluster. This is particularly true for "reduced" forms of iron-sulfur clusters (which have at least one iron in the Fe(II) formal oxidation state), since delocalization of the highest d electrons can interact strongly with spin coupling effects.

0097–6156/89/0394–0366$06.00/0

Considerable progress has been made in the past few years in the application of "vector coupling" or spin Hamiltonian models to iron sulfur clusters. These follow the classic application by Gibson and co-workers (3) of the Heisenberg Hamiltonian

$$\hat{H} = J\hat{S}_1.\hat{S}_2 \qquad (1)$$

to describe Fe_2S_2 clusters. Here $\hat{S}_1$ and $\hat{S}_2$ describe spins associated with each iron atom, and have magnitudes 5/2 and 2 for the mixed-valence Fe(III)/Fe(II) cluster. In isolated two-iron clusters, antiferromagnetic coupling is favored (J>0) and the clusters have a trapped valence character, with identifiable Fe(II) and Fe(III) sites. In three- and four-iron clusters, however, parallel spin alignment often occurs, in which Fe(II) and Fe(III) sites couple to a subsystem with spin S'= 9/2, and in which electron delocalization is sufficiently strong to make the two sites spectroscopically indistinguishable (4). These seemingly conflicting results can be rationalized by recognizing that spin interactions and electron delocalization are strongly related, so that the dimer energies are not given by just the Heisenberg formula E = (J/2)S'(S'+1), but are more closely approximated as

$$E = (J/2)S'(S'+1)\pm B(S'+\tfrac{1}{2}) \qquad (2)$$

Here B is a resonance parameter, and the proportionality factor, (S'+½) can be derived in a variety of ways, perhaps most transparently from the "double exchange" model of Anderson and Hasegawa (5). For the antiferromagnetic dimer (S'=½), the resonance splitting is quite small, so that minor static or dynamic distortions that break the strict equivalence of the iron sites will lead to trapped valence states. For the feromagnetic (S'=9/2) case, however, the resonance interaction predicted by Eq. (2) is much larger, and delocalized states are obtained. In a simple language, it is easier for the final d-electron to "hop" between iron sites when the spins are parallel than when they are opposed, since no net change in exchange interactions is involved in the former shift. (A detailed analysis along these lines for Fe_2S_2 dimers has been given by Noodleman and Baerends (6).) For these clusters J is positive, favoring low S', while the resonance interaction stabilizes states of high S', so that the final spin state often represents a balance of opposing forces.

The extension of this theory to three-iron clusters has been pioneered by Münck, Girerd and their co-workers (7). Here the total system spin S enters into the Hamiltonian along with the subdimer spin S'. The initial motivation came from analysis of the reduced three iron cluster in *D. gigas* ferredoxin, which showed clearly that one component of this cluster was a delocalized Fe(II)/Fe(III) dimer with S'=9/2. Later studies have confirmed that a spin Hamiltonian that leads to energies analogous to those given in Eq. (2) can explain many of the properties in a variety of clusters. In parallel work, we had been studying iron sulfur clusters for many years (6,8) and had shown how broken-symmetry molecular orbital calculations can be used to provide a physically consistent picture of many of their properties; this included an alternate derivation of the (S'+½) factor relating resonance energies to cluster spin (6). We recently extended this analysis to the oxidized forms of three iron clusters (with both linear and cubane-like geometries (9).) Here we consider the reduced forms of the three iron clusters, and analogous $ZnFe_3$ clusters, in the reduced and doubly-reduced oxidation states.

The systems we have chosen for illustration are themselves of considerable interest as models for the active sites in certain ferredoxins (from *D. gigas* and *Azobacter vinlandii*) and in aconitase (10). Many of these sites can be readily converted to four-iron clusters, and it has recently been shown that cobalt and zinc can be used in place of the fourth iron to create novel mixed-metal systems (11). The Zn complex is of particular interest because it stabilizes the more reduced forms of three-iron clusters but does not affect the spin coupling problem since it is diamagnetic.

Broken Symmetry Analysis for a Three-Iron Cluster.

The crux of our computational approach arises from the recognition that broken symmetry wavefunctions (in which otherwise equivalent metal sites have different spin populations) are relatively easy to compute and interpret, especially if local density functional methods are used. By contrast, the "correct" wavefunctions (which are eigenfunctions of S^2) are generally multiconfiguration states that are considerably more difficult to approximate and understand. Hence, we choose to fit an (assumed) spin Hamiltonian to energies computed from broken symmetry wavefunctions, and use the resulting parameters to estimate the locations of the pure spin states, including the pure-spin ground state. We have earlier applied this procedure to three S=5/2 spins, as in oxidized three-iron clusters (9). Here we extend the analysis to the reduced and doubly reduced species, where for the first time we consider the effects of electron delocalization in polynuclear clusters.

We assume that the true electrostatic interactions that couple iron spins together can be replaced by an interaction of the Heisenberg type:

$$\hat{H} = J_{12}\hat{S}_1\hat{S}_2 + J_{13}\hat{S}_1 \cdot \hat{S}_3 + J_{23}\hat{S}_2\hat{S}_3 \qquad (3)$$

and that the off-diagonal matrix elements connecting states where electron delocalization is "allowed" will be of the form $B(S'+\frac{1}{2})$, as discussed above. The essence of choosing a spin Hamiltonian model then consists of choosing the basis states to be included, diagonalizing the resultant spin matrices, and comparing the resulting eigenvalues to the computed $X\alpha$ energies in order to estimate B and J. In this paper, we will consider only the simplest models that have the important physical interactions; more complex spin Hamiltonian models may be necessary for quantitative interpretations of all the experimental data.

We consider first a three iron cluster with three equivalent metal sites in the "singly reduced", Fe(II)/Fe(III)/Fe(III) formal oxidation state. In the high spin configurations, the first five d-electrons on each site are aligned in a parallel fashion, say spin-up. We form three basis configurations by allowing the final d-electron (which must be spin-down) to reside in turn on each of the three sites. We have worked out the matrix elements of the

Heisenberg portion of the Hamiltonian in earlier work ($\underline{9}$); for the delocalization terms we assume that a single parameter B' characterizes resonance interactions between each pair of sites. Hence, the spin Hamiltonian matrix becomes:

$$H_{hs} = \begin{bmatrix} (65/4)J & 5B' & 5B' \\ 5B' & (65/4)J & 5B' \\ 5B' & 5B' & (65/4)J \end{bmatrix} \qquad (4)$$

Here and below, the diagonal elements represents the system energy in the absence of delocalization, and the off-diagonal elements give the specific resonance delocalization effects, recognizing that $(S'+\frac{1}{2})$ = 5 for parallel spin Fe(II)/Fe(III) dimers. We are assuming three equivalent sites so that $J_{12} = J_{23} = J_{13} = J$. The eigenvalues are E_1= $(65/4)J + 10B'$, and $E_{2,3} = (65/4)J - 5B'$ (doubly degenerate). For these clusters, we find $B'<0$ and, hence E_1 lies lowest. For the broken symmetry state, the first five d-electrons of one of the iron atoms (which we call "a") is of opposite spin to that of both equivalent atoms in the pair "b". There are still three basis configurations, corresponding to the three possible locations of the last d-electron. Following Papaefthymiou *et al.* ($\underline{7}$), we will adopt the simplest delocalization hypothesis, that resonance interaction is important only between the two irons of the same spin, pair "b". Here the spin matrix is:

$$H_{bs} = \begin{bmatrix} -(25/4)J & 0 & 5B \\ 0 & -(15/4)J & 0 \\ 5B & 0 & -(25/4)J \end{bmatrix} \qquad (5)$$

Here we have allowed the delocalization parameter in the broken symmetry state, B, to differ from that in the high spin state; the $X\alpha$ results reported below show that this indeed happens. Eigenvalues for the broken symmetry case are $E_{1,2} = -(25/4)J \pm 5B$ and $E_3 = -(15/4)J$.

The J's and B's can, thus, be estimated by comparing the energy differences arising from these formulas with those computed from a broken symmetry molecular orbital approach, and estimates of the pure spin state energies (including the ground state energy and its spin value) are then made from the resulting parametrized spin Hamiltonian. (For the simple model used here, the eigenstates for various values of B and J are given in Ref. 7.) In the language of our previous papers, we are using the Heisenberg Hamiltonian as a tool to carry out an approximate spin projection on the broken symmetry wavefunctions.

For the doubly reduced complexes, the spin algebra is only slightly different from that considered above. The high spin matrix becomes:

$$H_{hs} = \begin{bmatrix} 14J & 5B' & 5B \\ 5B' & 14J & 5B' \\ 5B' & 5B' & 14J \end{bmatrix} \qquad (6)$$

where the columns of the matrix now identify the site that does not have a sixth electron added to it. Eigenvalues are $E_1 = 14J + 10B'$, $E_{2,3} = 14J - 5B'$ (doubly degenerate). The broken symmetry matrix is:

$$H_{bs} = \begin{bmatrix} -4J & 0 & 5B \\ 0 & -6J & 0 \\ 5B' & 0 & -4J \end{bmatrix} \qquad (7)$$

with $E_{1,2} = -4J \pm 5B$, $E_3 = -6J$. The resonance formulas used here are based on the model of a single atomic orbital on each site to accept the sixth d-electron. As we show below, this model begins to break down, especially for the doubly reduced complexes, where at least two d-orbitals on each site are close in energy. We plan in the future to pursue more complex spin Hamiltonians that incorporate such features; for qualitative purposes, however, the present simple Hamiltonians incorporate most of the essential features of the spin interactions.

Scattered Wave Calculations

To apply these ideas, we have carried out scattered wave calculations (12) on models for Fe_3 and $ZnFe_3$ clusters, using the Xa approximation for exchange and correlation effects. Structures of "cubane" 3-Fe clusters are not known from high resolution X-ray crystallography, so we prepared a geometry for $[Fe_3S_4(SH)_3]^{2-}$ by removing the $Mo(SH)_3$ "corner" from an $MoFe_3$ cluster recently studied by Cook and Karplus (13) without changing the remaining geometry or sphere radii. This structure is in accord with EXAFS data on 3-Fe proteins in solution (14) and is also consistent with the 3.0 Å resolution X-ray model of aconitase (15). For the Zn cluster, Zn was placed in the Cook and Karplus Mo position, and a linear SH attached to it along the threefold axis. This 180° Zn-S-H angle is certainly unrealistic, but allows us to maintain the overall C_{3v} symmetry of the nuclear geometry; it seems likely that the effect of this less than optimal ligand geometry on the spin coupling among the irons will be slight.

Coordinates and sphere radii for the linear three-iron cluster are given in our earlier paper (9) based on synthetic model clusters (16).

The singly reduced Fe_3 cluster was computed in the following manner. A high-spin unrestricted self-consistent wavefunction (with 15 spin-up d-electrons and one spin-down d electron) was computed in the usual manner from standard programs (17). Then, the spin-up and spin-down potentials of one iron atom were interchanged, and the solution re-converged to a broken symmetry state with a large spin-up population on two irons and a large spin-down population on the third. As in our previous calculations, significant spin populations are also found on sulfur centers. Figures 1 and 2 show the resulting molecular orbital diagrams. As in all our previous calculations, the highest occupied, orbitals (except for the last one) are primarily of sulfur p- character, and the majority-spin Fe d-orbitals are below these. For each wavefunction, the very highest occupied orbital is primarily of minority-spin Fe d-character. For the high spin case, the two potential orbitals for this last electron (of a_1 and e symmetry) lead to total energies that are separated by 450 cm^{-1}, which can be identified with $15B'$ through Eq. (4). Hence, the effective resonance interaction B' is fairly small, about -30 cm^{-1}. In the broken symmetry case, where delocalization can occur over only two centers (the "b" pair, which have the same spin) we find g and u states (of

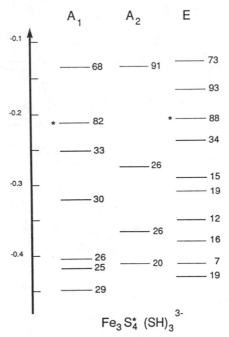

Figure 1. Orbital energies (in Rydberg atomic units) for the high spin form of the cubane three-iron cluster. Solid lines show spin-up orbital energies, dashed lines show spin-down. Numbers beside each line give the percent iron d-character in that orbital. The two orbitals marked with an asterisk (of a_1 and e symmetry) are the potential locations for the final electron. The plot corresponds to the state with the final electron in the a_1 orbital marked with an asterisk.

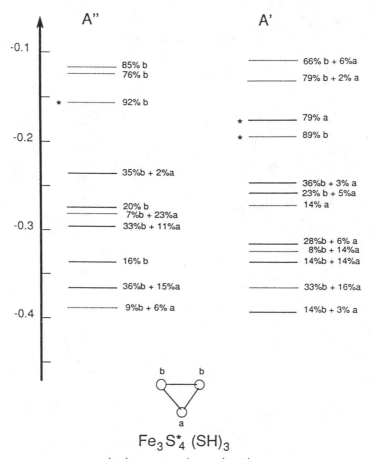

$Fe_3 S^*_4 (SH)_3$

broken symmetry, reduced

Figure 2.   As in Figure 1, but for the broken symmetry state.   Iron "a" is the unique iron, and "b" labels the equivalent pair. The three orbitals marked with an asterisk show the g and u spindown orbitals on the "b" irons (whose energy difference determines B), and the spin-up orbital mostly localized on iron "a".   The plot corresponds to the state with the final electron in the lower a' orbital marked with an asterisk.

a' and a" symmetry) which are split by a much larger amount, 4060 cm^{-1} (which by Eq. (5) is 10B), so that B is -406 cm^{-1}. The J value for this complex is determined primarily by the difference in the high spin and broken symmetry energies; fitting the Xα energies to the eigenvalues given above yields J = 297 cm^{-1}. Other calculations were carried out in a similar fashion.

Results.

Table I summarizes the results of our calculations, along with previous work on two- and three-iron clusters. As we have noted before, the general tendency of the Xα scattered wave method is to overestimate J: for example, the experimental estimate of J for the oxidized linear three-iron cluster (18) is 300 cm^{-1}, about 45% lower than the value we estimate. Estimates for J in oxidized two-iron clusters are smaller than our estimates by similar amounts (6). We expect the behavior to be about the same for the reduced species, but that the qualitative trends should be correct. We have discussed the results of the oxidized three-iron clusters in our previous paper (9), showing that the Xα calculations predict a S=½ ground state for the cubane-like geometry and S=5/2 for the linear cluster, in agreement with experiment.

Table I. Spin Hamilton Paramters (cm^{-1})

Compound	J_{ox}	J_{red}	B'	B	J_{red}/J_{ox}	$|B|/J_{red}$[a]
$Fe_2S_2^{+2/+1}$	530	346		-516	0.65	1.5
$Fe_3S_4^{+1/0}$ linear	544[b]	407[b]		-568	0.75	1.4
$Fe_3S_4^{+1/0}$ cubane	369	297	-30	-406	0.81	1.4
$ZnFe_3S_4^{+3/+2}$	360	282	-28	-407	0.78	1.4
$ZnFe_3S_4^{+1}$		45[c]	-64	-426	0.16[d]	9.5

[a]Since the computed values of J are systematically higher than those observed, we expect $|B|/J$ to be larger than indicated above.
[b]J value for neighboring irons; J values connecting the two "outside" iron atoms are very small, see Ref. 9.
[c]Based on a model of a single orbital on each iron atom, which breaks down significantly for this cluster; see text.
[d]Ratio for the +1 and +2 oxidation states.

Several interesting features emerge from Table I. First, the magnitude of J decreases as the complexes are reduced. This happens even when the geometries are unchanged, as we have assumed here. It is likely that reduction of complexes will in addition lead to slight expansions of the core, which would reinforce this drop in J as the complexes are reduced. The magnitudes of J appear to be correlated with geometry: the linear two- and three-iron complexes (with nearly identical bridging geometries) have higher values for J than do other cubane-like clusters. Second, the resonance parameters B are negative, indicating that the symmetric combination of atomic orbitals

on the "b" irons lies below the antisymmetric combination.   As with
J, the magnitudes of B appear to be related to the geometry of the
bridge, with larger absolute values for the linear than for the
cubane geometry.   Computed magnitudes of $|B|/J$ are around 1.5 (for
all clusters except the doubly Zn complex, discussed below); since,
however, we expect the computed J's to be too large, we likewise
expect the true values of $|B|/J$ to be larger than those shown in the
Table, probably in the range 2.0 to 3.0.   For the cubane-like
three-iron cluster, with all J's equal, the ground state is predicted
to be S=2 (as is observed) when $|B|/J>2.0$ (7).   It is worth
remembering, though, that the experimental geometries of the
three-iron clusters are not known, nor do we have much knowledge
about the sensitivity of J and B to changes in structural parameters.

    The doubly reduced $ZnFe_3$ cluster has some special features,
arising from stabilization of the iron d-orbitals by the zinc ion.
Figure 3 shows the molecular orbitals near the Fermi level for this
cluster, for our lowest-energy broken symmetry state, with one extra
electron localized on the "a" iron (in orbital 57a'), and the other
delocalized over the "b" pair (in the g orbital 55a').   This
corresponds to antiferromagnetic coupling of an S'=9/2 Fe(II)/Fe(III)
dimer and a S=2 Fe(II) monomer, yielding a net spin of 5/2, as
observed (11).   (An alternative occupation, in which both of the "b"
irons are reduced to Fe(II) and the unique "a" iron remains Fe(III)
is computed to be higher in energy by 1,300 cm^{-1}).   However, several
other low-energy orbitals are now possible accepters of these extra
electrons.   In addition to the u combination of the "b" pair (36a"),
there are now additional low-lying orbitals both there (e.g., 56a'
and 37a") and on the unique, "a" site (38a" and 60a').   These extra
orbitals imply the existence of many more states than in our model
of Eq. (7), and suggest that more complicated spin Hamiltonian models
may be necessary to describe the molecular orbital calculations.
Hence, the estimate of 45 cm^{-1} for J in the doubly reduced complex is
rather approximate; nevertheless, the qualitative conclusion of a
much smaller J for the doubly reduced complex should be reliable.
    The results reported here use the Xα exchange-correlation
function, which has historical interest and can be compared to past
calculations.   Within   the   local   density   approximation,
parametrizations that include the correlation effects found in a
uniform electron gas often give a better account of spin-dependent
properties (19).   Since correlation effects generally stabilize low
spin species more than high-spin states (20), one would expect
correlation effects to increase J over the values reported here, and
this was indeed found in our earlier studies of oxidized three-iron
clusters (9).   Calculations on the reduced species using improved
exchange-correlation potentials are in progress.

Conclusion

The calculations present here among the first to give serious
consideration to the competing effects of spin coupling and electron
delocalization in polynuclear transition metal   clusters.   The
qualitative features support the "double exchange" model put forth
to describe experimental spectra (7), although we do not postulate
a particular mechanism in these calculations--the results arise
solely from analysis of computed total energies for various states.

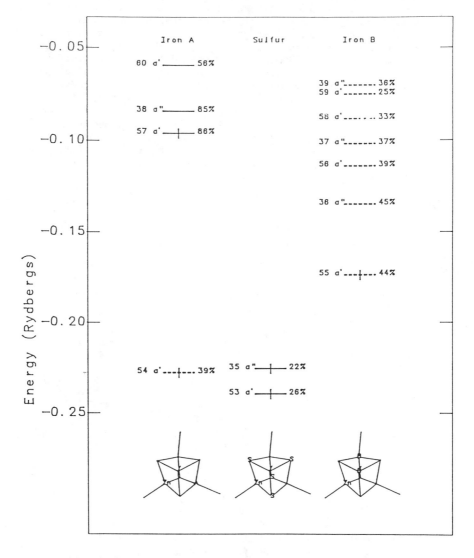

Figure 3. As in Figure 1, for the lowest energy broken symmetry state of the doubly-reduced ZnFe$_3$ cluster. Valence orbitals are numbered within each symmetry and have been placed in three columns depicting the location of their primary charge distribution. Vertical bars indicate occupied orbitals.

Even though the magnitudes of J we predict are often too high, the general trends seen in Table I are likely to be reliable. Even such qualitative information can be quite useful: one of us (L.N.) shows elsewhere (21) that a simple extension of these ideas can explain spin coupling in $Fe_4S_4^{+3}$ clusters, where the experimental results are at first sight perplexing. Although there are clear limitations to the $X\alpha$ and scattered wave models, they offer a clear and relatively straightforward path to obtaining useful models that incorporate essential physics of exchange interactions and electron delocalization in complex clusters, and the method is being adapted by others (22). Furthermore, the local exchange and muffin-tin approximations are not essential to the method, and one can look forward (with faster computers) to *ab initio* studies along the same lines (23). Thus, the scheme outlined here should enable connections to be made between spin Hamiltonian models and practical calculations for a wide range localized and delocalized polynuclear transition metal complexes.

Acknowledgments

We thank Eckard Münck for many useful discussions, and The National Institutes of Health (GM39914) for financial support.

Literature Cited

1. Spiro. T.G., ed. "Iron-Sulfur Proteins", Vol. 4; New York, John Wiley, **1982**.
2. a. Anderson, P.W. Phys. Rev. **1959**, 115; 2. b. Hay, P.J.; Thibeault, J.C.; Hoffman, R. J. Am. Chem. Soc. **1975**, 97, 4884; c. Ballhausen, C.J. "Molecular Electronic Structures of Transition Metal Complexes", New York, McGraw-Hill, **1979**; Section 3-6.
3. Gibson, J.F.; Hall, D.O.; Thornley, J.H.M.; Whatley, F.R. Proc. Natl. Acad. Sci. USA **1966**, 56, 987.
4. For a review, see Münck, E.; Kent, T.A. Hyp. Int. **1986**, 27, 161.
5. Anderson, P.W.; Hasegawa, H. Phys. Rev. **1955**, 100, 675. See also Borshch, S.A.; Kotov, I.N.; Bersuker, I.B. Sov. J. Chem. Phys. **1985**, 3, 1009. Borshch, S.A. Sov. Phys. Solid State **1984**, 26, 1142.
6. Noodleman, L.; Baerends, E.J. J. Am. Chem. Soc. **1984**, 106, 2316.
7. Papaefthymiou, V.; Girerd, J.-J.; Moura, I.; Moura, J.J.G.; Münck, E. J. Am. Chem. Soc. **1987**, 109, 4703; Münck, E.; Papaefthymiou, V; Surerus, K.K.; Girerd, J.-J., ACS Symposium Series, L. Que, ed. (in press).
8. a. Noodleman, L. J. Chem. Phys. **1981**, 74, 5737; b. Norman, J.G., Jr.; Ryan, P.B.; Noodleman, L. J. Am. Chem. Soc. **1980**, 102, 4279; c. Aizman, A.; Case, D.A. *ibid.* **1982**, 104, 3269. d. Noodleman, L.; NormanJ.G., Jr.; Osborne, J.H.; Aizman, A.; Case, D.A. *ibid.* **1985**, 107, 3418; e. Noodleman, L.; Davidson, E.R. Chem. Phys. **1986**, 109, 131.
9. Noodleman, L.; Case, D.A.; Aizman, A. J. Am. Chem. Soc. **1988**, 110, 1001.
10. a. Beinert, H.; Thomsom, A.J. Arch. Biochem. Biophys. **1983**, 222, 333, and references therein. See also Ref. 4.
11. Moura, I.; Moura, J.J.G.; Münck, E.; Papaefthymiou, V.; LeGall,

J. J. Am. Chem. Soc. 1986, 108, 349; Surerus, K.K.; Münck, E.; Moura, I.; Moura, J.J.G.; LeGall, J. *ibid.* 1987, 109, 3805.

12. a. Johnson, K.H. Annu. Rev. Phys. Chem. 1975, 26, 39. Case, D.A. *ibid.* 1982, 33.151.

13. Cook, M.; Karplus, M. J. Chem. Phys. 1985, 83, 6344.

14. Antonio, M.R.; Averill, B.A.; Moura, I.; Moura, J.J.G.; Orme-Johnson, W.H.; Teo, B.-K.; Xavier, A.V. J. Biol. Chem. 1982, 257, 6646. Beinert, H.; Emptage, M.H.; Dreyer, J.-L.; Scott, R.A.; Hahn, J.E.; Hodgson, K.O.; Thomson, A.J. Proc. Natl. Acad. Sci. USA 1983, 80, 393. Stephens, P.J.; Morgan, T.V.; Devlin, F.; Penner-Hahn, J.E.; Hodgson, K.O.; Scott, R.A.; Stout, C.D.; Burgess, B.K. *ibid.* 1985, 82, 5661.

15. Robbins, A.H.; Stout, C.D. "Iron-Sulfur Cluster in Aconitase at 3.0 Å Resolution", (submitted for publication).

16. Hagen, K.S.; Holm, R.H. J. Am. Chem. Soc. 1982, 104, 5496.

17. Cook, M.; Case, D.A. Quantum Chemistry Program Exchange #465, Bloomington, Indiana.

18. Girerd, J.J.; Papaefthymiou, G.C.; Watson, A.D.; Gamp, E.; Hagen, K.S.; Edelstein, N; Frankel, R.B.; Holm, R.H. J. Am. Chem. Soc. 1984, 106, 5941.

19. Vosko, SN.; Wilk, L.; Nusair, M. Can. J. Phys. 1980, 58. 1200.

20. Salahub, D.R. In: other *ab initio* Methods in Quantum Chemistry-II. K.P. Lawley, ed. (John Wiley, 1987), p. 447.

21. Noodleman, L., Inorg. Chem. (in press).

22. Bencini, A.; Gatteschi, D. J. Am. Chem. Soc. 1988, 108, 5763; S. Mattar, personal communication.

23. Yamaguchi, K.; Tsunekawa, T.; Toyoda, Y.; Fueno, T. Chem. Phys. Lett. 1988, 143, 371. For a recent overview of *Ab initio* calculations, see de Loth, P; Karafiloglou, P.; Daudey, J.-P.; Kahn, O. J. Am. Chem. Soc. 1988, 110, 5676. See also Ref. 6.

RECEIVED October 24, 1988

Chapter 27

Role of High- and Low-Spin Electronic States in the $Co(NH_3)_6^{2+/3+}$ Exchange Reaction

Marshall D. Newton[1]

Institute for Molecular Science, Myodaiji, Okazaki 444, Japan

Ab initio SCF and Möller Plesset calculations with flexible valence basis sets including 4f orbitals are carried out for the ground and first excited spin states of the $Co(NH_3)_6^{2+}$ and $Co(NH_3)_6^{3+}$ complexes. The results of the calculations in conjunction with a first-order spin-orbit coupling model yield an estimate of 10^{-2} for the electronic transmission factor in the $Co(NH_3)_6^{2+/3+}$ exchange reaction using an apex-to-apex approach of reactants, thus providing a mechanism characterized by only a modest degree of non-adiabaticity, consistent with the experimental kinetic data.

The mechanistic analysis of the kinetics of electron transfer processes involving transition metal complexes in solution continues to stimulate intense theoretical activity (1-17). In terms of the conventional transition state expression for the rate constant for activated electron transfer,

$$k_{et}^{act} = \kappa_{e\ell} \nu_n \Gamma_n \exp(-\beta E^{\dagger}) \qquad (1)$$

it is of particular importance to assess the various contributions to the activation energy ($E^{\dagger}$) and to the electronic transmission factor ($\kappa_{e\ell}$), which is a measure of the probability of successful reaction once the reactants have been activated (ν_n and Γ_n are, respectively, the effective harmonic frequency associated with the reaction coordinate, and the nuclear tunnelling factor). In the context of the present volume, it is to be emphasized that the techniques of computational quantum chemistry have proven to be valuable tools for estimating and analyzing the relevant activation energies and electronic transmission factor (2-11). Since electron transfer between transition metal complexes generally involves the nominal exchange of valence d-electrons, it is seen that the particular challenge to quantum chemistry is that of treating "d-electrons" -- including the energetics of the various possible electronic states associated with d-electron manifolds, and the

[1]Permanent address: Department of Chemistry, Brookhaven National Laboratory, Upton, NY 11973

0097–6156/89/0394–0378$06.00/0

nature of the coupling between metal d-orbitals and the valence
orbitals of their coordinated ligands. The density of low-lying
electronic states may be high for transition metal complexes, and it
is not always obvious which is the kinetically dominant state --
either from the point of view of minimizing the activation energy or
maximizing the electronic transmission factor. A balanced treatment
of the energies of different states is important for determining the
latter quantity since it is often dominated by various intermediate
"virtual" states through super-exchange mechanisms (7-10,16).

Although the Hartree Fock model can yield valuable information,
the reliable calculation of the properties of transition metal
complexes often necessitates the inclusion of electron correlation.
Electron correlation is necessary in some cases even to obtain
reasonable metal-ligand bond lengths (18-25), and is certainly
required in general for reliable excitation energies, especially for
states with different spin multiplicities or different electron
pairing schemes (11). Among the various types of electron
correlation effects arising in transition metal complexes, angular
(via available 4f orbitals) as well as radial (via relatively diffuse
d-orbitals) correlation of d-electrons is expected to be important
(26-33).

With this background we now consider a specific process of
interest, the $Co(NH_3)_6^{2+/3+}$ exchange reaction:

$$Co(NH_3)_6^{2+} + Co(NH_3)_6^{3+} \rightarrow Co(NH_3)_6^{3+} + Co(NH_3)_6^{2+} \quad (2)$$

$$[t_{2g}^5 e_g^2/\ ^4T_{1g}] \quad [t_{2g}^6/\ ^1A_{1g}] \quad [t_{2g}^6/\ ^1A_{1g}] \quad [t_{2g}^5 e_g^2/\ ^4T_{1g}]$$

(the lowest energy electronic configuration and state for each
species at equilibrium is indicated in brackets). While the
experimental data (34) do not indicate a strongly nonadiabatic
reaction (i.e., one where $\kappa_{el} \ll 1$), the reaction as written
corresponds formally to a "three-electron" process (a $t_{2g} \leftrightarrow e_g$
interchange within each reactant, as well as the inter-complex e_g
exchange), and thus would be expected to have a very low probability
relative to a conventional "1-electron" process (16). Accordingly,
one seeks to invoke low lying ligand-field excited states so as to
provide "1-electron" pathways for e_g exchange (11,17):

$$Co(NH_3)_6^{2+} + Co(NH_3)_6^{3+} \rightarrow Co(NH_3)_6^{3+} + Co(NH_3)_6^{2+} \quad (3)$$

$$[t_{2g}^6 e_g^1/\ ^2E_g] \quad [t_{2g}^6/\ ^1A_{1g}] \quad [t_{2g}^6/\ ^1A_{1g}] \quad [t_{2g}^6 e_g^1/\ ^2E_g]$$

$$Co(NH_3)_6^{2+} + Co(NH_3)_6^{3+} \rightarrow Co(NH_3)_6^{3+} + Co(NH_3)_6^{2+} \quad (4)$$

$$[t_{2g}^5 e_g^2/\ ^4T_{1g}] \quad [t_{2g}^5 e_g^1/\ ^3T_{1g}] \quad [t_{2g}^5 e_g^1/\ ^3T_{1g}] \quad [t_{2g}^5 e_g^2/\ ^4T_{1g}]$$

and the two "cross reactions",

$$Co(NH_3)_6^{2+} + Co(NH_3)_6^{3+} \rightarrow Co(NH_3)_6^{3+} + Co(NH_3)_6^{2+} \quad (5)$$

$$[t_{2g}^6 e_g^1/\ ^2E] \quad [t_{2g}^5 e_g^1/\ ^3T_{1g}] \quad [t_{2g}^6/\ ^1A_{1g}] \quad [t_{2g}^5 e_g^2/\ ^4T_{2g}]$$

$$Co(NH_3)_6^{2+} \quad + \quad Co(NH_3)_6^{3+} \quad \to \quad Co(NH_3)_6^{3+} \quad + \quad Co(NH_3)_6^{2+} \quad (6)$$

$$[t_{2g}^5 e_g^2 / \,^4T_{1g}] \qquad [t_{2g}^6 / \,^1A_{1g}] \qquad [t_{2g}^5 e_g^1 / \,^3T_{1g}] \qquad [t_{2g}^6 e_g^1 / \,^2E_g]$$

These excited states, which have been analyzed in previous theoretical studies (11,17), may participate in the reaction either through thermal population (as determined by appropriate Boltzmann factors) or via spin-orbit mixing into the ground electronic states. In either case, reliable estimates of the high-spin/low-spin energy separations are required: i.e., for direct estimation of the Boltzmann factors in the former case, and for estimates of spin-orbit mixing through first-order perturbation theory in the latter case. With spin-orbit coupling, we may combine Equations 2-6, to yield (17)

$$Co(NH_3)_6^{2+} \quad + \quad Co(NH_3)_6^{3+} \quad \to \quad Co(NH_3)^{3+} \quad + \quad Co(NH_3)_6^{2+} \quad (7)$$

$$[^4T_{1g} + c\,^2E_g] \quad [^1A_{1g} + c'\,^3T_{1g}] \quad [^1A_{1g} + c'\,^3T_{1g}] \quad [^4T_{1g} + c\,^2E_g]$$

where c and c' are the spin-orbit mixing coefficients for the 2+ and 3+ states, respectively. The overall electron-transfer matrix element, $H_{if} \equiv \int \Phi_i^* H \Phi_f$ (1-17), which couples the initial and final states (corresponding to the left and right hand sides of Equation 7, respectively), will be a sum of the various "1-electron pathways" and will vary quadratically with the spin-orbit coefficients. Since κ_{el} is proportional to the square of H_{if} if the degree of adiabaticity is not too great (e.g., via the Landau-Zener model) (16), and since the spin-orbit coefficients vary inversely with the high-spin/low-spin energy separations (via first-order perturbation theory), we expect κ_{el} to have an inverse fourth-order dependence on the energy separations, thus underscoring the importance of having reliable estimates for these quantities. Furthermore, these energy separations must be known at the transition state, where their magnitudes may be substantially smaller than for equilibrium geometries (11). This situation arises because low-spin states become relatively less favorable at larger Co-N bond lengths due to reduced ligand-field strength.

The vertical $^1A_{1g} \to \,^3T_{1g}$ energy for the 3+ complex is known from experiment (35) to be ~ 13,700 cm^{-1}. A variety of estimates for the analogous $^4T_{1g} \to \,^2E_g$ vertical separation in the 2+ complex are available (11,17,36), ranging from the recent estimate of Larsson et al. (~ 3300-3800 cm^{-1}) (11) to an earlier estimate of 9000 cm^{-1} (17). A conventional ligand-field model yields an estimate of 7000 cm^{-1} (36). Larsson et al. carried out CI calculations using a modified INDO model, and concluded that the transition state associated with the $^2E_g/^1A_{1g}$ reactants may be no higher in energy than the $^4T_{1g}/^1A_{1g}$ transition state.

In the present work, we evaluate the high-spin/low-spin energy separations for various Co-N bond lengths, using ab initio molecular orbital theory (RHF and UHF), including electron correlation via perturbation theory. The known energy separation for the 3+ complex at its equilibrium geometry is used to calibrate and adjust the raw calculated results.

Computational Details

The <u>ab initio</u> calculations were carried out at the SCF and second-order Möller-Plesset level (MP2) ($\underline{37}$), using a version of the Gaussian 82 computer program ($\underline{38}$) adapted for the HITAC computers at the Institute for Molecular Science. For the Co atoms, a Wachters basis ($\underline{39}$) (14s, 9p, and 5d primitives) was supplemented with two additional p-GTO's (contracted as a 2-GTO least-squares fit to an STO 4p orbital with orbital exponent 1.0 a_0^{-1}), an additional, diffuse 3d GTO (after Hay) ($\underline{40}$), and a single 4f GTO (with GTO orbital exponent 2.58 a_0^{-2}, based on optimization of the Co^{2+} and Co^{3+} ions at the MP2 level, using UHF orbitals). The full contracted basis for Co consists of Wachters' original (8/4/2) contracted basis (contraction no. 1 in Table VI of Reference $\underline{39}$, except that we employed a [4/1] 3d contraction instead of [3/2]), extended as noted above, thus yielding an (8/5/3/1) basis. For the NH_3 ligands, the 4-31G basis ($\underline{41}$) was employed.

The parameters of the contracted basis described above for Co were optimized directly for the Co^{2+} and Co^{3+} ions at the SCF level, using the computer program of Huzinaga, Obara, and Tatewaki ($\underline{42}$). The parameters so obtained differed little from the original parameters. Accordingly, the latter parameters were employed in all subsequent calculations, except for the outer two Co s-functions and the outer p-function, whose GTO exponents were scaled by 2.0 in the $Co(NH_3)_6$ calculations, based on minimization of the SCF energy for the $^1A_{1g}$ state.

The above basis set corresponds to a total of 45 contracted functions for the Co atom and 135 contracted functions for the $Co(NH_3)_6$ complexes.

The Cr atom calculations reported in the next section employed a basis analogous to that defined for the Co atom, with a 4f exponent of 1.14 a_0^{-2}, based on the results of Botch et al. ($\underline{26}$).

For the $Co(NH_3)_6^{2+/3+}$ complexes, the Co-N bond lengths of the ^1A_g (3+), $^4T_{1g}$ (2+), and ^2E_g (2+) complexes were optimized subject to the constraint of octahedral symmetry for the CoN_6 framework. Although the $Co(NH_3)_6$ structures employed in the calculations correspond strictly to D_{3d} symmetry, the departure from octahedral symmetry induced by the NH_3 protons is quite small (splittings in the t_{2g} levels $\lesssim 10^{-3}$ au). Accordingly, octahedral notation is used, even though the actual calculations reflect the small departures from O_h symmetry. For the $t_{2g}^6 e_g^1 /^2E_g$ state, we ascertained that the choices of x^2-y^2 or z^2 for the singly-occupied e_g orbital yield total energies within 10^{-5} au (cf. cautions noted in Reference $\underline{43}$). Representation of the $t_{2g}^5 e_g^1 /^3T_{1g}$ state as a single determinant required the use of an xy hole in the t_{2g}^5 manifold and an occupied x^2-y^2 e_g orbital. An xy hole was also employed for the $t_{2g}^5 e_g^2 /^4T_{1g}$ state.

The MP2 calculations were intended to provide an estimate of the differential correlation energy associated with various valence-state configurations. Accordingly, the 1s-2p(Co) and 1s(N) cores were not "active" in the MP2 calculations. Extension of the Co core to include the 3s/3p shell was found to yield results similar to those based on the 1s-2p core, and some of the results reported below were obtained at this level.

Calculations for Atomic Ions

Before proceeding to the $Co(NH_3)_6^{2+/3+}$ complexes, we consider the
effect of ionic charge on electron correlation by examining various
atomic species. We first compare the $3d^6/^5D$ electronic states in the
isoelectronic Cr^0 and Co^{3+} species (Table I). The d^6 configuration
is of interest since it is the nominal configuration of the Co atom
in the $Co(NH_3)_6^{3+}$ complex. Of course, in the latter case one has a
low-spin d^6 configuration in the ground state. However, so as to
allow comparison with other calculations, we consider the high-spin
^5D states. Botch et al. (26) have previously demonstrated the
importance of diffuse d-orbitals for radial correlation energy in
neutral first transition row elements in configurations of the
$3d^{n+1}4s$ and $3d^{n+2}$ type, and the utility of MCSCF and
multi-reference CI calculations in accounting for their correlation
energy. The tendency to place at least one of the $Cr(d^6)$ electrons
in a diffuse d-orbital is manifested in the present calculations in
the appreciable difference in the UHF and RHF energies, and in the
$\langle S^2 \rangle$ value of 6.5, which corresponds roughly to five "tight" 3d
orbitals of α spin and one orbital of β spin which can be viewed as
approximately a 50-50 mixture of a tight 3d orbital (similar to those
for the α-spin electrons) and an orthogonal d orbital more diffuse in
character. Thus relative to the RHF reference, the UHF wavefunction
already captures nearly 1 eV of correlation energy. The MP-level in
the present study yields a correlation energy (3.4 eV) similar to the
value of 3.7 eV obtained in the previous variational study (26),
which employed a similar basis set. The final entry for the Cr atom
in Table I offers an estimate of the limit of $d^2 \rightarrow d^2$ and $d^2 \rightarrow f^2$
correlation energy, based on variational calculations employing very
large basis sets and a natural orbital expansion. In all of these
studies, angular $d^2 \rightarrow f^2$ correlation is found to account for ~ 20%
of the total calculated correlation energy.
    Turning to the isoelectronic Co^{3+} ion, we find indications of
the reduced importance of diffuse d-orbitals and the associated
radial d-electron correlation (manifested in the similarity of UHF
and RHF results). Furthermore, while the magnitude of the
correlation energy is seen to decrease with increasing (positive)
atomic charge (for a given electronic configuration), the angular
contribution associated with the f-orbital is seen to increase in
magnitude. Similar behavior has been observed by Jankowski et al.
(33) in their studies of the $(3d)^{10}$ isoelectronic series.
    Additional comparisons of calculated correlation energies are
provided in Table II. Once again, the increasing importance (both
absolute and relative) of angular correlation ($d^2 \rightarrow f^2$) with
increasing positive charge, and the relative unimportance of diffuse
d orbitals, are observed. The calculated correlation energies do not
depend strongly on the order of the MP calculation (MP2 - MP4).
    The valence shell correlation energies in Tables I and II were
based on a frozen 1s-3p UHF core. Additional calculations with only
a 1s or 1s-2p UHF core indicate that correlation energy differences
for different d-electronic configurations can change typically by
< 0.1 eV when correlation of the 3s/3p shell is included in the MP
model. While these changes are rather small, they can correspond to
appreciable relative changes (typically 10-20%). So as to eliminate
this source of uncertainty, we have employed a frozen 1s-2p core for

Table I. Valence Shell Correlation Energy (eV) for the

$3d^6/^5D$ States of Cr and Co^{3+} [a]

Species	Correlated Wave Function / SCF Reference		Basis Set[a]	
	Correlated Wave Function	SCF Reference	(8/5/3/1)	(8/5/3)
Cr				
present results				
	UMP3 /	UHF $(\langle S^2 \rangle = 6.5)$[c]	2.6	1.9
	UMP3 /	RHF	3.4	2.7
previous results				
	RHF+1+2 /	RHF[d]	3.7	3.1
	RHF+1+2 /	RHF[e]	4.6	3.5
Co^{3+} present results				
	UMP3 /	UHF $(\langle S^2 \rangle = 6.0)$[c]	2.0	0.9
	UMP3 /	RHF	2.2	1.1

a) The correlation energy is the energy of the correlated wavefunction relative to the energy of the reference SCF wavefunction. In the present work, this corresponds to the UMP3 energy relative to the UHF or RHF energy. Both the spin and spatial constraints of the RHF wavefunction were relaxed in the UHF calculations. The UMP3 results are similar (to within 0.2 eV) to those from UMP2 and UMP4 (see also Table II). The MP calculations employed a frozen 1s-3p core.

b) For the present results, the basis sets are as defined in the previous section, either with or without the 4f function (see also footnotes d and e).

c) The expectation values of the S^2 operator at the UHF level are to be compared with the exact value of 6.0.

d) Variational singles and doubles CI relative to a RHF reference (26), using a basis set similar to that used in the present work. A multi reference singles and double CI, using configurations obtained from a MCSFC calculation, and with a 4f virtual, yielded a correlation energy of 4.2 eV (relative to the RHF energy).

e) CI calculation using a very large atomic orbital basis set (M. Sekiya, private communication).

Co in the $Co(NH_3)_6$ calculations. The highest five unoccupied
orbitals were also excluded ("virtual" 1s-2p orbitals), along with
the N atom 1s orbitals, from the MP active space.

From this brief consideration of atomic correlation energies, we
conclude that for the positively charged complexes of interest here,
the single-reference MP2 level, with a split d-level basis, and a
single set of 4f primitives, provides a suitable electronic structure
model. This is to be contrasted with some notable cases of
zero-valent ground state transition metal systems (26-32), where the
reduced Coulombic field from the nucleus and the formal occupation of
the valence s-level increase the importance of including a
multi-configuration reference state and more diffuse d and f
orbitals. Even though it did not appear to be crucial, we included
the additional diffuse Co d-orbital (40) in the $Co(NH_3)_6$
calculations.

Properties of $Co(NH_3)_6^{2+/3+}$ Complexes at the SCF Level

As a partial measure of the quality of the SCF calculations, we
compare in Table III calculated and experimental Co-N bond lengths
for the $^1A_{1g}$ and $^4T_{1g}$ species, the ground electronic states for
the 3+ and 2+ complexes, respectively. In addition, we have obtained
the bond length of the octahedrally-constrained ^2E (2+) complex (the
equilibrium bond lengths disproportionate about this center of
gravity, as noted in the earlier work of Larsson et al.) (11). As
expected (18-25), the calculated bond lengths are longer than the
experimental values. However, this bias is uniform (0.10 Å for each
species) and relatively modest when compared to cases of major
exaggeration observed for some transition metal complexes. Perhaps
consistent with the present results for positive oxidation states, it
has been noted previously (21) that calculated metal-ligand bond
lengths for metallocene positive ions are in significantly better
agreement with experiment than those for the neutral parents.

In the context of activated charge-transfer reactions, one of
the most important quantities is the <u>difference</u> in bond lengths of
the oxidized and reduced species (16), and we find that the
calculations give the same value for this quantity (0.22 Å) as the
experimental data. The "inner-shell" component of the transition
state geometry for the exchange reaction is defined by that CoN bond
length ($r^\dagger$), common to both the oxidized and reduced reaction
partners, which minimizes the total energy of both species (16). If
each species had the same symmetric breathing force constant, then
$r^\dagger$ for the $^1A_{1g}$ (3+)/$^4T_{1g}$ (2+) system would simply be the
arithmetic mean of the two equilibrium Co-N distances. Since the
force constant for the 3+ species is about twice that for the 2+
species (14), we estimate a value of 2.15 Å for $r^\dagger$, based on the
equilibrium SCF bond lengths (the corresponding experimental value is
thus estimated to be ~ 2.05 Å), with the value of the calculated
activation energy ~ 15 kcal/mole (SCF) and ~ 16 kcal/mole (MP2).
The experimental inner-shell activation energy is estimated to be
about 17 kcal/mole (14,15). Thus aside from the uniform horizontal
shift of 0.10 Å, the calculations give encouraging account of
important features of the potential energy surface pertaining to
variations in CoN bond length.

Table II. Valence Shell Correlation Energy for Co^{2+} and Co^{3+} Ions

Ion	Level[a]	Basis		
		(8/5/3/1)	(8/5/2/1)	(8/5/3)
$Co^{2+}(3d^7/^4F)$				
	MP2	3.24	3.19	1.86
	MP3	3.26	3.22	1.84
	MP4[b]	3.26-3.35	----	----
$Co^{3+}(3d^6/^5D)$				
	MP2	1.91	1.90	0.83
	MP3	2.01	2.00	0.87
	MP4[b]	2.01-2.04	----	----

a) Based on UHF orbitals. UHF spin contamination shows up only in the fourth significant figure of the $\langle S^2 \rangle$ values. The MP calculations employed frozen ls-3p UHF core.

b) The range corresponds to presence or absence of single and triple excitations.

Table III. Equilibrium Co-N Bond Lengths (Å)[a]

Species	Present Results[b]	Experiment[c]	Δr[d]	Results of Larsson et al.[e]	
				ab initio	INDO
$^1A_{1g}$ (3+)	2.07	1.97[f]	0.10	2.03	2.02
$^4T_{1g}$ (2+)	2.29	2.19[g]	0.10	2.27	2.14
^2E_g (2+)	2.23	[2.13][h]	[0.10][h]	2.20	2.09

a) Based on an octahedral CoN_6 framework.
b) RHF or UHF.
c) Taken from crystal and aqueous solution diffraction data.
d) Calculated minus experimental value.
e) Reference 11.
f) References 14 and 45.
g) References 14 and 44.
h) Estimate based on the assumption that the same Δr value applies to all three species.

Energy Splittings of the Co(NH$_3$)$_6^{2+/3+}$ Complexes at the MP2 Level

The salient results of the MP2 calculations are displayed in
Table IV. The results of primary interest are the corrected MP2
state separations. Since the low-spin to high-spin transition for
each charge state corresponds to a $t_{2g} \rightarrow e_g$ process in which a
t_{2g}^2 pair is broken, it seems reasonable to correct the calculated
(MP2) transition energy for each process by the same constant term --
namely, that which brings the $^1A_{1g} \rightarrow {}^3T_{1g}$ energy separation at
the calculated $^1A_{1g}$ Co-N bond length (2.29 Å) into coincidence with
the known vertical transition energy for the 3+ complex (13,700 cm^{-1})
(35). Accordingly, the raw MP2 splitting for the vertical
$^1A_{1g} \rightarrow {}^3T_{1g}$ excitation has been <u>increased</u> by 6000 cm^{-1}, and the
$^4T_{1g} \rightarrow {}^2E_g$ excitation energy has been <u>decreased</u> by the same
amount. A similar procedure was adopted in Reference 11.
     Convenient expressions for estimating the 3+ and 2+
high-spin/low-spin splittings over a range of Co-N bond lengths are
provided by linear interpolation of results obtained for 2.15 Å and
2.05 Å, and 2.15 Å and 2.25 Å, respectively. This linear procedure
is exact to the extent that the variation of energy with Co-N bond
length is harmonic (with a common force constant) for each spin
state, differing only in the magnitude of the equilibrium bond
length. In fact, the force constant for the state of higher spin is
expected to be somewhat smaller than that for the lower spin state,
due to a higher population of anti-bonding (e$_g$) electrons.
Nevertheless, the linear expression should provide useful estimates
of state separations. For purposes of estimating the electronic
transmission factor, $\kappa_{e\ell}$, dealt with in more detail in the next
section, the splittings at r† ($\sim$ 2.15 Å) are of greatest
interest. As in Reference 11 (where the ^2E state was estimated to
lie slightly below the $^4T_{1g}$ state at r†), we find the two spin
states for the 2+ complexes at r† to be very close in energy
(ΔE^{2+} = 2200 cm^{-1}). The corresponding calculated state splitting at
the $^4T_{1g}$ equilibrium geometry is appreciably larger (by
$\sim$ 4200 cm^{-1}) than that found at r†, although still well below the
earlier estimate of 9000 cm^{-1} (17).
     All MP2 results in Table IV include the Co 3s/3p shell in the
active space. Freezing this shell as part of the UHF core increases
the $^4T_{1g}/^2E_g$ splitting by 990 cm^{-1} and reduces the $^1A_{1g}/^3T_{1g}$
splitting by 510 cm^{-1} at r_{CoN} = 2.15 Å. While excitations to the
4f virtual space increase the magnitude of both the $^4T_{1g}$ and ^2E_g
correlation energies by $\sim$ 0.06 au, the effect on the $^4T_{1g}/^2E_g$
splitting is much smaller (e.g., a net reduction of only 440 cm^{-1} at
r_{Co-N} = 2.25 Å).
     The state splittings discussed above for r† are based on
individual reactants which, aside from the bond length constraint
(r†), are otherwise non-interacting. If we now consider the full
initial and final states associated with the exchange reaction
(Equations 2-7), we note that the relative energies of high-spin and
low-spin states in the bimolecular transition-state complex may be
affected by differences in initial and final state coupling, as
represented by the electron-transfer matrix element, H_{if}, discussed
in the introduction. While H_{if} for the "three-electron"
$^1A_{1g}/^4T_{1g}$ exchange (Equation 3) is expected to be quite small, as

Table IV. High-Spin/Low-Spin Energy Separations (10^3 cm^{-1})

for $Co(NH_3)_6^{2+/3+}$ Complexes

$$\Delta E^{3+} \equiv E(^3T_{1g}) - E(^1A_{1g})$$

$$\Delta E^{2+} \equiv E(^2E_g) - E(^4T_{1g})$$

A) Results at $r^\dagger$

ΔE^{3+} (2.15 Å) = 10.0 (corrected MP2 result)[a]

$\left\{\begin{array}{l} 4.0 \text{ (uncorrected MP2 result)} \\ 2.4 \text{ (SCF result)} \end{array}\right\}$[b]

ΔE^{2+} (2.15 Å) = 2.2 (corrected MP2 result)[a]

$\left\{\begin{array}{l} 8.1 \text{ (uncorrected MP2 result)} \\ 13.8 \text{ (SCF result)} \end{array}\right\}$[b]

B) Dependence on Co-N Bond Length[c]

$\Delta E^{3+}(r_{CoN})$ = $109.3 - 46.2\ r_{CoN}$

$\Delta E^{2+}(r_{CoN})$ = $-61.6 + 29.7\ r_{CoN}$

C) Results at Equilibrium[d]

ΔE^{3+} (2.07 Å) = 13.7 (present results)

ΔE^{2+} (2.29 Å) = 6.4 (present results)
3.3-4.7 (INDO)[e]

a) ΔE^{2+} is shifted <u>downward</u> by the same amount (5900 cm^{-1}) needed to shift the $^3T_{1g}$-$^1A_{1g}$ value (ΔE^{3+}) <u>upward</u> so as to match the experimental separation (13,700 cm^{-1}) at the calculated $^1A_{1g}$ equilibrium CoN bond length (2.07 Å), as described in the text.

b) The total SCF and uncorrected MP2 (1s-3p core) energies are, respectively (relative to -1717 au), -0.068346, -1.259428 ($^1A_{1g}$ (3+)); -0.079181, -1.241193 ($^3T_{1g}$ (3+)); -0.707071, -1.840157 ($^4T_{1g}$ (2+)); and -0.644179, -1.803265 (^2E_g (2+)).

c) Linear interpolation of results based on 2.15 Å and either 2.05 Å (3+) or 2.25 Å (2+).

d) Based on calculated (SCF) equilibrium r_{CoN} values. The MP2 splittings have been corrected (see footnote <u>a</u> and the text).

e) Reference 11.

noted above, the coupling for the "1-electron" $^1A_{1g}/^2E_g$ process
(Equation 4) may be as high as ~ 1000 cm^{-1}, depending on the
orientation of reactants in the transition state (6,7,11). We recall
(16) that the zeroth-order activation energy associated with the
crossing of the diabatic initial and final states will be reduced by
$|H_{if}|$ when the 2-state secular equation is solved (i.e., the
crossing is avoided). The maximum barrier reduction of ~ 1000 cm^{-1}
corresponds to an apex-to-apex configuration in which the two
reactants attain contact along a common 4-fold axis of their
octahedral frameworks (7). Various semiempirical methods have
provided estimates of the smaller H_{if} magnitudes for other
orientations (7,11), suggesting that the rms value of H_{if} is
~ 25% of the apex-to-apex value (7), assuming equal weight for each
orientation. While an accurate determination of the relevant
distribution of orientations in the encounter complex would require
detailed computer simulations, it is nevertheless clear that
suppression of the $^1A_{1g}/^2E_g$ barrier due to $|H_{if}|$, as well as an
expected Jahn-Teller relaxation energy (11), may possibly place the
energy of the $^1A_{1g}/^2E_g$ couple somewhat below that of
$^1A_{1g}/^4T_{1g}$ couple at the $^1A_{1g}/^4T_{1g}$ transition state
($r^\dagger$ ~ 2.15 Å), in spite of the fact that the energetics of the
"non-interacting" reactants favors the $^1A_{1g}/^4T_{1g}$ couple. Further
calculations (currently underway) are required to ascertain if the
$^1A_{1g}/^2E_g$ activation energy (based on the appropriate $r^\dagger$ value)
is less than the $^1A_{1g}/^4T_g$ value of ~ 16 kcal/mole (see previous
section). In the meantime, we proceed in the next section to
evaluate κ_{el} at the $^1A_{1g}/^4T_{1g}$ transition state.

Electronic Transmission Factor (κ_{el})

For cases of electron transfer between relatively weakly coupled
reactants, the 2-state Landau-Zener model leads to the following
expression for the electronic transmission factor, κ_{el} (as in
Equation 1):

$$\kappa_{el} = 2H_{if}^2(\pi^3/E_\lambda RT)^{1/2}/h\nu_n \qquad (8)$$

where E_λ is the total reorganization energy and ν_n is the
effective harmonic frequency associated with the initial or final
state system (15,16) (the expansion of exponential factors implicit
in Equation 8 is accurate to better than ~ 10% when κ_{el} is less
than ~ 0.2). For the $Co(NH_3)_6^{2+/3+}$ exchange reaction we take
$\nu_n = 409$ cm^{-1} and $E_\lambda = E_{in} + E_{out} = 97.7$ kcal/mol (for
exchange reactions, $E_\lambda \sim 4E^\dagger$; to include a small effect from
tunnelling, 97.7 kcal/mol should be replaced by 87.6 kcal/mol when
evaluating E_λ in Equation 8) (6,15). To complete the evaluation of
κ_{el} at room temperature we must evaluate H_{if} for the process
schematically represented by Equation 7. Following Buhks et al.
(17), we express c and c' as,

$$c = -(\int \Psi_{4T_{1g}} H_{so} \Psi_{2E})/\Delta E^{2+}(r^\dagger) \qquad (9)$$

$$c' = -(\int \Psi^*_{3T_{1g}} H_{so} \Psi_{1A_{1g}})/\Delta E^{3+}(r^\dagger) \qquad (10)$$

In the O_h double group (46), spin orbit coupling via H_{so} mixes the totally symmetric Γ_1 component of $^3T_{1g}$ with $^1A_{1g}$. The $^4T_{1g}$ state yields two different four-fold degenerate Γ_8 states which mix with ^2E_g, and accordingly, Equation 9 yields two distinct coefficients, denoted c_1 and c_2. At the level of first-order perturbation theory, these two states lie at $\xi^{2+}/2$ and $4\xi^{2+}/3$ with respect to the Γ_6 ground state, where ξ^{2+} is the effective spin-orbit coupling integral for the Co^{2+} 3d orbital (17).

Because two low-lying Γ_8 levels are available for the $Co(NH_3)_6^{2+}$ species in both the initial and final states of the reaction, the usual 2-state rate constant model for activated electron transfer (i.e., one initial and one final state) must be generalized as follows:

$$k_{et}^{act} = \sum_{j,k=1}^{2} y_j k_{jk} \qquad (11)$$

where the normalized Boltzmann factor y_j is given by

$$y_j = g_j \exp - (\beta E_j)/(\sum_{k=0}^{3} g_k \exp(-\beta E_k)) \qquad (12)$$

In Equation 12, $E_0(\Gamma_6) = 0$, and $E_3(\Gamma_7) = E_2(\Gamma_8)$ since the doubly-degenerate Γ_7 state is accidentally degenerate with the second Γ_8 state (through first order in spin-orbit coupling). The g_j are the degeneracy factors (2 or 4), and $\beta \equiv (k_B T)^{-1}$.

Consistent with the form of Equation 1, we find it convenient to reexpress Equation 11 as follows:

$$k_{et}^{act} = \bar{\kappa}_{el} \bar{\nu}_n \bar{\Gamma}_n \exp(-\beta \bar{E}^\dagger) \qquad (13)$$

where $\bar{\nu}_n$, $\bar{\Gamma}_n$, and $\bar{E}^\dagger \equiv E_\lambda/4$ are taken to be common to all four "paths" involved in Equation 11, and where (using Equation 8),

$$\bar{\kappa}_{el} = \sum_{j=1}^{2} y_j \kappa_{ij} = 2[\sum_{j,k=1}^{2} y_j (H_{if}^{jk})^2](\pi^3/\bar{E}_\lambda RT)^{1/2}/h\bar{\nu}_n \qquad (14)$$

We have evaluated the transfer integrals, H_{if}^{jk} (where the j,k denote the four possible combinations of initial and final Γ_8 states of the $Co(NH_3)_6^{2+}$ complex) assuming an effective 1-electron Hamiltonian, h, with relevant Γ_1 (3+) and Γ_8 (2+) states represented by appropriate multi-determinantal wavefunctions, and with the coefficients c_1, c_2, and c' evaluated using $\xi^{2+} = 515$ cm^{-1}, $\xi^{3+} = 600$ cm^{-1} (the same values employed in Reference 17). The values for $\Delta E^{2+}(r^\dagger)$ and $\Delta E^{3+}(r^\dagger)$ are taken from Table IV (2200 cm^{-1} and 10,000 cm^{-1}, respectively). The numerators in Equations 9 and 10 have magnitudes $1/5^{1/2}$ (c_1), $3/5^{1/2}$ (c_2), and $6^{1/2}$ (c') in units of ξ.

For simplicity, we construct the many-electron wavefunctions using a one-electron basis of 3d spin-orbitals on each Co site. Although these model wavefunctions nominally contain only the thirteen electrons of predominantly 3d-character (i.e., the d^7/d^6 and d^6/d^7 configurations corresponding, respectively, to the initial and final states), with the ligand electrons confined to a rigid core, the influence of the metal-ligand mixing is implicitly contained in the various coefficients and energy parameters defined above, and also in the effective 1-electron transfer integrals defined below.

   The foregoing model is somewhat analogous to that employed in Reference 17, but differs considerably in the detailed specification of the wavefunctions and evaluation of the H_{if} integrals.

   The results of the calculations depend on the relative orientation assumed for the reacting species in the transition state encounter complex. To illustrate the method, we have adopted the apex-to-apex configuration, which has been found to be especially favorable for transfer of an e_g electron (6,7,11). In this encounter geometry, the inner-shell reactant complexes achieve contact along a common 4-fold axis. Taking this 4-fold axis to be the z-axis, we assume that the effective 1-electron Hamiltonian matrix element between $3d_{z^2}$-type orbitals on the two Co atoms, $h_{3d_{z^2}3d_{z^2}}$, is the same for each of the four possible 1-electron "pathways" (see Equations 3-6). We assign a value of 1000 cm^{-1} to this matrix element, based on previous calculations (6,7), and neglect all other 2-center off-diagonal matrix elements of $\underline{h}$, as well as all off-diagonal overlap elements. We finally obtain for the bracketed quantity in Equation 14,

$$[\sum_{j,k=1}^{2} \gamma_j (H_{if}^{jk})^2] = (0.030\ h_{3d_{z^2}3d_{z^2}})^2 \tag{15}$$

Equation 14 in turn yields

$$\bar{\kappa}_{e\ell} = 1.0 \times 10^{-2} \tag{16}$$

   This result suggests that at least to the extent that the apex-to-apex orientation is important in the overall kinetics, the spin-orbit coupling mechanism is capable of yielding a rate constant characterized by only a moderate degree of non-adiabatic character, consistent with the most recent kinetic data (34) and in contrast to previous estimates which suggested a significantly greater degree of non-adiabaticity (typically, $\kappa_{e\ell} < 10^{-4}$). While the 10^{-2} value calculated for the apex-to-apex orientation should be considered as an upper limit, the mean value based on orientational averaging is likely to be within an order of magnitude of this value, a topic which will be dealt with in greater detail in future studies.

   The possible role of an alternative mechanism involving participation of the thermally excited $Co(NH_3)_6^{2+}$ state of predominantly low spin character (^2E_g) is currently under investigation and will be the subject of a future publication.

Acknowledgments

Most of the work reported herein was completed while the author was on sabbatical leave at the Institute for Molecular Science as a Visiting Scientist (11/87 - 3/88). The author is very grateful to

Prof. Morokuma and his associates in the Department of Theoretical Studies for their hospitality, and to the staff of the IMS Computer Center for providing generous access to the HITAC S-810/10 and S-820/20 computers, and for making available Library Program ATOMHF (42). Special thanks are offered to Dr. N. Koga for assistance with the use of the IMS version of the Gaussian 82 computer program, and to Mr. M. Sekiya (IMS and Hokaido University) and Prof. D. Geselowitz (Haverford College) for providing results prior to publication.

This research was supported in part (including all work done at Brookhaven National Laboratory) under contract DE-AC02-76CH00016 with the U. S. Department of Energy and supported by its Division of Chemical Sciences, Office of Basic Energy Sciences.

Literature Cited

1. Kestner, N. R.; Logan, J.; Jortner, J. J. Phys. Chem. 1974, 78, 2148.
2. Newton, M. D. Int. J. Quantum. Chem., Quantum Chem. Symp. 1980, No. 14, 363.
3. Jafri, J. A.; Logan, J.; Newton, M. D. Israel J. Chem. 1980, 19, 340.
4. Logan, J.; Newton, M. D. J. Chem. Phys. 1983, 78, 4086.
5. Logan, J.; Newton, M. D.; Noell, J. O. Int. J. Quantum. Chem., Quantum. Chem. Symp. 1984, 18, 213.
6. Newton, M. D. J. Phys. Chem. 1986, 90, 3734.
7. Newton, M. D. J. Phys. Chem. 1988, 92, 3049.
8. Larsson, S. J. Am. Chem. Soc. 1981, 103, 4034.
9. Larsson, S. Chem. Phys. Lett. 1982, 90, 136.
10. Larsson, S. J. Phys. Chem. 1984, 88, 1321.
11. Larsson, S.; Stähl, K.; Zerner, M. C. Inorg. Chem. 1986, 25, 3033.
12. Siders, P.; Marcus, R. A. J. Am. Chem. Soc. 1981, 103, 741.
13. Brunschwig, B. S.; Logan, J.; Newton, M. D.; Sutin, N. J. Am. Chem. Soc. 1980, 102, 5798.
14. Brunschwig, B. S.; Creutz, C.; Macartney, D. H.; Sham. T.-K.; Sutin, N. Faraday Discuss, Chem. Soc. 1982, 74. 113.
15. Sutin, N. Prog. Inorg. Chem. 1983, 30, 441.
16. Newton, M. D.; Sutin, N. Ann. Rev. Phys. Chem. 1984, 35, 437.
17. Buhks, E.; Bixon, M.; Jortner, J.; Navon, G. Inorg. Chem. 1979, 18, 2014.
18. Taylor, T. E.; Hall, M. B. Chem. Phys. Lett. 1985, 114, 338.
19. Williamson, R. L.; Hall, M. B. Int. J. Quantum Chem., Quantum Chem. Symp. 1987, 21, 503.
20. Lüthi, H. P.; Ammeter, J. H. J. Chem. Phys. 1982, 77, 2002.
21. Almlöf, J.; Faegri, K., Jr.; Schilling, B. E. R.; Lüthi, H. P. Chem. Phys. Lett. 1984, 106, 266.
22. Lüthi, H. P.; Siegbahn, P. E. M.; Almlöf, J.; Faegri, K. Jr.; Hedberg, A. Chem. Phys. Lett. 1984, 111, 1.
23. Luthi, H. P.; Siegbahn, P. E. M.; Almolf, J. J. Phys. Chem. 1985, 89, 2156.
24. Antonovic, D.; Davidson, E. R. J. Am. Chem. Soc. 1987, 109, 5828.
25. Antonovic, D.; Davidson, E. R. J. Chem. Phys. 1988, 88, 4967.
26. Botch, B. H.; Dunning, T. H., Jr.; Harrison, J. F. J. Chem. Phys. 1981, 75, 3466.

27. Sunil, K. K.; Jordan, K. D.; Raghavachari, K. J. Phys. Chem.
     1985, 89, 457.
28. Sunil, K. K.; Jordan, K. D. J. Chem. Phys. 1985, 82, 873.
29. Raghavachari, K.; Sunil, K. K.; Jordan, K. D. J. Chem. Phys.
     1985, 83, 4633.
30. Sunil, K. K.; Jordan, K. D. Chem. Phys. Lett. 1986, 128, 363.
31. Martin, R. L. Chem. Phys. Lett. 1980, 75, 290.
32. Rohlfing, C. M.; Martin, R. L. Chem. Phys. Lett. 1985, 115, 104.
33. Jankowski, K.; Malinowski, P.; Polasik, M. J. Chem. Phys. 1985,
     82, 841.
34. Hammershoi, A.; Geselowitz, D.; Taube, H. Inorg. Chem. 1984, 23,
     979.
35. Wilson, R. B.; Solomon, E. I. J. Am. Chem. Soc. 1980, 102, 4085.
36. Geselowitz, D., private communication.
37. Krishnan, R.; Frisch, M. J.; Pople, J. A. J. Chem. Phys. 1980,
     72, 4244.
38. Binkley, S.; Whiteside, R. A.; Krishnan, R.; Schlegel, H. B.;
     Seeger, R.; DeFrees, D. J.; Pople, J. A. Quantum Chemistry
     Program Exchange, Indiana University, Bloomington, Indiana,
     1982.
39. Wachters, A. J. H. J. Chem. Phys. 1970, 52, 1033.
40. Hay, P. J. J. Chem. Phys. 1977, 66, 4377.
41. Ditchfield, R.; Hehre, W. J.; Pople, J. A. J. Chem. Phys. 1971,
     54, 724.
42. Computer Program ATOMHF, developed by S. Huzinaga, S. Obara, and
     H. Tatewaki, and provided by the Program Library of the IMS
     Computer Center.
43. Davidson, E. R.; Borden, W. T. J. Phys. Chem. 1983, 87, 4383.
44. Kummer, S.; Babel, D. Z. Naturforsch. 1984, 39b, 1118.
45. Beattie, J. K.; Moore, C. J. Inorg. Chem. 1982, 21, 1292.
46. Herzberg, G. Electronic Spectra of Electronic Structure of
     Polyatomic Molecules; van Nostrand: New York, NY, 1967.

RECEIVED October 24, 1988

Author Index

Affiliation Index

Subject Index

A

Ab initio calculations of electronic structure
  advantages and disadvantages, 6
  applications, 3,5
  basis set, 5–6
  classes, 3,4t,5
  model core potential techniques, 7
  role in chemistry, 7–8
  technical choices, 5
  transition metal complexes, 6
Ab initio configuration interaction methods,
  examples, 106
Ab initio electronic structure techniques,
  transition metal dihydrogen
  chemistry, 92–104
Ab initio pseudopotential method of Durand
  basis set, 107
  basis set superposition errors, 107
  comparison to other methods, 108
  competing effects for H_2 activation by
  copper, 118,120f
  configuration analysis, 107–108
  convergence criterion, 107
  H_2 activation by copper, 116
  interactions of excited platinum atom
  states, 113,114–115f
  limitation, 107
  nitrogen matrix hindrance of H_2 activation
  by copper, 119,120–121f
  palladium–ethylene interaction, 108,109f,110
  photoexcited copper activation of
  H_2, 116,117f,118
  use of CIPSI algorithm, 107
Ab initio theory of bonding, transition
  metal complexes, 153–163
Actinide(s)
  electron correlation, 2
  electronic structure calculation, 7
Actinide compounds, relativistic corrections
  to bond energies, 335t,337
Agostic hydrogen
  definition, 18
  distortion of geometry, 18
  flattening of methyl hydrogens, 18
Alkali chemisorption on nickel surfaces
  background, 191
  energy, 191,192t,193f
  geometry, 192,194–195t,196
  Mulliken populations, 194–195t,196
  properties of clusters, 191,192t
  structure of clusters, 188f
Alkali electron affinities, calculation
  using effective potential quantum Monte
  Carlo simulations, 312t

α density, definition, 55
Antibonding orbitals, definition, 54
Atomic chemisorption, cluster
  requirements, 130,131t,132
Atomic correlation problems for transition
  metal complexes
  systematic errors in CI, 154
  systematic errors in relative SCF
  energies, 154t
Atomic spectra, transition metal
  clusters, 127–128
AuH, relativistic configuration
  interaction, 294–296,297t
Autocorrelation function, definition, 203

B

Basis sets
  LCGTO–Xα method, 185t,186–187
  titanium, 19
β density, definition, 55
Binuclear d^8 metal complexes
  atom-transfer mechanism for halocarbon
  photooxidative addition, 357,359f,360
  electrochemical oxidations, 364
  photochemical two-hydrogen transfer
  catalytic cycle, 362,363f,364
  pictorial representation of M_2-localized
  hole, 357,358f
  $S_{RN}1$ mechanistic scheme for halocarbon
  photooxidative addition, 357,359f
  Stern–Volmer quenching rate
  constants, 360,361t
  structural representations, 357,358f
Bis(π-allyl)nickel
  electronic structure, 233,236
  orbital diagram, 233,235f
  photoelectron spectrum, 233,234t,236
Bis(pyridine)(meso-tetraphenylporphinato)-
  iron(II)
  d electron orbital populations, 43t
  deformation electron density, 43,45f
  perspective drawing, 43,44f
Bis(tetrahydrofuran)(meso-tetraphenylporphinato)iron(II)
  deformation density, 46,47f
  iron atom d-orbital populations, 46t
Bloch states
  configurations, 281,283f
  definition, 281
Bond distances, TiCl$_4$, 19,21t,27
Bonding, relativistic effects, 296

Production: Paula M. Befard
Indexing: Deborah H. Steiner
Acquisition: Cheryl Shanks

Elements typeset by Hot Type Ltd., Washington, DC
Printed and bound by Maple Press, York, PA

Other ACS Books

Biotechnology and Materials Science: Chemistry for the Future
Edited by Mary L. Good
160 pp; clothbound, ISBN 0–8412–1472–7, paperback, ISBN 0–8412–1473–5

Chemical Demonstrations: A Sourcebook for Teachers
Volume 1, Second Edition by Lee R. Summerlin and James L. Ealy, Jr.
192 pp; spiral bound; ISBN 0–8412–1481–6
Volume 2, Second Edition by Lee R. Summerlin, Christie L. Borgford, and Julie B. Ealy
229 pp; spiral bound; ISBN 0–8412–1535–9

The Language of Biotechnology: A Dictionary of Terms
By John M. Walker and Michael Cox
ACS Professional Reference Book; 256 pp;
clothbound, ISBN 0–8412–1489–1; paperback, ISBN 0–8412–1490–5

Cancer: The Outlaw Cell, Second Edition
Edited by Richard E. LaFond
274 pp; clothbound, ISBN 0–8412–1419–0; paperback, ISBN 0–8412–1420–4

Chemical Structure Software for Personal Computers
Edited by Daniel E. Meyer, Wendy A. Warr, and Richard A. Love
ACS Professional Reference Book; 107 pp;
clothbound, ISBN 0–8412–1538–3; paperback, ISBN 0–8412–1539–1

Practical Statistics for the Physical Sciences
By Larry L. Havlicek
ACS Professional Reference Book; 198 pp; clothbound; ISBN 0–8412–1453–0

The Basics of Technical Communicating
By B. Edward Cain
ACS Professional Reference Book; 198 pp;
clothbound, ISBN 0–8412–1451–4; paperback, ISBN 0–8412–1452–2

The ACS Style Guide: A Manual for Authors and Editors
Edited by Janet S. Dodd
264 pp; clothbound, ISBN 0–8412–0917–0; paperback, ISBN 0–8412–0943–X

Personal Computers for Scientists: A Byte at a Time
By Glenn I. Ouchi
276 pp; clothbound, ISBN 0–8412–1000–4; paperback, ISBN 0–8412–1001–2

Chemistry and Crime: From Sherlock Holmes to Today's Courtroom
Edited by Samuel M. Gerber
135 pp; clothbound, ISBN 0–8412–0784–4; paperback, ISBN 0–8412–0785–2

For further information and a free catalog of ACS books, contact:
American Chemical Society
Distribution Office, Department 225
1155 16th Street, NW, Washington, DC 20036
Telephone 800–227–5558